Implementing Machine Learning with SAP S/4HANA®

SAP PRESS

Siar Sarferaz, Raghu Banda

Implementing Machine Learning with SAP S/4HANA®

Rheinwerk Publishing

Editor Megan Fuerst
Acquisitions Editor Hareem Shafi
Copyeditor Melinda Rankin
Cover Design Graham Geary
Photo Credit iStockphoto.com: 510391350/© wsfurlan
Layout Design Vera Brauner
Production Graham Geary
Typesetting III-satz, Husby (Germany)
Printed and bound in the United States of America, on paper from sustainable sources

ISBN 978-1-4932-2011-3

1st edition 2020

Library of Congress Cataloging-in-Publication Data
Names: Sarferaz, Siar, author. | Banda, Raghu, author.
Title: Implementing machine learning with SAP S/4HANA / Siar Sarferaz, Raghu Banda.
Description: 1st edition. | Bonn ; Boston : Rheinwerk Publishing, [2020] | Includes index.
Identifiers: LCCN 2020028040 (print) | LCCN 2020028041 (ebook) | ISBN 9781493220113 (hardcover) | ISBN 9781493220120 (ebook)
Subjects: LCSH: Machine learning. | SAP HANA (Electronic resource)
Classification: LCC Q325.5 .S27 2020 (print) | LCC Q325.5 (ebook) | DDC 006.3/1--dc23
LC record available at https://lccn.loc.gov/2020028040
LC ebook record available at https://lccn.loc.gov/2020028041

Contents at a Glance

Dear Reader,

I make a lot of predictions on a given day.

It begins with my alarm, when I reach in its general direction to turn it off. I make more predictions as I go about my morning routine: stepping around the bench in the hallway, wearing a seasonally-appropriate outfit, finding my cat in her usual morning hideaway, and so on.

These autopilot tasks don't seem like predictions in the moment, but each relies on established, daily patterns. For instance, because my cat always chooses to hide under the coffee table at seven in the morning (for reasons unknown), I can predict where to look for her. These patterns become so engrained that if something changes them—say, my cat decides to hide under the bed—it can cause quite a disruption; in this case, an extended search party.

Machine learning uses algorithms to find established patterns (a more reliable prediction method than depending on the whims of my cat). As a result, data analysis improves and organizations can make better decisions. But how do you implement machine learning in your SAP S/4HANA system? With this guide, you'll get the expertise you need to set up machine learning and apply it to your business needs.

What did you think about *Implementing Machine Learning with SAP S/4HANA*? Your comments and suggestions are the most useful tools to help us make our books the best they can be. Please feel free to contact me and share any praise or criticism you may have.

Thank you for purchasing a book from SAP PRESS!

Megan Fuerst
Editor, SAP PRESS

meganf@rheinwerk-publishing.com
www.sap-press.com
Rheinwerk Publishing · Boston, MA

Contents

Preface

SAP S/4HANA operates as a central system of record for many organizations, integrating data and business processes from sales, marketing, finance, supply chain, manufacturing, services, procurement, and human resources. SAP S/4HANA efficiently enables day-to-day operations, business decision-making, and tracking of enterprise-critical data. Enterprises are driven by data and insights. It's not just about having the right data at your fingertips, but having the insights from that data linked to the core business processes. This is where machine learning can help, as it unlocks knowledge from structured and unstructured data to enable intelligent enterprise resource planning (ERP). Machine learning makes sense of raw data and discovers hidden insights and relationships by learning from data, rather than programming explicit rules. Integrating machine learning capabilities across relevant business processes helps identify untapped opportunities, expose hidden risks, and automate repetitive tasks or knowledge-based work.

What Is the Impact of Machine Learning?

The foundation and drivers for machine learning are increased computer processing power, better algorithms, and the evolution of big data. Machine learning enables disruptive innovation in many domains and has a major impact on the enterprise software market. Today, most organizations have enterprise software that uses rule-based processing to automate tasks, enable compliance, and ensure data consistency. Rule-based automation has resulted in significant efficiency improvements over the years, but the software can't learn and improve with experience as humans do. Machine learning helps to close this gap.

Self-learning algorithms take enterprise software to a new level. Machine learning allows for much faster improvement of processes with less human intervention or expertise. The intelligent enterprise utilizes both machine learning and rule-based techniques to the best effect, freeing up humans to focus on high-value tasks. Integrating machine learning into business processes results in tremendous efficiency gains, driving innovation and enabling new business models. Market analysts agree that machine learning is a core driver for economic development. As of the time of writing (summer 2020), McKinsey & Company estimates that the global machine learning applications market will reach a value of $127 billion by 2025. PricewaterhouseCoopers predicts that machine learning could potentially contribute almost $16 trillion to the global economy and boost GDP growth by up to 26% in 2030. According to Gartner, by 2023 machine learning will automate 60% of manual tasks. Benefits will be gleaned in almost all industries and sectors.

However, the benefits go beyond cost savings. They also include improved forecasting, optimized operations, more personalized customer services, and enhanced user experiences. Examples of using machine learning in the context of ERP solutions include the following:

- For manufacturers, failure of high-tech equipment is a business-critical issue. By combining sensor data with business information in ERP systems and applying machine models, the health of these machines can be foreseen. Thus, maintenance scheduling and logistics planning for spare parts and repair crew management can be transformed into a proactive business process.
- Early detection of local market trends and offering exactly the products that the market demands is key to success. Machine learning algorithms can analyze ordering trends to recommend products to add to/eliminate from a company's offerings. They can help in answering questions about which variants are currently nonsellers and which ones will be future top sellers, segmented by specific markets.
- Validating that a product has been produced exactly to its specifications and configuration is an important step in final quality assurance and readiness checks for product shipments. Image-recognition algorithms support this via visual product quality checks. Thus, the accuracy and automation level of quality processes in production can be increased, which leads to fewer returns, higher customer satisfaction, and better profitability.
- High-quality master data is crucial as the basis for ERP business processes. Machine learning models can automate the identification and implementation of validation rules to safeguard data consistency. Machine learning-based autocompletion of attribute values while maintaining master data simplifies the interaction with end users and saves on costs thanks to reduction of manual activities.
- Intercompany reconciliation is a very time-consuming activity in the financial close. Machine learning algorithms help in automating and controlling the end-to-end, intragroup invoicing, payment, and settlement scenarios to improve and speed the local and group closes. Machine learning models support identifying and resolving mismatches in reconciliations, which results in accurate data for month-end corporate group reporting and enhanced transparency of intercompany management.

We conclude that early adopters of machine learning can disrupt entire industries and formulate completely new business models.

What Ethical Aspects Will Be Considered?

Although machine learning positively impacts the economy and our society, at the same time, it raises public concerns, such as negative effects on jobs or loss of human control in automated decision-making. In practice, however, most of these concerns are more fiction than fact. Machine learning is evolutionary rather than revolutionary, giving us time to prepare for the changes and challenges ahead. Nevertheless, these concerns must be taken seriously.

Certainly, machine learning impacts the labor market as additional tasks are automated. Employees are liberated from routine and repetitive activities, allowing them to shift to higher-value tasks that depend on capabilities unique to human beings, like creativity or emotionality. Thus, existing jobs are transformed through machine learning and new jobs are created for designing, operating, and using intelligent systems. Conversely, the future of work will be marked by human-machine collaboration.

Automated decision-making based on machine learning entails consumer protection issues in terms of neutrality and the objectiveness of algorithms. Good software design can mitigate this. The design of intelligent systems will have humans in control as a guiding principle. In a business environment, actions triggered in a system have a tangible outcome in the real world that impacts the goals and profits of the company. As the responsibility and accountability for these actions still lies with the human user, humans must always remain in control of the outcome. Intelligent systems will augment human capabilities so that users can better understand and control the software. Providing sufficient information about the underlying machine learning model and explaining the reasoning behind the results of algorithms should be an essential part of designing intelligent applications in order to build trust between humans and machines. Furthermore, machine learning development must respect ethical values and legal standards. The success of machine learning depends on its broad social acceptance. SAP defined guiding principles to steer the development and deployment of machine learning/artificial intelligence (AI) software to help the world run better and improve people's lives. The people at SAP are guided by the following principles:

- They are driven by values
- They design for people
- They enable business beyond bias
- They strive for transparency and integrity in all that they do
- They uphold quality and safety standards
- They place data protection and privacy at the company's core
- They engage with the wider societal challenges of AI

You can learn more about these values at *http://s-prs.co/v515800*.

In addition, there are further challenges (e.g., economic and social impacts, normative aspects) that require a much broader discourse across industries, disciplines, borders, and cultural, philosophical, and religious traditions.

What Is the Objective of This Book?

Improved processing power, better algorithms, and the availability of big data are facilitating the implementation of machine learning for infusing intelligence into back-office processes and to provide intelligent ERP. SAP S/4HANA's underlying in-memory database SAP HANA increases speed, combines analytical and transactional data, and brings innovation with embedded machine learning capabilities. Thus, machine learning is natively integrated into SAP S/4HANA and can be easily used across the entire organization to optimize business operations, improve employee job satisfaction, and create better customer services. With conversational AI, a natural language experience is provided by SAP S/4HANA that revisits the way users interact with the system by enabling hands-free applications. However, incorporating machine learning into ERP solutions and using it effectively is a challenging task, one that this book addresses.

The objectives of this book are as follows:

- Explain the machine learning and predictive analytics tools, services, and architecture of SAP S/4HANA
- Teach how to implement your own machine learning and predictive analytics applications based on the SAP S/4HANA programming model
- Describe the machine learning business applications that have been already delivered with SAP S/4HANA

We organized the structure of the book around those objectives. First, the tools and services for machine learning and predictive analytics are described. Second, the solution architecture for systematically implementing and integrating machine learning and predictive analytics into business processes is explained. This architecture ensures that intelligence is provided to the right person, in the right place, and at the right time. In context of ERP, fused machine learning functionality must be enterprise-ready. This covers qualities like compliance, security, scalability, robustness, extensibility, configurability, operations, supportability, globalization, and auditing, and this is the third area we cover. Fourth, the machine learning applications that have been already delivered with SAP S/4HANA are described from functional scope, business process, and configuration points of view.

We recommend reading this book sequentially from Chapter 1 onward. However, if you prefer, you can directly go to any chapter and start reading about that chapter's topic. If you're interested in learning about how to implement your own machine learning and predictive analytics applications, for example, you can start by reading Chapter 5 on the technical implementation without reading the previous chapters. If you want to explore the machine learning applications delivered with SAP S/4HANA, just jump to Chapter 6 on the business implementation.

Let's review what's covered in each chapter of this book:

- **Chapter 1: Introduction to Predictive Intelligence**
  This chapter introduces predictive intelligence as a component of the SAP intelligent enterprise. You will discover how predictive intelligence—the application of machine learning and predictive analytics—is evolving at SAP and get an overview of how intelligent technologies are being infused into SAP S/4HANA business processes.
- **Chapter 2: The Evolution of Predictive Analytics and Machine Learning at SAP**
  This chapter discusses how predictive analytics and machine learning were implemented in SAP landscapes before SAP S/4HANA, and it helps explain how those technologies, methodologies, and best practices were adapted for SAP S/4HANA systems.
- **Chapter 3: Tools, Technologies, and Services**
  This chapter teaches you about the tools, technologies, and services they can use to consume predictive analytics and machine learning in SAP S/4HANA. It also includes a brief introduction into topics like intelligent robotic process automation and the Internet of Things.
- **Chapter 4: Architecture**
  In this chapter, you'll learn about the technical challenges of both implementing machine learning and predictive analytics in SAP S/4HANA and embedding intelligence in SAP S/4HANA business processes. We'll cover the solution architecture of SAP S/4HANA with machine learning and predictive technologies.
- **Chapter 5: Technical Implementation**
  This chapter walks you through the technical implementation of machine learning and predictive analytics for SAP S/4HANA, which includes setting up the services and training models and managing machine learning and predictive analytics applications.
- **Chapter 6: Business Implementation**
  In this chapter, you'll learn how machine learning and predictive analytics—embedded, side-by-side, or a combination of the two—are applied to specific business processes and how these technologies solve business problems within various lines of business.
- **Chapter 7: Services on SAP Cloud Platform**
  This chapter explores the complete SAP Cloud Platform services at a high level. You'll learn about the full scope of services that relate to machine learning, including peripheral services like Internet of Things (IoT).
- **Chapter 8: The Road Ahead and Further Learning**
  This chapter explains how you can access support resources and where you can go if you want to learn more about machine learning and SAP S/4HANA. You'll also learn about the roadmap for machine learning in SAP S/4HANA.

What Is the Target Audience for This book?

This book is for everyone interested in applied machine learning and predictive analytics based on SAP S/4HANA. We kept the level of detail appropriate for all interested readers. However, the main target groups for this book are as follows:

- Architects who want to understand the solution approach for incorporating machine learning and predictive analytics into their software designs
- Developers who want to implement their own machine learning and predictive analytics applications based on the SAP S/4HANA platform
- Consultants who want to configure the delivered SAP S/4HANA business applications for machine learning and predictive analytics
- End users who want to consume SAP S/4HANA business applications based on machine learning and predictive analytics capabilities

This book doesn't include details on data science processes and data exploration techniques. However, it does help you build a solid foundation from which you could also deepen your knowledge in those areas.

Now, let's begin our journey with an overview of the predictive intelligence concept in Chapter 1 .

Chapter 1
Introduction to Predictive Intelligence

To get started with machine learning in SAP S/4HANA, first you must familiarize yourself with the foundations of predictive analytics, machine learning, and the intelligent enterprise, which we'll discuss in this chapter.

In a digital world where the consumer is in charge, companies need to adapt faster to changes and catch-up with their customers' needs. Value creation is the need of the hour in a smart, intelligent, and digital economy.

In this chapter, we'll unpack the predictive intelligence concept and how it fits into our digital world and SAP. In the first section, we'll briefly discuss the need for an intelligent enterprise and how it benefits the customer. We'll then discuss how predictive intelligence has developed within SAP technologies, the workings of an intelligent suite, and how analytics are infused within SAP business processes.

1.1 The Intelligent Enterprise

Algorithms have changed how we interact with the world in the last 10 to 15 years. As a consumer, you are constantly looking for better ways of doing transactions or communicating and interacting with the people and systems around you. Whether you want to take a flight, buy an item, eat at a restaurant, or drive to a place, you are constantly guided by intelligent applications to do a certain set of things. The human mind is evolving to take basic guidance on mundane tasks and focus on more intelligent and productive tasks.

In the same way, enterprises are trying to understand how to increase the top-line revenues and cut the bottom-line costs. This is where enterprises are evolving to do better.

You do not want to access various different systems to perform your daily tasks at an enterprise; instead, you want to perform all the tasks in a connected enterprise. This is where the concept of intelligent enterprise evolves. SAP's history of helping companies harness the power of technology for competitive advantage goes back more than 45 years. SAP has leveraged emerging technologies, from mainframes and client/server architectures, to the internet, cloud computing, and mobile, and now to the rise of artificial intelligence. With intelligent cloud ERP systems, SAP is helping companies of all sizes in every industry around the world manage the incredible complexity of today's global, digital marketplace.

Figure 1.1 articulates it best. Notice how SAP's technologies have transitioned through the journey of the intelligent enterprise.

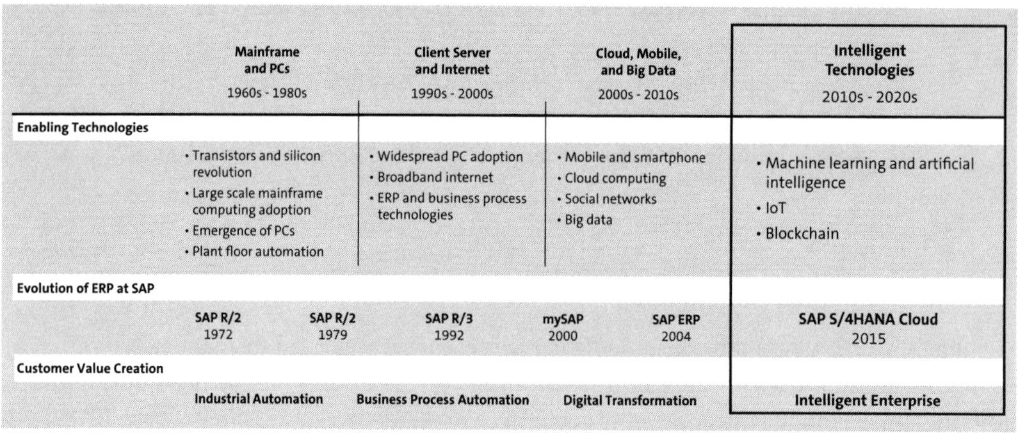

Figure 1.1 Evolution of Intelligent Enterprise

At each stage of the technology innovation, SAP has responded to customers' needs by developing appropriate enterprise solutions. During the mainframes and PCs era at the time of industry automation, SAP started the journey with SAP R/2 and then moved to SAP R/3 once client-server architectures took enterprises by storm. Later, with the invention of the internet, SAP offered the mySAP solutions for web-based architectures during the era of business process automation. Now, with the rise of mobile, big data, and cloud innovations, we have seen SAP entering into the digital transformation phase with SAP ERP. With technology disruption happening all over with the incubation of intelligent technologies such as machine learning and artificial intelligence, SAP is riding the intelligent enterprise wave with the inception of the flagship product SAP S/4HANA. A lot of technology disruption has just started, and SAP is at the center of it all while addressing customers' pain points with SAP S/4HANA.

Often customers come to SAP with the goal to stop buying point solutions. They want to manage their businesses in a holistic way so that different solutions can talk to each other, and they want to reduce the total cost of ownership. Customers typically want to get out of the IT business of building and maintaining software.

Managing key aspects of the business in an autopilot mode helps increase top-line revenue and reduce bottom-line costs. SAP achieves this through the intelligent enterprise in the following ways:

- Rationalizing foundational data by managing each person irrespective of what system he or she logs in from
- Transactional integration by doing an activity once and removing redundancy and duplicate entries
- Streamlined look and feel of the systems across the different solutions and offerings

- Commercial alignment by paying only once for a transaction even if it travels through multiple channels or systems

Let's now briefly look into SAP's intelligence strategy. Figure 1.2 gives a big-picture overview of SAP's intelligent enterprise strategy in the context of SAP's intelligent applications for every line of business (LoB) and how these different products and applications are connected to the digital core, SAP S/4HANA Cloud. Figure 1.2 also explains the various integration points, the user interface (UI) technologies, and so on, and how they all fit into the big picture.

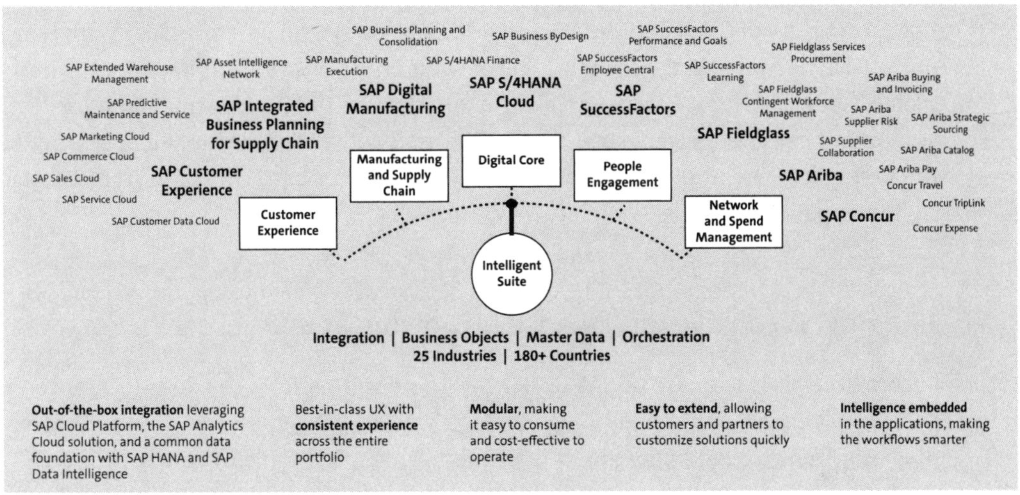

Figure 1.2 Intelligent Enterprise

As shown in Figure 1.2, SAP's intelligent enterprise strategy focuses on outcomes for customers, helping them remove repetitive tasks and move their efforts to higher-value activities. Giving better tools to SAP customers will enable them to engage better with their own customers and to develop their next-generation customer experience.

Some of the key methods include the following:

- Concentrating on integration. SAP can now bring together all LoB offerings on a standard platform using SAP Cloud Platform and next-generation data management, with the purpose of providing a seamless user experience from end to end in a united modular suite. Customers will no longer need to develop their own strategies to integrate our products into their environments.
- Building intelligence by leveraging intelligent technologies. We can use artificial intelligence (AI)/machine learning, robotic process automation (RPA), Internet of Things (IoT), and analytics delivered not just in an innovative model with design thinking, but also into our applications. Embedding these functionalities inside applications and business processes will radically simplify how our customers consume and apply innovations.

- Return on investment for the execution of a digital transformation strategy. Customers are investing heavily in their digital transformations—for example, with cloud offerings or big data solutions. So far, they have not been able to take full advantage of these investments and derive their real meaning. An intelligent enterprise can gain considerably more insights, improve predictive capabilities, and inject more intelligence into their business processes.

So you might ask, what is the difference between our digital transformation strategy and our intelligent enterprise strategy? The intelligent enterprise strategy is an evolution of our digital business transformation. An intelligent enterprise is able to automate a majority of the repetitive tasks in its day-to-day business processes by applying innovative technology like machine learning and AI so that its workforce can shift to higher-value activities. For example, after automating the invoice-matching process, accountants will not need to spend their time manually assigning incoming payments to outstanding invoices and instead can focus on other, better tasks to improve the customer payment terms.

1.2 How Predictive Intelligence Is Evolving at SAP

Even before the inception of SAP S/4HANA, there were efforts to add intelligence to SAP applications by leveraging AI technologies. SAP's journey started with leveraging the basic predictive analytics algorithms and running them side by side with SAP ERP. *Predictive analytics* is the analysis of historical information and existing external data to find patterns and make informed decisions about future events.

During the pre-SAP S/4HANA days, we leveraged predictive analytics technologies to augment intelligence in the SAP ERP applications. This started with enhancing or appending the rule-based analysis via additional parameters and variables that take into consideration the historical aspects of a situation. Because data plays a huge part in this, the algorithms have been designed to extract the data based on past events and to identify patterns to provide different outcomes or simulations of a future situation.

Let's now dive into the details of how AI capabilities are evolving in SAP applications, as shown in Figure 1.3. In each phase, SAP applications are adapting with AI technologies:

1. **Phase 1: Basic algorithms for predictions**
   Before the SAP S/4HANA days, basic predictive algorithms were used from the SAP HANA Predictive Analysis Library (PAL) in conjunction with the data in the SAP ERP applications to simulate predictions. We will look into the details of this in Chapter 2 with some examples.

   During this phase, other SAP visual tools and technologies, such as SAP BusinessObjects and SAP Lumira, have been used to display these predictions. The SAP HANA Automated Predictive Library (APL) has also been used to build more predictive models. Depending on the required functionality, PAL also has been used. In this context, most of the work was accomplished by data scientists to build these predictive models. Later, end users such as operational users and business users have used the resulting predictions with regular visual tools such as SAP Lumira. We have also leveraged open-source tools and scripts such as the R language to create additional flavors of predictions and analysis for some customer scenarios.

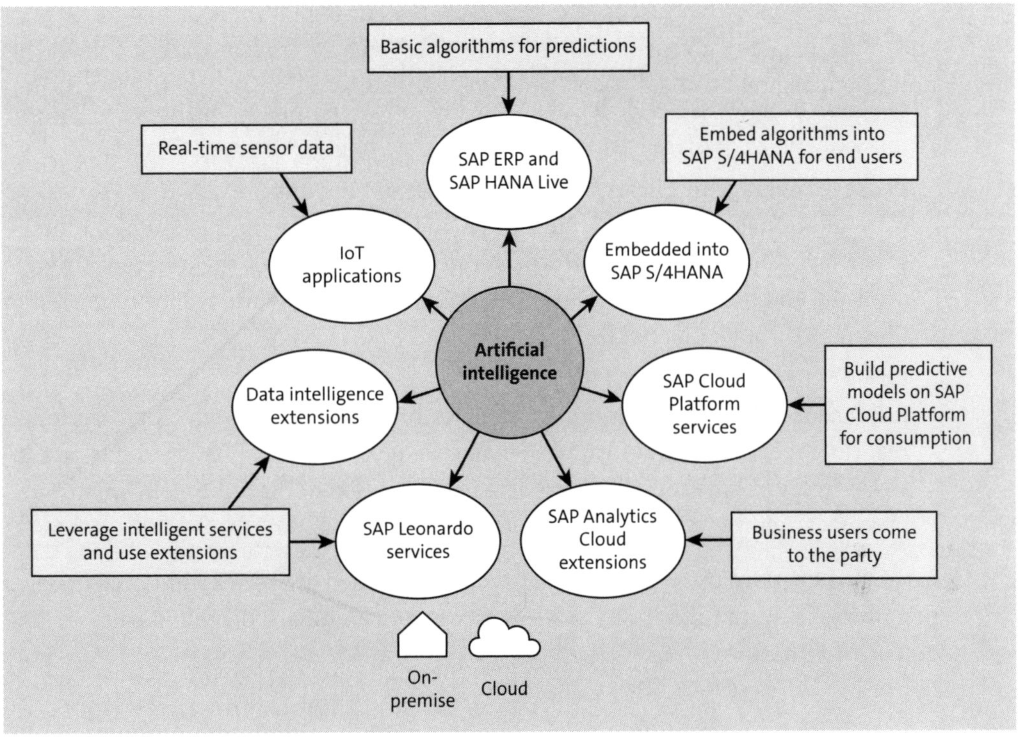

Figure 1.3 Evolution of AI at SAP

2. **Phase 2: Embedded algorithms in SAP S/4HANA for end users**
   After the inception of SAP S/4HANA, the basic algorithms leveraging both PAL and APL have been used to inject these predictive models into the SAP S/4HANA business processes. In this phase, some of the basic SAP S/4HANA business processes are incubated with these predictive technologies to perform better simulations and predictions with out-of-the-box models and the like, which can help customers realize more value for their money while using these processes. We will be discussing this at length in Chapter 3 through Chapter 6 on the architecture and will see some examples of how this is leveraged.
3. **Phase 3: Build predictive models on SAP Cloud Platform for consumption**
   Later, additional machine learning services were built on SAP Cloud Platform for performing deep learning functions, and these services are consumed by the SAP

S/4HANA business processes. In this phase, we extended some of the complicated customer requirements by building machine learning services that leverage deep learning techniques and are consumed by the SAP S/4HANA processes, helping the customer to realize more benefits. Detailed commentary and discussion with architecture and examples is given in Chapter 3 through Chapter 6.

4. **Phase 4: Business users come to the party**
   With the inception of SAP Analytics Cloud, the APL algorithms could be even leveraged by the business user and create quick simulations and predictions. In this phase, we extended the beauty of leveraging the predictive analytics models to the business users with a wide variety of ready-to-use templates. Again, we will give a quick overview of these basics and examples in Chapter 3 through Chapter 6.
5. **Phase 5: Leverage intelligent services and use extensions**
   Now additional intelligent services are being built leveraging the SAP Data Intelligence layer on SAP Cloud Platform. A brief overview about the different services available and the journey into the SAP Data Intelligence world will be discussed in Chapter 3 through Chapter 6.
6. **Phase 6: Real-time sensor data**
   IoT sensor data provides more meaningful predicted results, which gives a more real-time feel to the data insights. Finally, we give additional flavor to the data insights by leveraging the IoT sensor data in the context of the SAP S/4HANA business processes. Additional details are again provided in Chapter 3 through Chapter 6.

Except for the development in phase 1, which has ended, there are additional updates pertaining to all the other phases. This is where we embark on the journey of SAP S/4HANA with AI.

1.3 Connected End-to-End Scenarios

Now, let's connect the intelligent enterprise and AI technologies in the context of the end-to-end processes so that you can understand the complete story and can realize the true potential of connected enterprises in the context of an intelligent enterprise.

SAP was a suite company for its first 35 years of existence, but in the last few years it has been more LoB- and cloud-focused in terms of market share capture and has made a number of acquisitions and investments to build out the capabilities for these LoBs. In addition, SAP has been relentlessly moving to a cloud model, made huge strides leveraging in-memory computing with SAP HANA, and supported its customers with digital transformations through cloud, mobile, and big data. This gives customers a complete end-to-end picture of how the intelligent suite is built and its true potential can be leveraged.

In this section, we'll take a closer look at potential end-to-end scenarios.

1.3.1 Modular Approach

Many customers may not be ready to adapt or may feel overwhelmed by what they need to do to implement the whole intelligent enterprise framework, so there is no one single template to start with. Those just starting this journey can focus on immediate needs; the beauty of the intelligent suite is that it is modular, so customers don't need to change everything from day one but can start anywhere and get to the end state. It's important to drive long-term vision and value, but we can start in the most relevant areas for each customer to show them an immediate return on investment.

Some examples for this modular approach could also include the following:

- Dramatically changing the customer experience with SAP Customer Experience to create an end-to-end customer experience that includes fulfillment, supply chain, real-time inventory management, and next-generation field service.
- Reducing costs with total workforce management by bringing in SAP Ariba, SAP Fieldglass, and SAP Concur.
- Changing the business model with SAP S/4HANA: moving from being a product-based company to an outcome-based company with next-generation supply chain and manufacturing systems.
- Solving data problems by bringing in SAP HANA and SAP Data Intelligence to search for data outside the SAP data stores. By taking unstructured data into Hadoop, Spark, Amazon, or Google we can bring these insights into SAP HANA to deliver the next level of data management.
- Gaining access to intelligent tools by moving to SAP S/4HANA Cloud, which can also help boost the digital transformation. A much higher level of intelligence can be embedded in the cloud, data access is simplified because it is all in one place, and we can train and produce algorithms in the cloud.

The idea is to design an end-to-end process that helps solve a business problem rather than building cool bots that leverage leading-edge algorithms or skills. If the customers' pain points are not addressed, any number of solutions developed, however cool they might be, will end up sitting on a shelf.

1.3.2 Example Scenarios

Now that you understood the benefits of leveraging the intelligent suite and the modular approach, let's dive into some of the end-to-end scenarios.

Here, we'll discuss some of the intelligent scenarios available as part of the SAP S/4HANA business processes. The intelligent suite is built based on these business processes, though not all the processes have AI technologies. Like we discussed earlier, intelligence is the vision; to get there, you follow the roads of AI and machine learning while building an intelligent suite.

Figure 1.4 shows the 21 focused, end-to-end processes that define the intelligent ERP system or the intelligent suite. We'll see how they interact in the following sections.

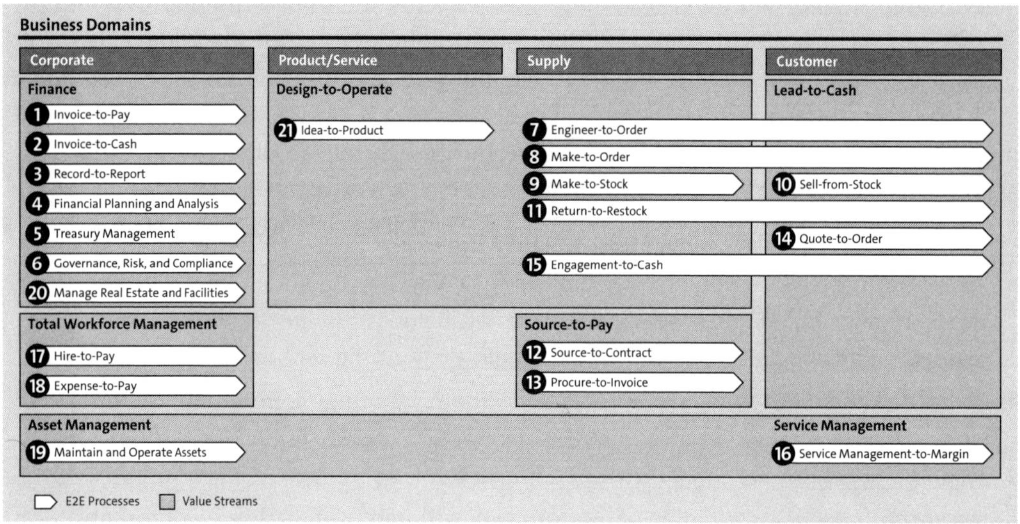

Figure 1.4 End-to-End Processes

Intelligent Procure-to-Invoice Scenario

As shown in Figure 1.5, let's now look at the different steps involved in this scenario and how the intelligent technologies have been applied in some of these steps:

1. Process purchase requisition
2. Assign sources
3. Process purchase order
4. Receive products and services
5. Evaluate supplier performance

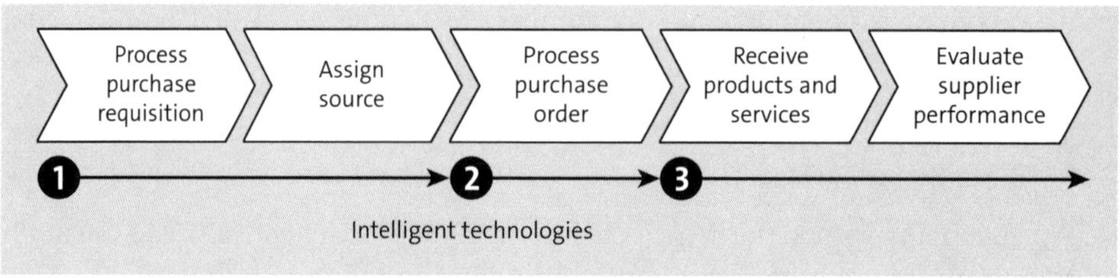

Figure 1.5 Steps in Intelligent Procure-to-Invoice Scenario

Table 1.1 shows different business outcomes and the corresponding intelligent technologies embedded for the three main steps (❶, ❷, and ❸ in Figure 1.5).

Business Outcomes	Outcomes Enabled by Intelligent Technologies
Process Purchase Requisition	
■ Accurate ordering ■ Streamlined activity	Improve efficiency: ■ Intelligent RPA automates creation of purchase requisition from Excel Capture savings: ■ Situation handling prevents lost savings by efficiently notifying of contract availability and assigning purchase requisition items ■ Machine learning proposes matching material group for required free text item to reduce wrong assignment to operational purchaser
Process Purchase Order	
■ Centralized workstream ■ On-time delivery	Improve efficiency: ■ Intelligent RPA automates down payment requests and email extraction of purchase order confirmations to eliminate manual updates ■ Machine learning streamlines approval processes, enabling accounts payable (AP) teams to focus on exceptions Manage risk: ■ Situation handling informs purchaser of missing order confirmation for upcoming scheduled deliveries to reconcile information and avoid delays ■ Machine learning proposes matching material group for required free text item to reduce wrong assignment to operational purchaser
Receive Products and Services	
■ Fast resolution of delays ■ Up-to-date accounts	Increase planning accuracy: ■ Machine learning predicts supplier delivery delays by evaluating deviations from plan to prompt appropriate action Minimize errors: ■ Situation handling prompts owner of an overdue purchase requisition to confirm receipt of goods to validate entry

Table 1.1 Business Outcomes in Procure-to-Invoice Scenario

Intelligent Source-to-Contract Scenario

As shown in Figure 1.6, let's now look at the different steps involved in this scenario and how intelligent technologies have been applied in some of these steps:

1. Analyze spend
2. Plan spend
3. Create request for information (RFI)
4. Analyze demand
5. Process RFX (request for information, request for proposal, request for quote, request for bid, etc.)
6. Negotiate legal contract
7. Manage legal contract
8. Manage contracts

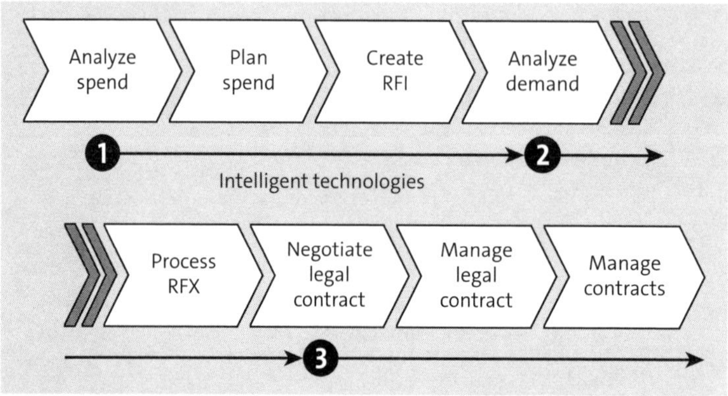

Figure 1.6 Steps Involved in Intelligent Source-to-Contract Scenario

Table 1.2 shows different business outcomes and the corresponding intelligent technologies embedded for the three main steps (❶, ❷, and ❸ in Figure 1.6).

Business Outcomes	Outcomes Enabled by Intelligent Technologies
Analyze Spend	
▪ Develop sourcing strategies	▪ Identify optimal sourcing opportunities via spend analysis ▪ Efficiently create requests for quotations (RFQs), leveraging spend history (transparency) ▪ Improve profitability by increasing percentage of on-contract spend
Analyze Demand	
▪ Select suppliers	▪ Improve supplier award decisions and collaboration ▪ Reduce cycle times for contract creation, negotiation, and approval ▪ Focus on knowledge tasks rather than repetitive rote tasks

Table 1.2 Business Outcomes in Source-to-Contract Scenario

Business Outcomes	Outcomes Enabled by Intelligent Technologies
Negotiate Legal Contract	
▪ Maintain contract	▪ Prevent contractual terms oversight ▪ Prompt purchasers to consistently renegotiate and secure better contract terms

Table 1.2 Business Outcomes in Source-to-Contract Scenario (Cont.)

Intelligent Quote-to-Order Scenario

As shown in Figure 1.7, let's now look at the different steps involved in this scenario and the intelligent technologies applied in some of these steps:

1. Receive request for proposal (RFP)/RFQ
2. Understand customer business and requirements
3. Develop solution and delivery
4. Configure product and services
5. Identify staffing requirements
6. Develop pricing and scheduling
7. Conduct profitability analysis
8. Manage legal content
9. Manage internal approvals
10. Negotiate quote with customer

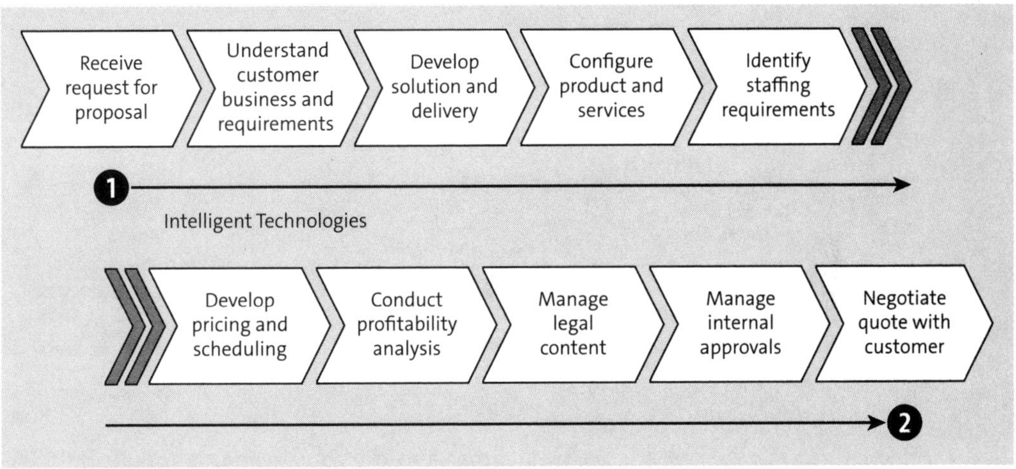

Figure 1.7 Intelligent Quote-to-Order Scenario

Table 1.3 shows different business outcomes and the corresponding intelligent technologies embedded for the two main steps (❶ and ❷ in Figure 1.7).

Business Outcomes	Outcomes Enabled by Intelligent Technologies
Receive RFP/RFQ	
▪ Streamlined process	Minimize errors: ▪ Intelligent RPA automates creation of sales inquiries
Negotiate Quote with Customer	
▪ Win rate ▪ Sales plan accuracy	Improve win rate: ▪ Machine learning predicts the probability that a quotation item will be converted into a sales order item, allowing sales managers to focus on high potential quotations Increase forecast accuracy: ▪ Machine learning predicts sales forecasts, enabling faster, more efficient decision-making to increase sales volume and optimize sales planning accuracy

Table 1.3 Business Outcomes in Quote-to-Order Scenario

Intelligent Sell-from-Stock Scenario

As shown in Figure 1.8, let's now look at the different steps involved in this scenario and the intelligent technologies applied in some of these steps:

1. Accept and validate sales orders
2. Determine availability
3. Enter orders into system
4. Process back orders and updates
5. Handle sales order inquiries
6. Invoice customer

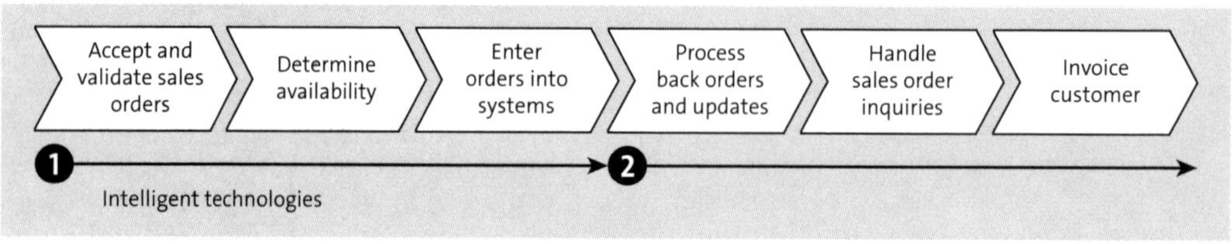

Figure 1.8 Intelligent Sell-from-Stock Scenario

Table 1.4 shows different business outcomes and the corresponding intelligent technologies embedded for the two main steps (❶ and ❷ in Figure 1.8).

Business Outcomes	Outcomes Enabled by Intelligent Technologies
Accept and Validate Sales Orders	
▪ Order management	Reduce manual effort: ▪ Intelligent RPA automates creation of down payments for sales orders with this requirement from customers ▪ Machine learning and situation handling work together to predict the conversion of sales quotes to sales orders and identifying sales orders with negative margin
Process Back Orders and Updates	
▪ Customer satisfaction ▪ Streamlined sales operations	Minimize delay: ▪ Machine learning predicts sales orders delivery delays to improve delivery performance and increase customer satisfaction

Table 1.4 Business Outcomes in Sell-from-Stock Scenario

Intelligent Invoice-to-Cash Scenario

As shown in Figure 1.9, let's now look at the different steps involved in this scenario and the intelligent technologies applied in some of these steps:

1. Evaluate manage credit
2. Invoice customer
3. Manage open receivables
4. Resolve disputes
5. Collect cash

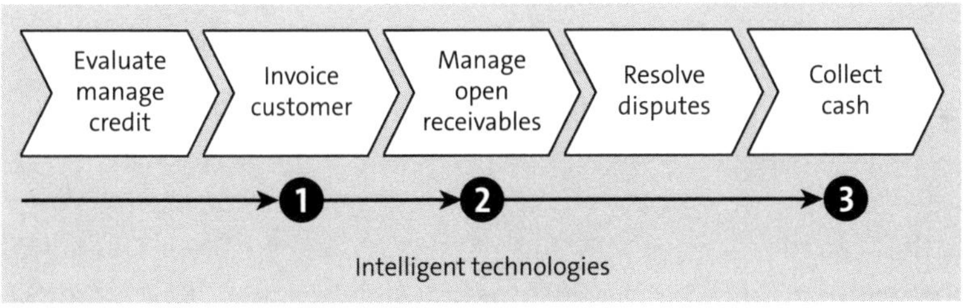

Figure 1.9 Intelligent Invoice-to-Cash Scenario

Table 1.5 shows different business outcomes and the corresponding intelligent technologies embedded for the three main steps (❶, ❷, and ❸ in Figure 1.9).

Business Outcomes	Outcomes Enabled by Intelligent Technologies
Invoice Customer	
▪ Accurate invoicing ▪ Effective relationship management	Improve customer service: ▪ Intelligent RPA reduces manual effort in payment plan generation by creating line items in billing plans of sales orders
Manage Open Receivables	
▪ Days sales outstanding (DSO) optimization ▪ Insights into doubtful accounts	Expedite dispute resolution: ▪ Intelligent RPA automates creation of emails to be sent to customers, monitors inboxes for replies, and attaches customers' replies to the content of a dispute for ease of reference
Collect Cash	
▪ Receivables matching ▪ Simplified remittance services	Minimize errors: ▪ Intelligent RPA automates payment advice file upload from email and triggers machine learning reconciliation of invoice payment information so that adjustments are matched and applied with accuracy ▪ Situation handling notifies if incoming payments do not cover invoices due so that accounts receivable (AR) accountants can follow up on payments for skipped invoices once payment advices are received Centralize workstream: ▪ Machine learning learns from past actions to capture and update customer behavior; incoming payments and open invoices are automatically matched and cleaned or proposed for review, minimizing manual efforts

Table 1.5 Business Outcomes in Invoice-to-Cash Scenario

These are some of the scenarios in which intelligent technologies are being leveraged. In Chapter 6, we'll discuss in detail how these use cases are realized in the SAP S/4HANA system.

1.4 Analytics of the Future

If data is fuel in a digital world, analytics is the real engine behind it. In leading organizations, executives agree on the following (Leaders 2020 Oxford Economics):

- 75% of decisions can be directly mapped to company strategies
- 78% of decisions are data-driven
- 62% of decisions are made in real time
- 62% of decision-making is distributed across the organization

SAP has come up with a best practices approach based on broad-based analytics solutions that span multiple platforms to help provide various data insights.

Figure 1.10 illustrates analytics in the context of SAP S/4HANA for the intelligent enterprise. It shows all the different components that the analytics portfolio comprises in the context of building an intelligent enterprise. You can see the three big pieces of the portfolio—SAP BW/4HANA, SAP S/4HANA, and SAP Analytics Cloud—and the corresponding tools and technologies required for these pieces.

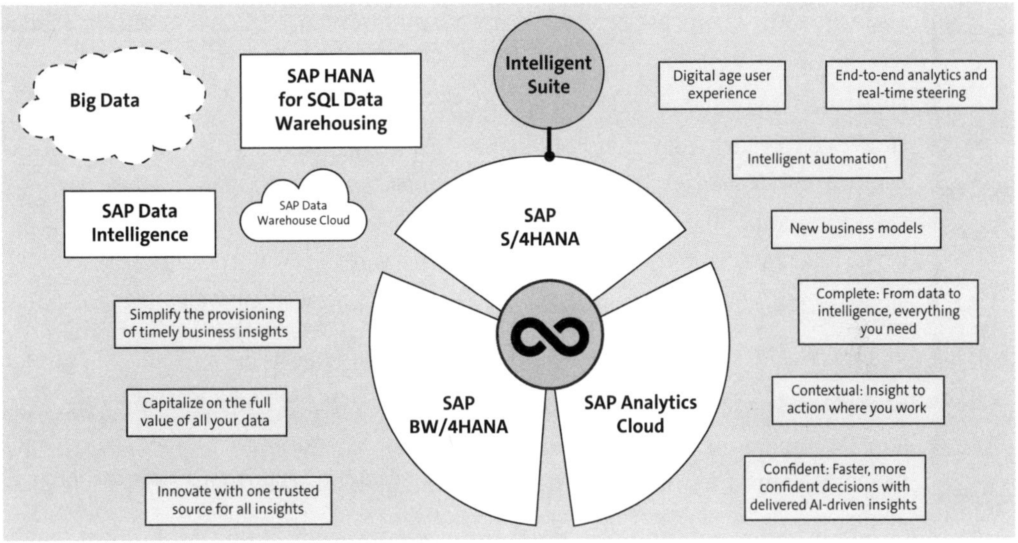

Figure 1.10 Analytics Portfolio in Context of SAP S/4HANA

Next, let's talk about unified analytics in the context of SAP, before discussing analytics-related business needs, trends, and user roles.

1.4.1 Unified Analytics

Enterprises are dealing with an unprecedented, exponential growth of data flowing from all sides of their organizations, as well as a multitude of outside sources. More than ever, companies need a modern solution to transform their growing mass of data into a truly strategic asset. Otherwise, they run the risk of missing out on tremendous insights when they make their next decisions.

To make informed decisions at the point of action, SAP S/4HANA provides embedded analytics to get real-time insights into all business operations. It includes machine learning and forecasting capabilities that help predict outcomes and simulate potential scenarios to determine the best course of action at the right moment.

To achieve even greater insights, many businesses are using the SAP Analytics Cloud solution as a single decision-making platform across their entire enterprise. The solution features best-in-class predictive analytics, planning, and business intelligence in a unified platform that is tightly integrated with SAP S/4HANA. Users can discover, visualize, plan, and predict in a single comprehensive tool, eliminating the need for inefficient, error-prone interfaces among various point solutions. SAP Analytics Cloud connects to many internal and external data sources, including enterprise data warehouses such as SAP BW/4HANA, to provide a 360-degree view of the business.

1.4.2 Business Problems and Business Needs

Let's consider the business problems that we are addressing here in the context of analytics:

- How to get real-time operational reports in SAP S/4HANA
- How to enhance productivity of business users by enabling them with insight-to-action scenarios within the context of a business process in SAP S/4HANA
- How to enable business users to better understand key performance indicators (KPIs) relevant for their business in an intuitive way
- How to enable business users to make better decisions for their businesses

Next, let's consider the business needs we are addressing here:

- Enable real-time operational reporting with SAP S/4HANA embedded analytics
- Rapidly integrate SAP S/4HANA with best-in-class analytical solutions provided by SAP and get up and running with prebuilt content in the SAP BusinessObjects BI platform
- Easily consolidate data from SAP S/4HANA core data services (CDS) views within SAP Business Warehouse (SAP BW)
- Extend SAP S/4HANA in SAP Analytics Cloud and build analytical use cases in SAP Cloud Platform
- Visualize and analyze SAP S/4HANA data in SAP Analytics Cloud

What's the solution? SAP Best Practices for analytics with SAP S/4HANA provides a comprehensive solution with guidance and prebuilt content so that customers can jump-start embedded analytics within SAP S/4HANA. Customers can build their analytical use cases by integrating with SAP BusinessObjects BI, SAP BW, SAP Analytics Cloud, and more.

SAP Best Practices for analytics with SAP S/4HANA is basically a collection of best practices analytics content and integration solutions. The content allows SAP S/4HANA customers to take advantage of the true power of SAP HANA.

1.4.3 Trends and Technologies

Users are empowered to uncover trends and opportunities with self-service analytics and reporting. SAP S/4HANA comes with advanced data modeling, powerful business simulation, and what-if analysis capabilities.

SAP Analytics Cloud and SAP S/4HANA work in perfect harmony with tight integration at all levels—including, but not limited to, the following:

- **Live data connectivity**
  Data replication from SAP S/4HANA to SAP Analytics Cloud is no longer required, so the highest level of data consistency and security can be achieved. Transactional data can be accessed from SAP Analytics Cloud in real time for live, up-to-date reports. In addition, information never leaves the corporate firewall, and IT always maintains the governance and security of the data.
- **Seamless navigation**
  A deep contextual link between solutions redefines user insights and efficiency. For example, stories generated in SAP Analytics Cloud can be opened directly from the SAP Fiori launchpad. In addition, users can directly jump from their analytics view in SAP Analytics Cloud to the associated transaction in SAP S/4HANA.
- **Industry and line-of-business content**
  With prebuilt industry and line-of business content using data from SAP S/4HANA and other applications, business users are empowered to produce real-time reports and technology adoption is accelerated.
- **Improved security**
  SAP Analytics Cloud uses existing security and authorization settings from SAP S/4HANA, without requiring backend modification.
- **Transform boardroom meetings**
  As an add-on to SAP Analytics Cloud, SAP Digital Boardroom leverages SAP S/4HANA to turn boardroom meetings into interactive discussions. Rendering enterprise information in a compelling board-level visualization gives executives a complete, real-time view of their company's current situation and KPIs across all departments.

1.4.4 User Roles with Analytics

Let's end this section by briefly discussing the various user roles that could leverage analytics in the context of SAP S/4HANA and SAP Analytics Cloud.

We won't go into the details of these various user roles; there are other books and blogs and SAP Help documentation that discusses them at length. We bring the concept up here briefly in the context of how analytics is realized for SAP S/4HANA and SAP Analytics Cloud offerings.

As you can see in Figure 1.11, there are broadly three kinds of user roles:

- *End user roles*, for realizing and visualizing the different kinds of reports and dashboards to analyze the data
- *Key user roles*, which enable end users and help visualize and interpret data for a more defined meaning, which could also help integrate it with other external sources
- *IT user roles*, to provide the raw data used in developing and designing many of the views for different analysis

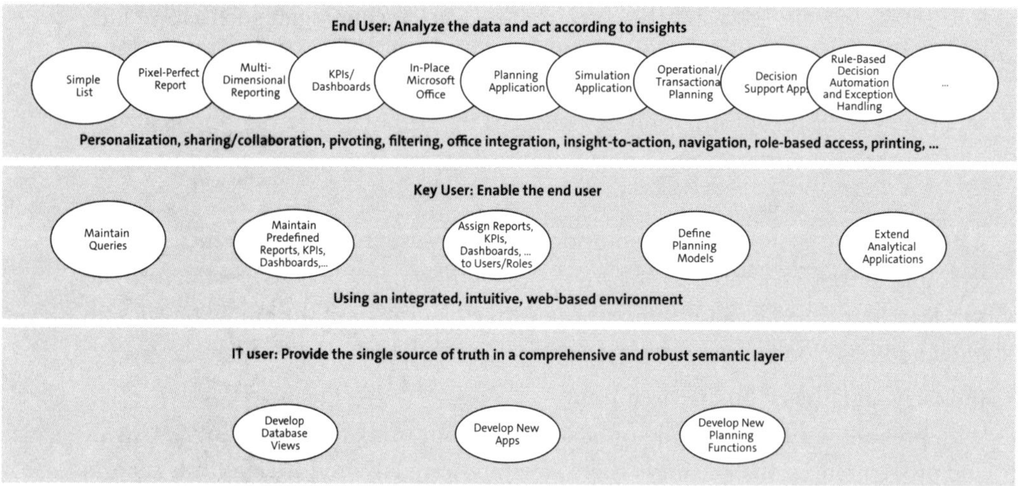

Figure 1.11 User Roles to Access SAP S/4HANA Apps

Let's now take a quick look at the different user types to complete our overview of how the different users of SAP S/4HANA can connect, extract, and visualize data. Figure 1.12 shows the different user types and analytics in the context of SAP S/4HANA and SAP Analytics Cloud.

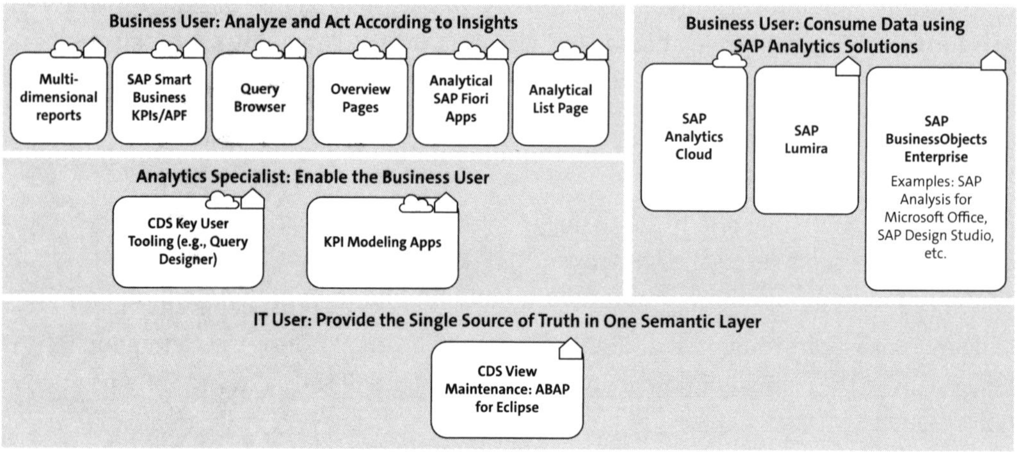

Figure 1.12 Different User Types Leveraging SAP S/4HANA

1.5 Summary

In this chapter, we began with a big picture definition of SAP's vision of the intelligent enterprise, coming from industrial automation, business process automation, and digital transformation and finally culminating in the intelligent enterprise and the evolution of SAP S/4HANA.

In the next section, we briefly discussed the journey of SAP in the context of leveraging the different artificial intelligence technologies, starting with predictive analytics, machine learning, and IoT, with the corresponding SAP technologies such as SAP ERP, SAP S/4HANA, SAP Cloud Platform, and SAP Analytics Cloud.

Then we discussed the evolution of the intelligent suite, reviewing the connected end-to-end scenarios and connecting the intelligent enterprise and the AI technologies.

Finally, we ended the chapter by looking at the various analytics tools and technologies that could be leveraged to realize the vision of the intelligent enterprise.

In the next chapter, we'll briefly discuss the evolution of AI technologies at SAP and how they have been applied before SAP S/4HANA, in the context of SAP ERP. This will illustrate SAP's strong background and focus on developing business processes in the context of innovative technologies.

Chapter 2
The Evolution of Predictive Analytics and Machine Learning at SAP

Before we dive into machine learning capabilities with SAP S/4HANA, it's important to review the underlying technologies, methodologies, and best practices that have evolved to support machine learning and predictive analytics at SAP.

In this chapter, we'll discuss the machine learning/predictive analytics evolution at SAP prior to SAP S/4HANA. We'll begin with a look at how SAP landscapes utilized predictive analytics and machine learning in the SAP ERP era. We'll then discuss key technologies that were adopted to build machine learning models. We'll end this chapter with a pathway to incorporating these technologies into SAP S/4HANA landscapes and systems.

2.1 Predictive Analytics and Machine Learning before SAP S/4HANA

American computer scientist Alan Kay said, "The best way to predict the future is to invent it." Quite a powerful quote, isn't it? Many businesses today also have a hard time inventing the future, even with large R&D budgets. This provokes a very interesting thought that's worth exploring. SAP asked the key question, is the opposite true as well? The journey into the AI space never remained the same.

In the early 2010s, SAP ERP was the single leading ERP application suite, with SAP HANA positioned as the real-time database engine for the enterprise customer space. SAP as a company has come a long way, and newer technologies like predictive analytics now add to SAP's HANA real-time database. This marked a new beginning, positioning SAP in a unique way with the combined strength of industry and the line of business (LoB) solutions, home-grown in-memory technology, and a robust predictive engine. This perfect marriage between SAP HANA, the real-time database, and the robust predictive engine leveraging predictive analytics technologies helped answer many aspects of handling the volume, velocity, variety, and veracity of data. Customers have huge sets of enterprise data from core SAP or legacy systems that need to be handled judiciously: time and data are what's really important for a customer.

This raises an interesting question: Can companies reinvent their businesses by analyzing their data to predict trends and find patterns? What if businesses have the ability to look into a crystal ball to explore what's in store for their future? SAP has already made this possible with some of the ERP business scenarios that were developed. Now, let's look into the past at how these scenarios evolved, what insights were gained from those experiences, and how these principles are applied with SAP S/4HANA business processes as well.

Figure 2.1 explains this in a nutshell, showing how companies have moved from a "sense and respond" model to a more engaging "predict and act" model. These companies have transitioned from a reactive approach to a more predictive approach, with intelligence being the key ingredient.

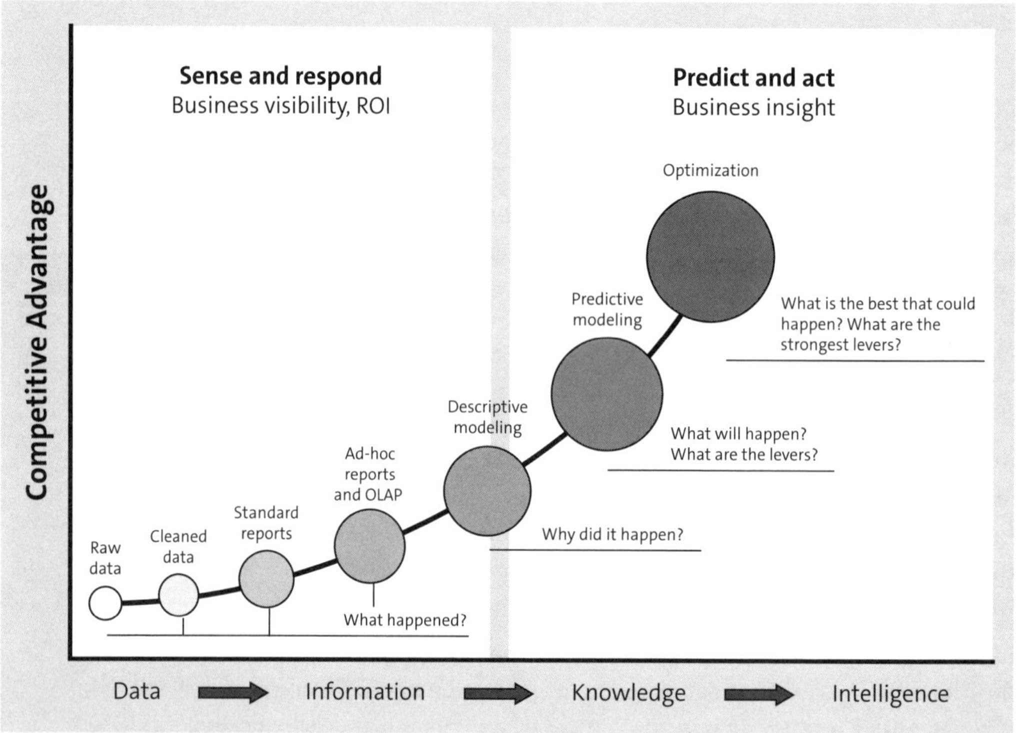

Figure 2.1 Predictive Analytics History and Competitive Advantage

2.2 Technologies and Methodologies

In this section, we'll focus on the tools, technologies, and methodologies that can build the predictive models that help customers. We'll look at the field of predictive analytics in detail, which encompasses techniques such as modeling, machine learning, and data mining to analyze current and historical facts to make predictions about future or otherwise unknown events. Using predictive models, businesses can discover patterns in data that allow them to identify risks and opportunities.

In this section, we'll briefly look at the concepts behind the following tools, which were leveraged in building customer use cases:

- Automated analytics
  - Classification model
  - Regression model
  - Clustering model
- Expert analytics

2.2.1 Automated Analytics

Automated analytics revolutionizes the way companies use predictive analytics to make better decisions on petabytes of big data. The unprecedented solution approach allows line of business users to solve common predictive problems without the need for highly skilled data scientists.

Some automated techniques involve building good models based on Russian mathematician Vladimir Vapnik's principles and methodologies. These principles and methodologies are called *structural risk minimization* (SRM), which leverages the Vapnik-Chervonekis theory, related to the statistical learning theory.

Figure 2.2 gives an idea of how this is conceptualized. The automated modeling process involves dealing with estimation techniques and the validation techniques that are applied, as well as how training is performed to build these models and arrive at predictions.

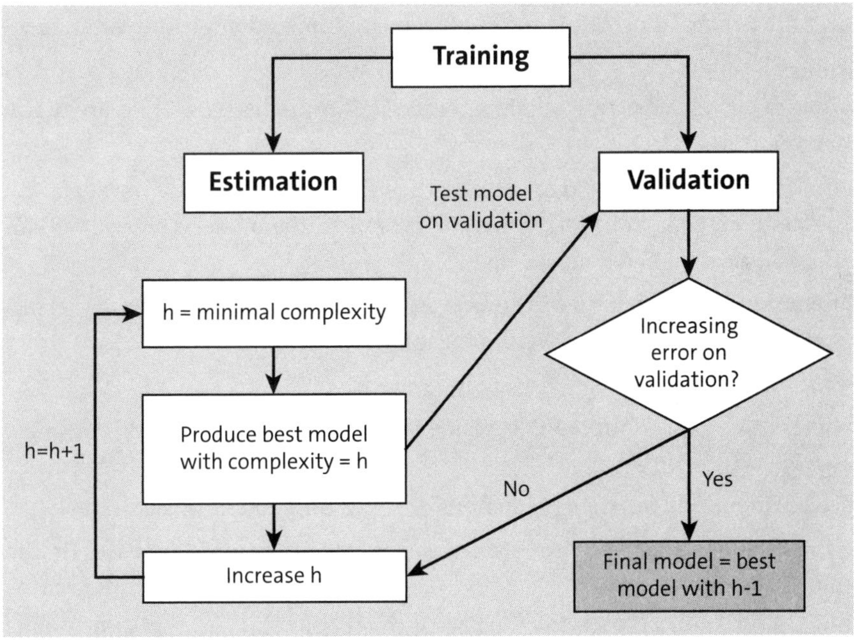

Figure 2.2 Automated Modeling Techniques

Data mining allows us to find patterns in the data with the following qualities:

- *Valid:* On new data with some degree of certainty
- *Novel:* Not previously known, and surprising in some way
- *Useful:* Answering a clear business question
- *Understandable:* Can lead to human insight

Here are some of the typical questions data mining should answer:

- *Classification:* Who will buy/fraud/churn next week/month/year?
- *Prediction:* What will the revenue/number of churners be next week/month?
- *Segmentation:* What are the groups of customers with similar behaviors/profiles?
- *Social network analysis:* How do you analyze interactions to identify events/communities/influencers?
- *Association rules:* How do you analyze transactions to identify events likely to occur together?
- *Forecasting:* What will the revenue/number of churners be for the next year on a monthly basis?

We would like to explore the Vapnik-Chervonenkis theory of statistical learning, the VC dimension, a little more as it's implemented in automated modeling techniques. The *VC dimension* is the capacity of a class of functions, or the largest number of observations that it can "shatter" (separate). Hence it can be defined as a measure of the complexity of the class of the functions.

Let's go over the four basic principles utilized by the automated modeling techniques:

1. **Consistency**
   Consistent results need to be obtained while doing predictions. Here are a few approaches:
   - If *n* (number of records) is large enough, then the error margin for new data will be the same as the learning error margin. Therefore, the model is robust and can be trusted when applied to new data.
   - Vapnik's theory says that this is the case provided you search for your model in a class of functions with a finite VC dimension:
     - Most classes have finite VC (e.g., polynomials).
     - Some classes have infinite VC (e.g., sine functions).
2. **Convergence speed**
   The speed of identifying correct predictions depends on a couple of factors:
   - The ability to generalize increases when the number of records increases (in the training dataset).
   - Automated analytics are sensitive to the size of learning dataset. Models will be robust if the number of records is high enough.

3. **Generalization capacity control**
   Here are a few hints for how to attain a model that can be utilized across various datasets:
   - If *n/h* (*n* = number of records; *h* = VC dimension) is large, the model generalizes well (R_{emp} (empirical risk – average cost on learning set) is minimized).
   - If *n/h* is small, the model doesn't fit.
   - For a given *n*, *h* must be large enough, but not too large.
   - There is an optimal balance between training and generalization errors.
4. **Strategy to obtain good algorithms**
   It is imperative to identify the correct algorithms to produce better predictions. Here are a few recommendations:
   - We have seen that the best compromise between accuracy (R_{emp} small) and robustness (generalization error (R_{gen}) small) is for that complexity h where R_{gen} is minimum.
   - SRM recommends progressively increasing h and building a succession of models until the minimum of R_{gen} is found.

Based on these principles, let's look at some examples of the various algorithms (e.g., classification, regression, and clustering) that are used with the automated modeling techniques.

Figure 2.3 shows an example based on classification modeling techniques. *Classification* is a data mining technique used to fit an equation to a dataset that can perfectly split a population between one value and another.

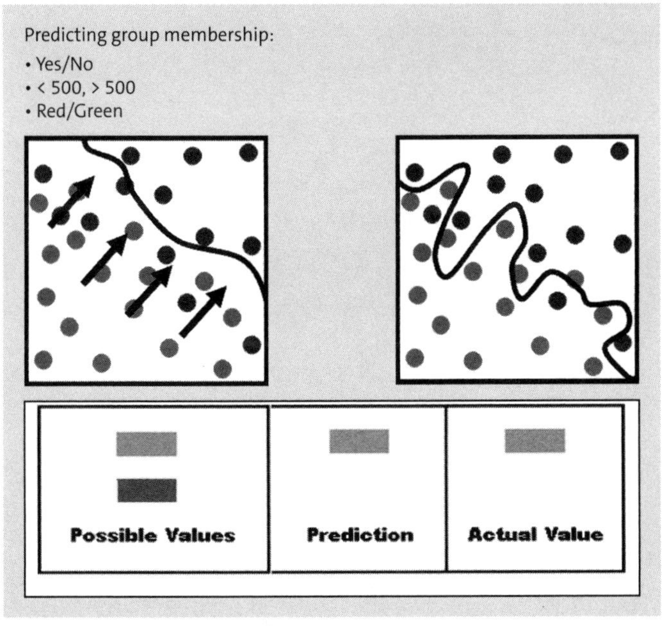

Figure 2.3 Classification Model Techniques

Figure 2.4 shows an example based on regression modeling techniques. *Regression* is a data mining technique used to fit an equation to a dataset that helps to predict the value of one variable or entity based on a few other factors.

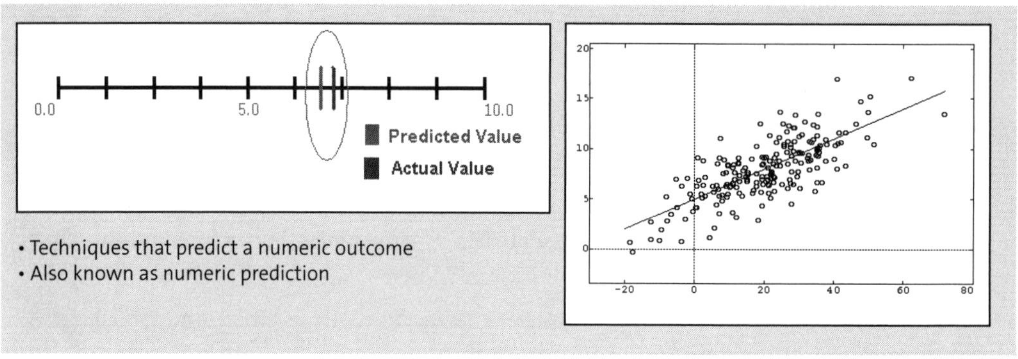

Figure 2.4 Regression Model Techniques

Figure 2.5 shows an example based on clustering modeling techniques. *Clustering* is a technique that helps to group data with similar distinctive behavior into clusters.

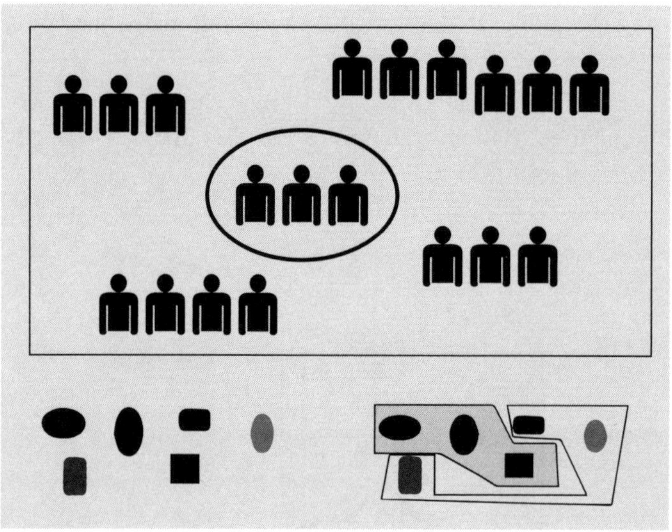

Figure 2.5 Clustering Model Techniques

2.2.2 Expert Analytics

Analytics techniques involve acquiring a dataset, then using the parameter settings to leverage additional transformational features, such as merge, append, and filter, to create and set new parameters that can be used in expert modeling.

Expert analytics provides flexibility to the user so that the best predictive model can be chosen while doing a visual comparative analysis. Most often, this function will be leveraged by the data scientist in picking a hybrid model to make the best use of the data at hand.

Additional algorithms leveraging the R integration feature can be utilized as well for the best results. Initially, four different classification algorithms were considered in the expert modeling flow:

1. **Neural network**
   This is a predictive analytics technique that builds the hidden network layer between the input and output nodes while making predictions.
2. **Auto classification**
   This includes a predictive analytics technique that is used for binary and categorical classification based on the target variable selected.
3. **Decision tree**
   This is an important predictive analytics technique used for making decisions following a particular flow of questions and answers.
4. **R-based logistic regression**
   This is a powerful technique that not only explores datasets but also utilizes the correct regression models to evaluate the results for accuracy.

We discussed some of these algorithms earlier, but there are many more flavors of these algorithms that have been leveraged to attain better results.

The following are the various phases involved in building a predictive model in the expert analytics modeling flow:

1. Acquire data
2. Enrich data
3. Predict insights
4. Build visualizations
5. Explore data
6. Share with others

2.3 Best Practices

Now that we've discussed the different tools, technologies, and methodologies leveraged, let's briefly touch on the best practices that were built around this framework. SAP has continued their predictive analytics and machine learning legacy, building similar best practices with SAP S/4HANA as well.

With the advent of SAP HANA, data movement between SAP ERP and SAP HANA must be addressed, and customized data models have been built to handle a multitude of datasets for different purposes. Many dynamic data models have also been built to feed this data appropriately to the predictive engine in SAP HANA to simulate the various ways of doing predictive analysis. To make this clear, SAP introduced the concept of packaged solutions. This is not a new approach, though it's engineered to provide customers with a proven way of handling various real-time scenarios. This is how SAP ventured into the AI space with the advent of the first set of tools leveraging real-time database access with SAP HANA and the robust predictive engine simulating predictions.

These packaged solutions for predictive analytics helped build complete end-to-end customer stories with quick templates that are ready to use for solving business challenges and providing predictions.

Figure 2.6 outlines one of those packaged solutions that helped provide customer solutions for about 30 different scenarios. We'll briefly look into some of these scenarios to better understand how these earlier technologies were used to enhance SAP ERP business processes and provide customer predictions. This helped to build a foundation to continue the AI journey after the inception of SAP S/4HANA.

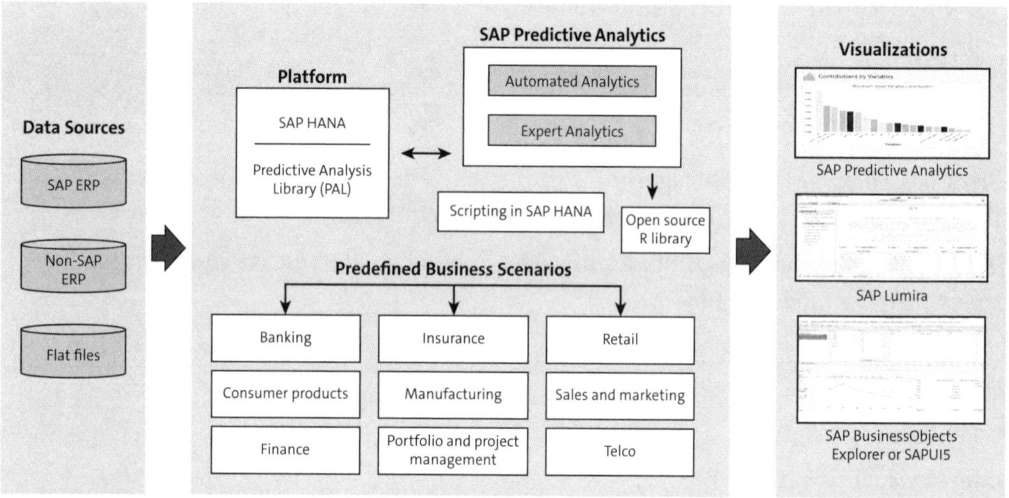

Figure 2.6 Predictive Analytics Rapid-Deployment Solution (RDS) Package Overview

As you can see, there about 30 different customer scenarios released across eight different domains (about five industries and three LoBs). Notice how they utilize different predictive algorithms in the context of SAP HANA, as well as SAP/non-SAP data sources to provide customers with different levels of basic predictions and simulations needed for some core SAP ERP processes.

Some of the algorithms used are as follows:

- K-means clustering
- Regression or classification
- Time series analysis
- Association a priori

- Decision tree analysis
- Outlier detection
- Neural network analysis
- Social network analysis
- Open-source R programming
- Hybrid models using SAP HANA scripting (combining more than one algorithm to arrive at predictions and results)

Now that many of these customer scenarios are built leveraging different tools and modeling techniques, SAP released SAP HANA Enterprise Cloud, which was hosted on Amazon Web Services (AWS). This helped customers quickly test different business scenarios, not only with the out-of-the-box SAP data but also with uploaded customer data.

Figure 2.7 explains how this was provided to benefit customers on a large scale, showcasing SAP ERP business processes leveraging the SAP HANA real-time database with the prebuilt predictive analytics use cases in an AWS-hosted environment. The SAP Predictive Analytics desktop client, SAP HANA studio, and the SAP HANA platform edition are leveraged and hosted on AWS to create the concept of a cloud image. SAP HANA Enterprise Cloud provides a very intuitive way for customers to try out the different example scenarios.

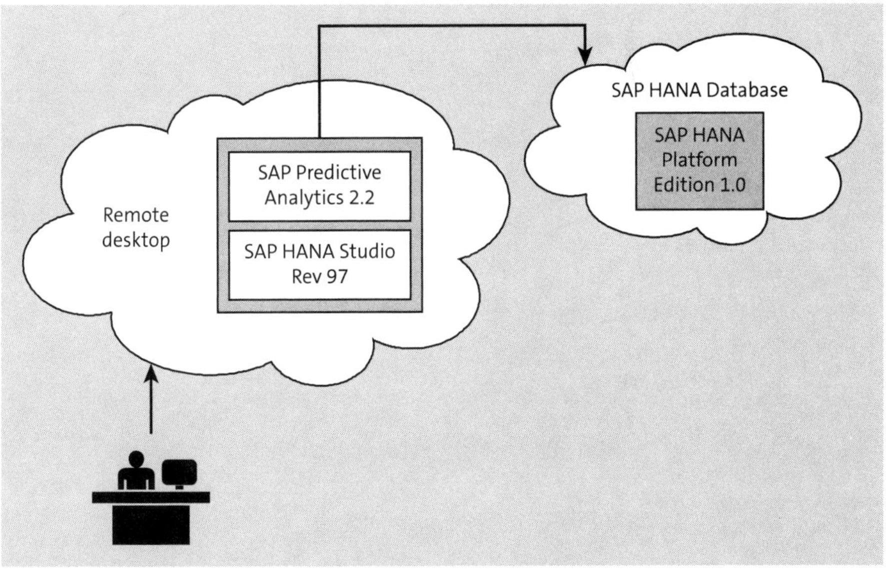

Figure 2.7 Architecture behind Hosting SAP HANA Enterprise Cloud

SAP is now taking the same concepts and building customer-centric use cases with SAP S/4HANA Cloud and SAP S/4HANA. A detailed walkthrough of all the available use cases can be found in Chapter 6.

2.4 Summary

In this chapter, we focused mainly on the background and history of SAP's journey into the AI world. We have taken you on a brief tour of the tools, technologies, and methodologies that were involved in building predictive intelligence capabilities into SAP ERP processes before SAP S/4HANA.

We briefly discussed the generic processes and approaches; most of these concepts continued to be implemented while building intelligence into SAP S/4HANA. This incremental approach not only grew the AI technology space but also built on the learning experiences gained from earlier ideas and datasets.

In the next chapter, we'll discuss the basic tools, technologies, and services that are leveraged in embedding predictive analytics and consuming machine learning services across SAP S/4HANA business processes.

Chapter 3
Tools, Technologies, and Services

In this chapter, you'll learn about the tools, technologies, and services you can use to realize and unlock the value of machine learning and predictive analytics with SAP S/4HANA.

We've discussed the predictive intelligence fundamentals and tracked how machine learning has evolved in SAP landscapes. Now it's time to explore current machine learning and predictive analytics with SAP—in particular, the tools, technologies, and services that you can use to consume them.

First, we'll establish the various ways you can implement machine learning and predictive analytics with SAP S/4HANA. Then, we'll walk through embedded machine learning technologies, SAP Cloud Platform, SAP Analytics Cloud, SAP Intelligent Robotic Process Automation (SAP Intelligent RPA), and SAP Internet of Things.

3.1 Machine Learning and Predictive Analytics Approaches

Let's first discuss the different approaches available with the current tools and technologies to realize machine learning and predictive analytics with SAP S/4HANA.

In this context, let's first briefly look into three different ways of using machine learning and predictive analytics:

1. Embedding machine learning into the SAP S/4HANA business processes
2. Consuming machine learning services built on SAP Cloud Platform with SAP S/4HANA business processes
3. Extending SAP S/4HANA business processes with predictions made via SAP Analytics Cloud

Note

We will be mainly focusing on the different approaches already available and those that are in the planning phase. We won't be diving into the architecture or technical details here, which are discussed in Chapter 4 and Chapter 5, respectively.

There are several different approaches available for the process of making business processes much smarter:

- **Building into the digital core, SAP S/4HANA**
  Embedding machine learning/predictive algorithms in SAP S/4HANA using the following technologies:
  - Using the SAP HANA Automated Predictive Library (APL) with the intelligent scenario lifecycle management (ISLM) framework
  - Using the SAP HANA Predictive Analysis Library (PAL) with the ISLM framework
- **Expanding around the digital core, SAP S/4HANA**
  Consuming predictive and machine learning services from SAP Cloud Platform using the following technologies:
  - Building hybrid models on SAP Data Intelligence
  - Consuming predictive services with Internet of Things (IoT) on SAP Cloud Platform (not yet available as of the time of writing, summer 2020)
- **Extending the digital core, SAP S/4HANA**
  Leveraging predictive analytics with SAP Analytics Cloud using the following technologies:
  - Using the smart assist, smart discovery, and smart predict features of SAP Analytics Cloud
  - Leveraging the ISLM framework with the smart predict features of SAP Analytics Cloud and integrating it into SAP S/4HANA
- **Extensibility of the models**
  Enhancing and extending the predictive and machine learning services using the following technologies:
  - Extending and modifying the embedded machine learning models (APL and PAL with the ISLM framework), available currently only in on-premise SAP S/4HANA
  - Extending and modifying the machine learning models on SAP Data Intelligence (not yet available as of the time of writing, summer 2020)
  - Leveraging the smart predict features of SAP Analytics Cloud to modify the embedded machine learning models in SAP S/4HANA (not yet available as of the time of writing, summer 2020)

Figure 3.1 illustrates the decision-making process for choosing predictive analytics and machine learning technology and the approach that will be leveraged accordingly. As you can see, the choices are to be made based on the complexity of the requirements, availability of the data, data volume, and best match with the algorithms needed in the specific scenarios. Figure 3.1 only gives an overview for analyzing the different approaches and choosing accordingly; this process is by no means set in stone and is subject to change with newer updates to the technologies and processes. We provided this decision tree only to provide some considerations for choosing an approach.

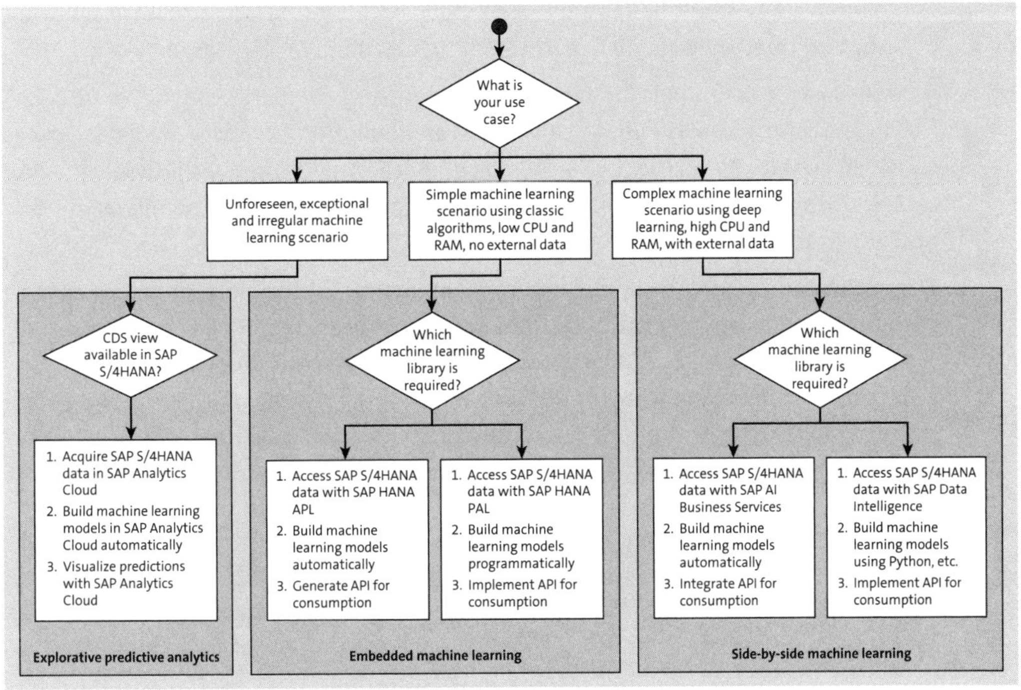

Figure 3.1 Decision Tree for Predictive and Machine Learning Approaches

Now let's discuss how to embed machine learning into SAP S/4HANA processes, consume machine learning services from SAP Cloud Platform, or extend the machine learning services that leverage SAP Analytics Cloud.

In the next section, we'll focus on the first approach: building into the digital core, SAP S/4HANA, by embedding predictive analytics algorithms.

3.2 Embedded Machine Learning and Predictive Analytics

In this section, we will mainly focus on how tools, technologies, and services are leveraged in the context of embedding predictive analytics into SAP S/4HANA business processes. This is where we discuss how predictive analytics and machine learning are built into the digital core of SAP S/4HANA.

3.2.1 Overview

At the core of the intelligent enterprise is *machine learning*, which enables computers to learn from large amounts of data without being explicitly programmed. One of the key enablers of this digital revolution is the ISLM framework, which was previously referred to as the Predictive Analytics integrator (PAi) framework. With ISLM, SAP S/4HANA is able to create and ship predictive use cases to customers that are tailored to their business needs.

Customers can manage and retrain these models within SAP S/4HANA, as well as adapt existing and create new predictive use cases to publish to the ISLM framework.

Now let's explore embedding machine learning and predictive analytics into SAP S/4HANA. This process is based on two main architectures: (1) using core data services (CDS) views and (2) making use of the machine learning techniques provided by SAP HANA. Chapter 4, Section 4.3 explains the concepts of embedding predictive models into SAP S/4HANA at a very detailed level.

The algorithms for embedding predictive analytics can be performance-intensive because high volumes of application data must be processed. For performance optimization, the algorithms should be processed closer to the data. SAP HANA provides the PAL and APL application function libraries to offer statistical and data mining algorithms, and there are additional algorithms that can be implemented as well. These algorithms require application data as input for model training. The trained models are then exposed to business processes by wrapping them up with CDS views. Figure 3.2 details how SAP S/4HANA business processes are embedded with a predictive model.

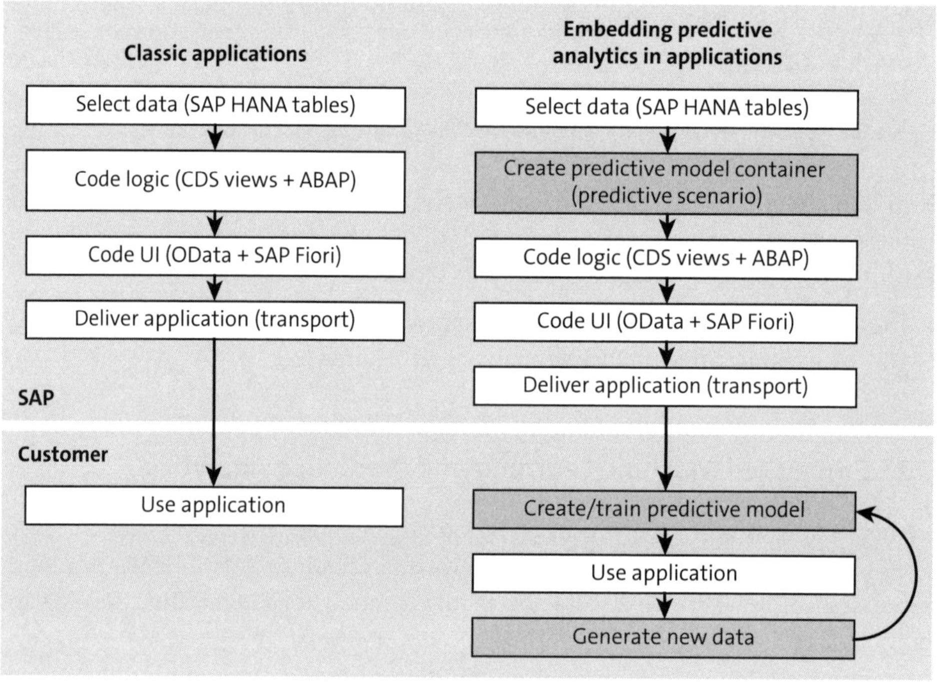

Figure 3.2 Predictive Analytics in SAP S/4HANA Processes

As shown in Figure 3.2, these CDS views can be combined with other virtual data model (VDM) CDS views and can then be exposed to consumers. By consuming these predictive models in CDS views, existing content and concepts are reused, which results in a simple but very powerful solution architecture. On the customer side, these predictive models can be created and trained with the application dataset and generate new predictions (known as *inferences* in the AI space) on a continuous basis. This is how the basic process flow works in the context of embedding predictive algorithms into SAP S/4HANA.

Let's now return to the ISLM framework, which provides a common interface for the embedded predictive models, leveraging either PAL or APL. Independent of the underlying predictive engine, the ISLM framework invokes either APL or PAL from SAP HANA by providing a repository for these predictive models. The ISLM framework also provides a pluggable infrastructure and adapters for the automated (APL) and expert (PAL, R, and machine learning) libraries. These models created in the automated or expert mode can be imported into the ISLM repository mode and executed.

The following are some of the criteria taken into consideration when embedded machine learning is applied:

- Business logic and machine learning logic both reside in the SAP S/4HANA platform.
- Use cases related to forecasting, trending, classification, clustering, and regression analysis use algorithms that might have a lower demand for data, RAM, and CPU time.
- Data located on the SAP S/4HANA platform is sufficient for model training with no need for a large amount of external training data.
- Required algorithms are provided by SAP HANA machine learning tools (e.g., PAL, APL, or text analysis) and handled by the ISLM framework in terms of machine learning model lifecycle management and ABAP integration.

3.2.2 Embedded Machine Learning with the SAP HANA Automated Predictive Library

In this section, we discuss how the APL is embedded into SAP S/4HANA business processes. Figure 3.3 gives an overview of the approach for doing so. This is discussed at length in Chapter 4, Section 4.3; we will only discuss the concepts briefly here for the current context.

As shown in Figure 3.3, the underlying application tables are retrieved via SQL views from the SAP HANA database models. The ISLM framework is a repository that holds the engine, the adapter, and the application programming interfaces (APIs) to access the predictive models from the SAP HANA database tables. Subsequently, SQL script procedures are written. These SQL script procedures are accessed via SAP S/4HANA ABAP classes and subsequent CDS table functions are created. The corresponding CDS view leverages these CDS table functions accordingly to access the inbuilt APL models.

Now that you understand the basic process flow behind embedding the APL algorithms into the SAP S/4HANA processes, let's see how these challenges are addressed. Lack of data science knowledge or little (or no) customer data might be some of the issues encountered while working on basic predictive algorithms like classification or regression analysis. Simplifying the model consumption and model management with the

ISLM framework addresses these challenges. Using the SAP HANA APL algorithms and integrating them into ABAP lifecycle management helps embed the predictive modeling into the SAP S/4HANA business processes. This approach has been leveraged by quite a few use cases and is available out of the box for SAP S/4HANA business processes (we'll discuss business adoption in Chapter 6).

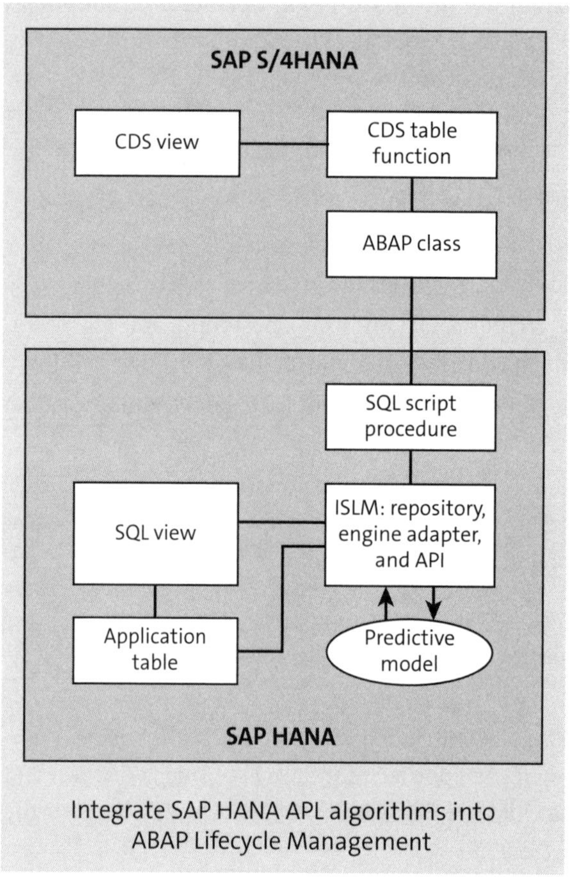

Figure 3.3 Embedding APL Algorithms into ABAP Lifecycle Management

Because we're discussing how the predictive analytics functionality is made available to the operational user in the context of SAP S/4HANA business processes, let's explain the steps involved:

1. **Accessing the system**
   The system is accessed via the SAP Fiori launchpad. The system administrator provides the URL to access the various apps assigned to your role.
2. **Role creation**
   The analytics specialist is a special role needed to first create the predictive model, then train the model, and finally activate the model.

3. **Preliminary steps needed**
   Business data must be created for the specified functionality and any prerequisite items for the same. The key step here is to identify and create the required business data.
4. **Model training**
   You must create and train the predictive model based on the business datasets available.
5. **Activating the model**
   You can create as many model versions with different sets of training parameters as needed. Activating the required model version lets you use the predictive model in the business process.
6. **Access the functionality**
   Finally, change the role to the corresponding operational user role and access the embedded SAP S/4HANA business process functionality to predict the business outcomes.

The implementation details and the mechanics behind the process of leveraging the ISLM framework used in the embedded predictive use cases are explained in Chapter 5, Section 5.2.

Let's now continue to explore the SAP HANA PAL embedding process.

3.2.3 Embedded Machine Learning with the SAP HANA Predictive Analysis Library

In this section, we discuss how the PAL is embedded into the SAP S/4HANA business processes. This process uses the SAP HANA embedded machine learning interface, which leverages the ISLM framework.

Figure 3.4 gives an overview of the approach for embedding PAL algorithms into SAP S/4HANA. This is discussed at length in Chapter 4, Section 4.3; we will only discuss the concepts briefly here for the current context.

As you can see, we are reusing SAP S/4HANA's transactional programming model to embed the PAL algorithms into the ABAP lifecycle management. The application data uses the PAL to access the machine learning model from the SAP HANA database tables and provide the details to the input CDS views in the SAP S/4HANA business process. The ABAP Managed Database Procedures (AMDP) class in the machine learning logic reads the input CDS views and is later consumed by the machine learning application for training and providing the predictive results (inferences) in the SAP S/4HANA business process.

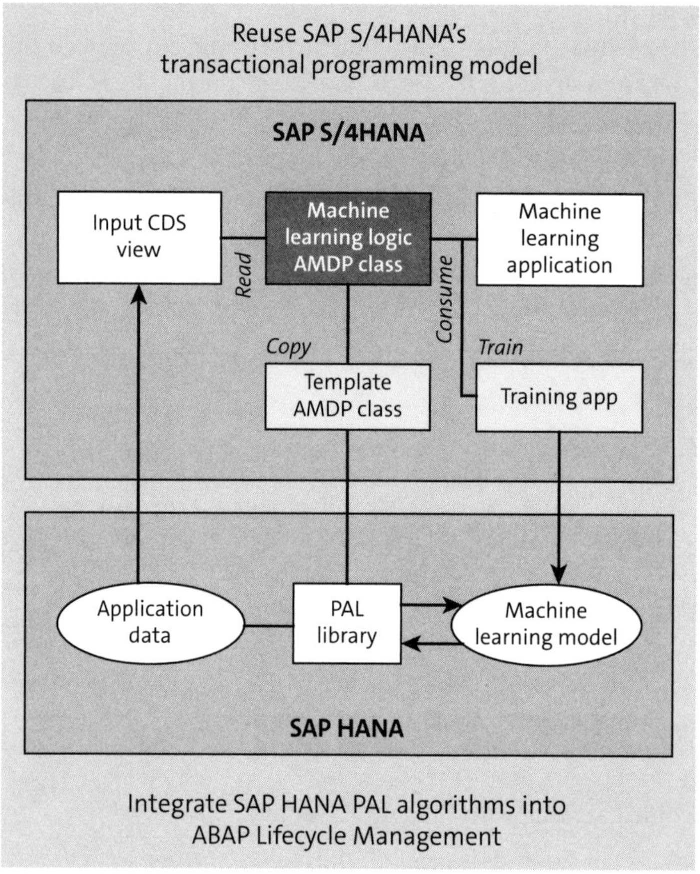

Figure 3.4 Embedding PAL Algorithms into ABAP Lifecycle Management

Some of the basic criteria that are considered to embed the predictive analytics models (PAL algorithms) into SAP S/4HANA are as follows:

- Seamless integration into the SAP S/4HANA business processes
- Reuse of the existing SAP S/4HANA functionality
- Powerful extensibility to reuse the transactional business processes
- Abundant availability of algorithms and open integration with external third-party algorithms

The concepts of embedding the predictive models into the SAP S/4HANA business processes is similar whether APL or PAL is used because we leverage the ISLM framework. Although APL is directly incorporated into the machine learning logic in SAP S/4HANA processes, PAL is specifically coded into the machine learning logic in SAP S/4HANA. PAL provides more than 100 machine learning algorithms and generally is used when APL is insufficient for the embedded scenarios.

3.3 SAP Cloud Platform

In this section, we'll mainly focus on how the tools, technologies, and services are leveraged in the context of consuming the machine learning services from the SAP S/4HANA business processes. In this approach, we focus on how we are expanding the digital core of SAP S/4HANA by consuming the machine learning services built on SAP Cloud Platform. We also call this approach *side-by-side machine learning* because the services are built on SAP Cloud Platform and leveraged by SAP S/4HANA.

Before getting ahead of ourselves, let's first introduce the concept of SAP Data Intelligence and how this solution is utilized in building the machine learning services.

3.3.1 SAP Data Intelligence

In this section, we'll discuss how machine learning services are leveraged from SAP Data Intelligence.

Let's dive into the consumption of these machine learning services. Figure 3.5 illustrates how machine learning services are consumed by SAP S/4HANA business processes.

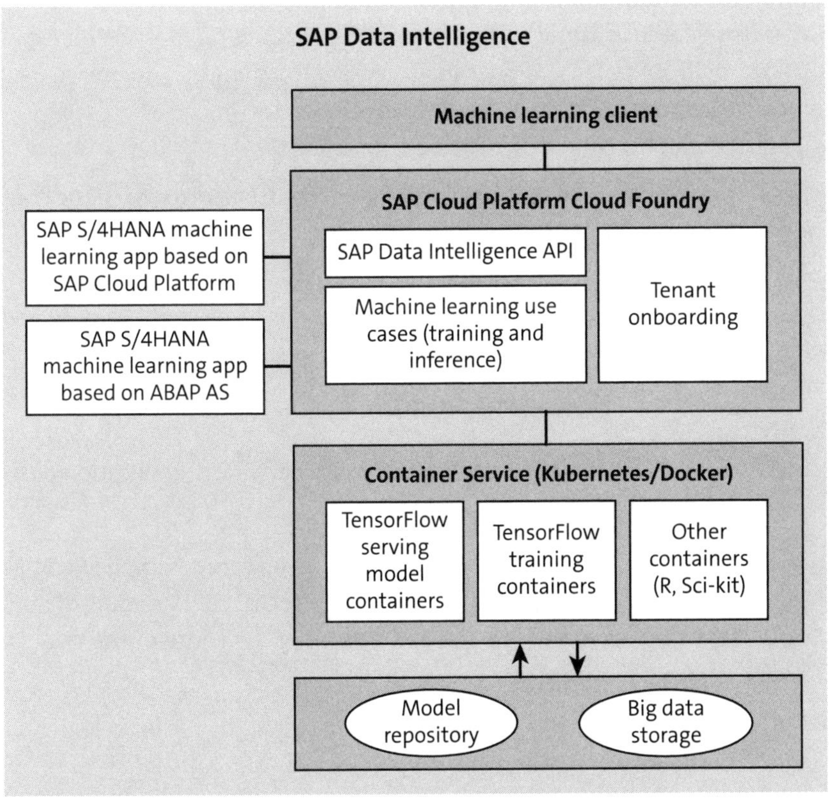

Figure 3.5 Side-by-Side Machine Learning Approach

As we discussed in Section 3.2, while embedded machine learning targets scenarios in which the business process and the machine learning logic reside in the SAP S/4HANA platform, side-by-side machine learning targets the following:

- **Business logic on SAP Cloud Platform**
  In this context, the application logic and the business logic reside on SAP Cloud Platform in addition to the machine learning logic that is leveraged from SAP Data Intelligence on SAP Cloud Platform. This follows the golden rule of bringing the algorithms to the data. The machine learning app is built on and runs on SAP Cloud Platform directly.
- **Business logic in SAP S/4HANA**
  In this context, the business logic and the application logic reside in SAP S/4HANA, while the machine learning logic stays on SAP Cloud Platform. Although the application logic and business logic are in SAP S/4HANA, the application data still resides on SAP Cloud Platform because the jobs are scheduled to pump data into SAP Cloud Platform. Data can also be leveraged from the external data sources into SAP Cloud Platform as needed to maintain the golden rule of data and algorithms staying together. The machine learning services built on SAP Cloud Platform are then accessed by the SAP S/4HANA apps via the communication protocols.

These side-by-side machine learning scenarios are usually based on lifecycle management in SAP Data Intelligence. This lifecycle management foundation is organized into three different development initiatives:

1. **Lifecycle management**
   Defines a framework for creating and executing "intelligent" algorithms in the context of the overall SAP ecosystem
2. **Functional services**
   Executed on top of the lifecycle management framework to provide ready-to-use implementations of a selected set of standard use cases, such as image classification, sentiment analysis, and so on
3. **Business services**
   Implements complex use cases that are tightly integrated with SAP application business processes on SAP S/4HANA

Let's also look in more detail at how side-by-side machine learning is applied and at the basic criteria that are considered. This will give an in-depth understanding of how the core machine learning services are built. The following are some of the criteria considered:

- Machine learning logic resides on SAP Cloud Platform, while the business logic can be either in SAP S/4HANA or SAP Cloud Platform.

- Use cases require deep learning techniques, such as complicated image recognition, video recognition, or natural language processing, along with other neural networks with a high demand for data, RAM, and CPU/GPU time.
- A huge volume of external data is required for model training, with the main focus on processing unstructured data.
- Required algorithms are not provided by SAP HANA machine learning but by other external libraries, such as TensorFlow, Scikit-Learn, and so on.

Let's continue into the process flow of how these machine learning services are consumed from SAP Cloud Platform by SAP S/4HANA business processes in the context of leveraging SAP Data Intelligence. Prior to the inception of SAP Data Intelligence, the SAP Leonardo Machine Learning Foundation framework was utilized to build the training models, deploy them, and create the inferences that would help configure communication scenarios with the SAP S/4HANA business processes. This helped the deployed machine learning services on SAP Cloud Platform to be consumed accordingly by the SAP S/4HANA business processes. SAP Data Intelligence is the new solution used for this process.

Because we're discussing how the machine learning functionality is made available to the operational user in the context of SAP S/4HANA business processes, let's articulate this better by explaining the steps involved here:

1. **Accessing the system**
   The system is accessed via the SAP Fiori launchpad. The system administrator provides the URL to access the various apps assigned to your role.
2. **Role creation**
   The specific role for the machine learning service needs to be assigned and used.
3. **Preliminary steps needed**
   You must create the master data, organizational data, and other data as needed.
4. **Basic business conditions**
   Any prerequisite functionalities must be implemented before the basic business conditions are met.
5. **Configuration**
   You must configure the machine learning service.
6. **Subscription**
   Subscribe to the corresponding application that uses the machine learning service.
7. **Communication**
   Create the communication system as per the administrator role (SAP_BR_ADMINISTRATOR), and then create the corresponding communication scenario assigned for the specific machine learning service.

8. **Training and inference**
   Schedule the training job. Infer the results from the prediction models by changing the role and logging in as a specific operational end user to access the functionality to predict the business outcomes.

Now that you know the steps involved in consuming the machine learning service, let's briefly look into the process flow while working on the side-by-side machine learning scenarios. Figure 3.6 highlights the steps involved in the creation of the training process flow and inference process flow with SAP Data Intelligence. The training process flow and inference process flow are the two most important steps here. The training process flow prepares the training model with the required datasets, and the inference process flow applies this training model to the new dataset.

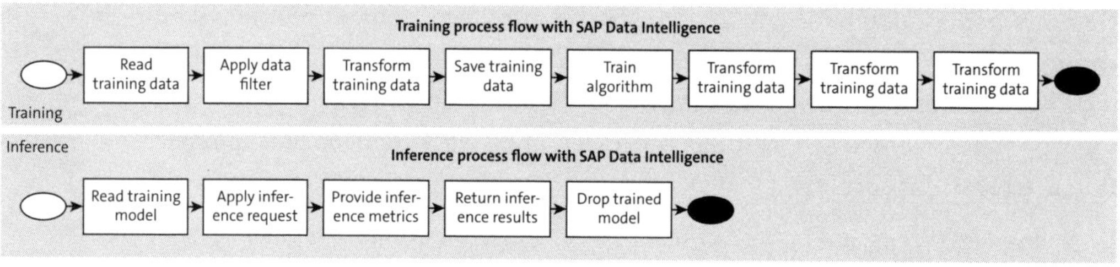

Figure 3.6 Side-by-Side Machine Learning Flow

Now, let's dive a bit deeper into SAP Data Intelligence and understand how customer needs are addressed in a simple and comprehensive way:

- End-to-end tooling support is available to set up and manage the machine learning lifecycle.
- You can accelerate and scale your machine learning projects as needed.
- You can automate retraining, maintenance, and retirement of machine learning artifacts.
- You can build agile, data-driven applications and benefit from enterprise visibility and governance.

Figure 3.7 gives a high-level understanding of the SAP Data Intelligence architecture and how the machine learning services are orchestrated. You can see the various components involved in the creation of the machine learning content and how the other internal and external services are leveraged. JupyterLab is built into the SAP Data Intelligence launchpad for building the machine learning content. The data governance component helps with governance and metadata management while preparing the data. The integration and orchestration component is helpful in pipeline modeling and data workflows and for connecting to the various pipelines to streamline the different sets of data. The systems management component focuses on the tenant containers of the machine learning models, monitoring and logging the actions as well as managing the lifecycle of the models in the repository.

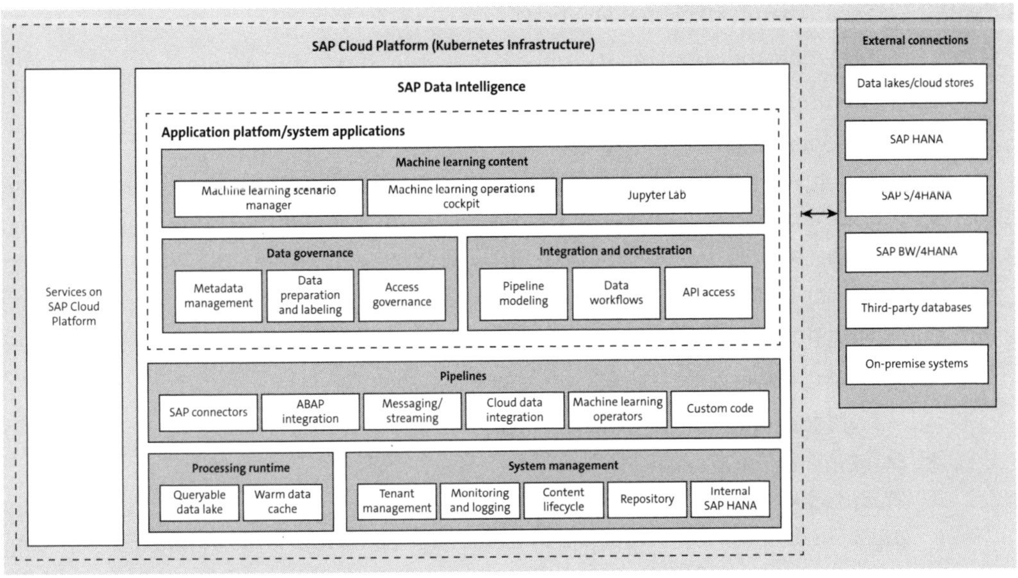

Figure 3.7 Data Intelligence: Architecture View

SAP Data Intelligence is built as a logical application offering enterprise-ready features to scale complex scenarios dealing with disparate data. It consists of four major building blocks:

1. **Data pipelining**
   This is the process of refining and enriching the raw data from a variety of sources by converting unstructured data to usable structures and formats. It creates a powerful data pipeline workflow application that cleanses, conforms, transforms, ingests, refines, orchestrates, prepares, and enriches information from a variety of sources across the organization. SAP Data Intelligence not only accelerates the modeling, execution, and management of data pipelines but also provides fast execution of pipeline activities by employing distributed processing while leaving the data where it resides so that you can accelerate and scale projects quickly. Key capabilities are as follows:
   - Modeling data pipelines and workflows
   - Data enrichment
   - Data preparation and quality
   - Distributed data pipeline processing
2. **Data landscape management**
   This includes understanding your end-to-end data landscape for proper orchestration. In addition, it ensures secured data sharing across the various users and systems. It also leverages existing connections and integration tools, with an open framework for adding new sources and destinations easily and flexibly. Key capabilities are as follows:

  - Security and policy management
  - Ingestion and connectivity
  - Monitoring and deployment
3. **Data governance**
   This helps to profile, view, and expose the data of all connected systems. The data models evolve more quickly in a highly visual environment, and this ensures consistent and proper handling of data quality, integrity, and security across the enterprise. Key capabilities are as follows:
   - Data profiling
   - Data discovery/centralized information catalog
   - Data preparation
4. **Machine learning content**
   The machine learning scenario manager represents a central entry point to easily organize data science artifacts and to manage all related tasks. A machine learning scenario contains SAP Data Intelligence pipelines and Jupyter Notebooks, and it's possible to do the following:
   - Version a scenario for reusability and add metadata such as a business question and scenario description.
   - Rate the pipelines and Jupyter Notebooks to identify those that are the best performing.

In addition, the machine learning operations cockpit is an application created to set up and manage the deployment of machine learning scenarios that have been released in the machine learning scenario manager. Among other things, it's possible to configure the underlying pipelines used for training and inference.

3.3.2 Hybrid Machine Learning Models

Let's now briefly discuss the different machine learning models that are developed as part of the machine learning services on SAP Cloud Platform. We've already focused on the *how* of SAP Data Intelligence in building the models. In this section, we briefly inspect some of the other aspects of these models.

Hybrid models are generally built by leveraging a mix of the TensorFlow algorithms, the Scikit-Learn algorithms, and the SAP HANA predictive algorithms, as shown in Figure 3.8.

As you can see, various functional services are created from these hybrid models, comprising of a mix of algorithms from different libraries. Based on the use case requirements, the business services are built out of these core functional services that are available and updated on a regular basis. These machine learning services are then made available for consumption into the SAP S/4HANA business processes. Furthermore, modifications and enhancements are made to these machine learning services as needed and are customized by various customers and partners. We have already discussed the other aspects of the training tools and lifecycle management behind these models in earlier sections, so we won't dive into those details again here.

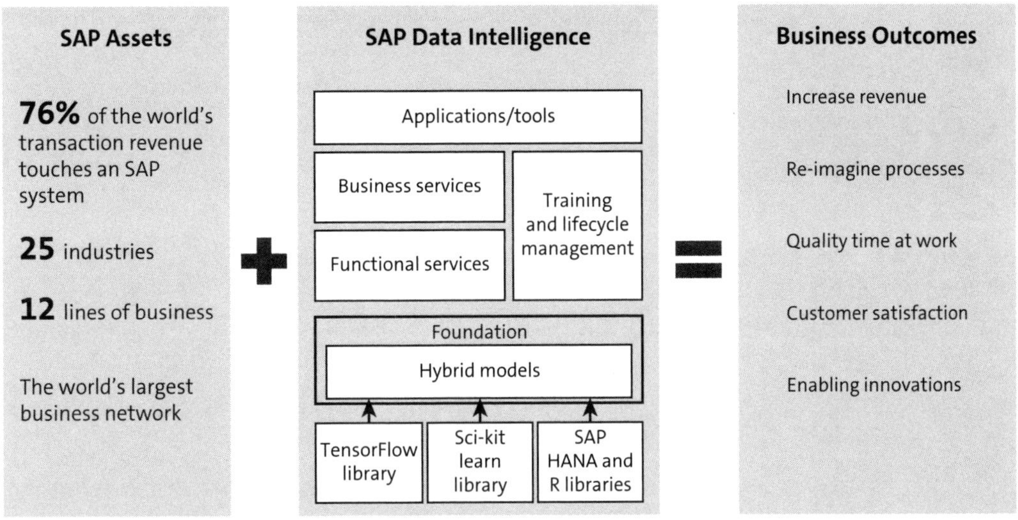

Figure 3.8 Hybrid Models for Machine Learning

3.4 SAP Analytics Cloud

In this section, we discuss how tools, technologies, and services are leveraged in the context of extending the predictive and machine learning services with SAP Analytics Cloud. In this approach, we focus on how we are extending the digital core of SAP S/4HANA by using the predictive services from SAP Analytics Cloud. We'll begin with an overview of the offerings, before diving into both the smart assist and smart predict services.

3.4.1 Overview

Table 3.1 illustrates the different SAP Analytics Cloud predictive services available in the context of the predictive functionality.

In Table 3.1, you can see that some of the core algorithms can be utilized from the APL for creating quick predictions and simulations as needed for different customer-centric scenarios. Choosing the right smart feature in SAP Analytics Cloud is key to performing the required functionality. There are different approaches in leveraging the predictive services functionality from SAP Analytics Cloud in the context of SAP S/4HANA. Although some functionality could be achieved using the live connection from SAP Analytics Cloud to the SAP S/4HANA CDS views, the other functionality would require a data acquisition approach to build additional VDMs in SAP Analytics Cloud. Depending on the available functionality and requirements at hand, the corresponding approach is leveraged.

	Search to Insight	Smart Insights	Smart Discovery	Time Series Forecasting	Smart Grouping	Custom Visualizations
Natural language query	Yes					
Natural language explanation		Yes	Yes			
Contribution analysis		Yes				
Automated story generation			Yes			
Key influencer analysis			Yes			
Outlier identification			Yes			
Scoring simulation			Yes			
Forecasting				Yes		
Clustering					Yes	
Custom visualizations						Yes

Table 3.1 Choosing Smart Features in SAP Analytics Cloud

Here is a quick overview of the available smart features in SAP Analytics Cloud:

- *Search to insight:* What is happening in my business?
- *Smart insights:* What is behind this number?
- *Smart discovery:* Why did this happen?
- *Time series forecasting:* What will happen?
- *Smart grouping:* Which data points are similar?
- *Custom visualizations:* What are the visualizations?

The following steps are involved in the creation of predictive models from SAP Analytics Cloud and publishing them back into SAP S/4HANA, which we'll explain further in Section 3.4.3:

1. Accessing the SAP S/4HANA application tables to retrieve the data.
2. The CDS view wrapped on top of the SAP HANA application tables provides the required functionality.

3. A custom CDS view might be created on top of the CDS views to fine-tune the required functionality from SAP S/4HANA.
4. Input datasets are created from the custom CDS views in SAP S/4HANA by acquiring these datasets in SAP Analytics Cloud.
5. Predictive scenarios are created with these input datasets and the results are made available in the output datasets.
6. The SAP Analytics Cloud data model is wrapped around these output datasets.
7. SAP Analytics Cloud user stories are created which embed the SAP Analytics Cloud data model and showcase the predictive functionality.
8. Each predictive scenario has a predictive model. These predictive models can also be published back into SAP S/4HANA with the ISLM framework.

3.4.2 Smart Assist Services

Let's now discuss the details of the smart assist services. We'll look into the features that are available with SAP Analytics Cloud.

Search to Insight

Search to insight provides a new way to interact with your data in order to quickly find usable information. This functionality helps you to quickly find usable information through questions in a natural language. It automatically generates charts that provide answers, allowing you to share info or move to the next question. Figure 3.9 gives an overview of the feature.

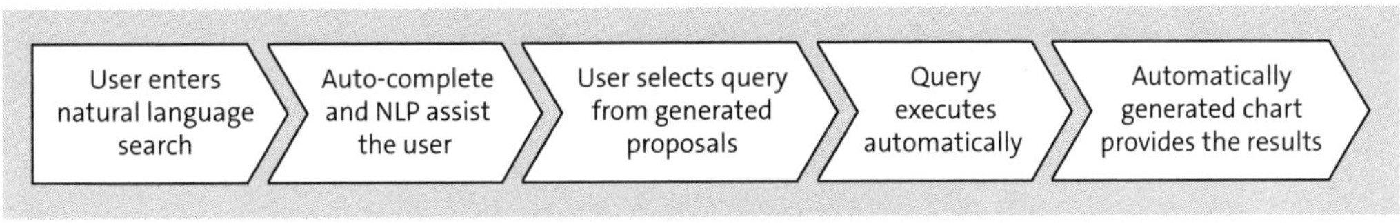

Figure 3.9 Search to Insight Feature in SAP Analytics Cloud Smart Assist

You can ask questions about your data in natural language and get instant answers. This feature helps you to interact naturally with your data through questions, such as "What are the current sales?" or "Can you show me the revenue for the last quarter?" Autocompletion of words and phrases is available as you type the measures and dimensions. This also allows you to quickly find the data that you are interested in. Answers are provided using charts or numeric values based on the question type, enabling you to make data-based decisions.

It is expected that by 2021, natural language processing (NLP) and conversational analytics will boost analytics and business intelligence adoption from 35% of employees to over 50%, including new classes of users, particularly front-office workers.

Some of the benefits are as follows:

- You can start with a question, asking for the data you want rather than relying on prebuilt stories or building your own charts manually.
- You can use questions to drill into the data and move from the data on the dashboard to more details.
- You can build stories more easily and quickly by using the automatically generated charts.

Smart Insights

Smart insights examines each of the dimensions in the data and detects members that have a significant influence on the data point you're interested in. This functionality automatically helps examine the variables in your data to determine what contributes to a value, and it embeds dynamic text tokens in charts to share the insights. A deep level of analysis is also provided by drilling down into the contributors.

Figure 3.10 gives an overview of the feature.

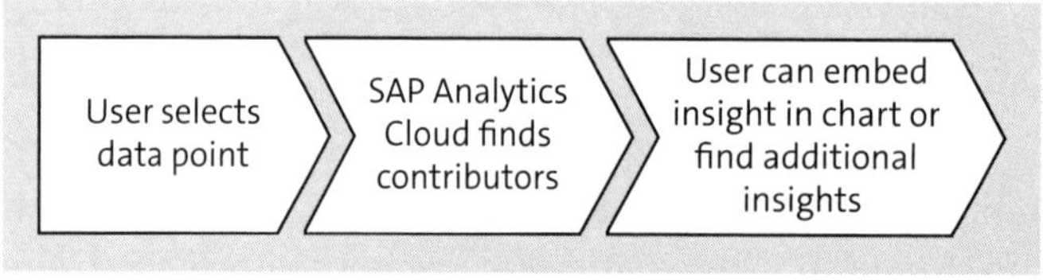

Figure 3.10 Smart Insights Feature in SAP Analytics Cloud Smart Assist

Smart insights uses proprietary rules and algorithms to do this in a performant way, whereas a user would need to manually investigate each dimension.

Some of the benefits are as follows:

- Quickly understand the reasons behind an interesting data point or variance and see what factors contribute to the value
- Add context to visualizations by embedding smart insights tokens

Smart Discovery

Smart discovery is an automated data exploration engine. Using machine learning algorithms, smart discovery reveals relationships and hidden patterns in your data model to drive accelerated insight discovery. This functionality acts as a digital business analyst by including visualizations in the overview page that determine the type and content of the data. The automated predictive engine builds a robust model without a data scientist's intervention. The key influencers, outliers, and simulation pages are derived from the predictive models created. Figure 3.11 gives an overview of the feature.

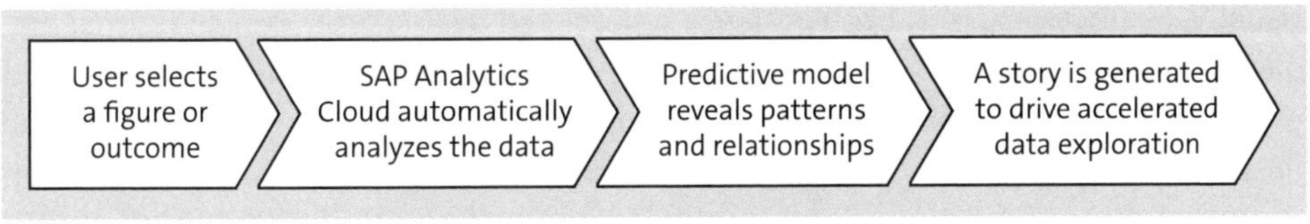

Figure 3.11 Smart Discovery Feature in SAP Analytics Cloud Smart Assist

Some of the benefits of smart discovery are as follows:

- Quickly create a story that guides you to interesting trends.
- Never need to start from scratch again!
- Spend less time on dashboard design and more time on insight discovery.
- Reduce human bias in analytics through machine-guided analysis.

3.4.3 Smart Predict Services

Let's now focus on the smart predict feature in the context of SAP Analytics Cloud and see how it can help benefit the SAP S/4HANA business processes.

SAP Analytics Cloud helps business users to make the right decisions by leveraging the augmented analytics features. A huge portfolio of augmented analytics features provides insights by utilizing machine learning and predictive services.

Data scientists are typically very good at understanding statistics and algorithms but do not have in-depth business knowledge. Therefore, most of the time, companies need to align data scientists with the right business experts to understand the business problem so that they can prepare the analytical datasets. There are a lot of questions to be considered while preparing these datasets: domain experts can help identify the right kind of data to be considered and where the data needs to be stored and organized. This is not a one-time task, so increased collaboration is needed between the business domain experts and the data scientists to provide realistic results. This not only takes a lot of time and effort but also adds to the companies' costs. Although some data scientists are experienced enough to provide real solutions while working on the machine learning algorithms, the much-needed trust from the business world might not be present, so their efforts might go to waste.

With the introduction of SAP Analytics Cloud's smart predict functionality, business experts and analysts can easily do data science even if they aren't comfortable with statistics and algorithms. The intuitive user experience and features of smart predict help business users and analysts to build predictive models in minutes, starting from data and creating meaningful insights. The science behind the algorithms is hidden in the smart predict functionality; business users/analysts just need to handle the data preparation while the wizard navigates them through the required steps. Data scientists can be freed from these mundane tasks and focus their efforts on building important machine learning functionality.

Usually, business users want to enhance reports and dashboards with additional predictive insights, and this is another area where smart predict proves very helpful. Smart predict extends traditional analytics by filling the gap between data preparation and business intelligence (BI) modeling/storytelling. To summarize, smart predict leverages machine learning that can be used by any end user with the click of a button and a few minimal steps for creating a BI story. Figure 3.12 explains this in a nutshell.

The focus is on the business questions, not algorithms, which helps speed the prediction and recommendation process. That is to say, the impact is driven by business outcomes, rather than data science know-how such as statistically balancing the dataset, removing outliers, encoding the data, picking the right algorithm, and selecting the right model. Smart predict users can train a predictive model in as little as three steps:

1. Select the business problem you are trying to solve.
2. Choose your analytical dataset.
3. Select your target (the column representing the historical answer to your business question).

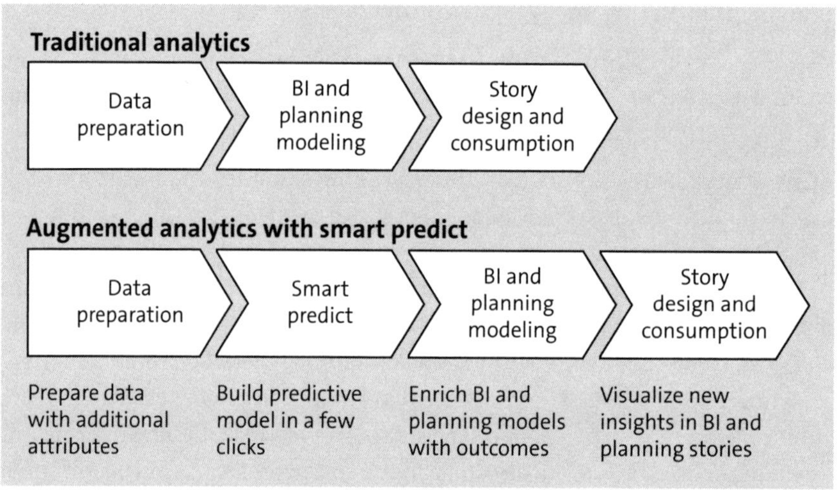

Figure 3.12 SAP Analytics Cloud Smart Predict Processes

Smart predict addresses most business problems with the following predictive scenarios:

- **Time series forecasting**
  The technique of predicting future values based on past observations.
- **Classification analysis**
  The technique of predicting groups that do not overlap with another group.
- **Regression analysis**
  The technique of identifying relationships among two or more variables while predicting another target variable.

- **Smart grouping**
  This is another form of classification analysis, which predicts and groups values based on a common theme or formula.
- **R-visualizations**
  Data visualization of turning numbers into insights using R programming.

Figure 3.13 shows how to select a particular predictive algorithm and then create a predictive scenario. We'll walk through the process of creating a predictive scenario in the steps ahead.

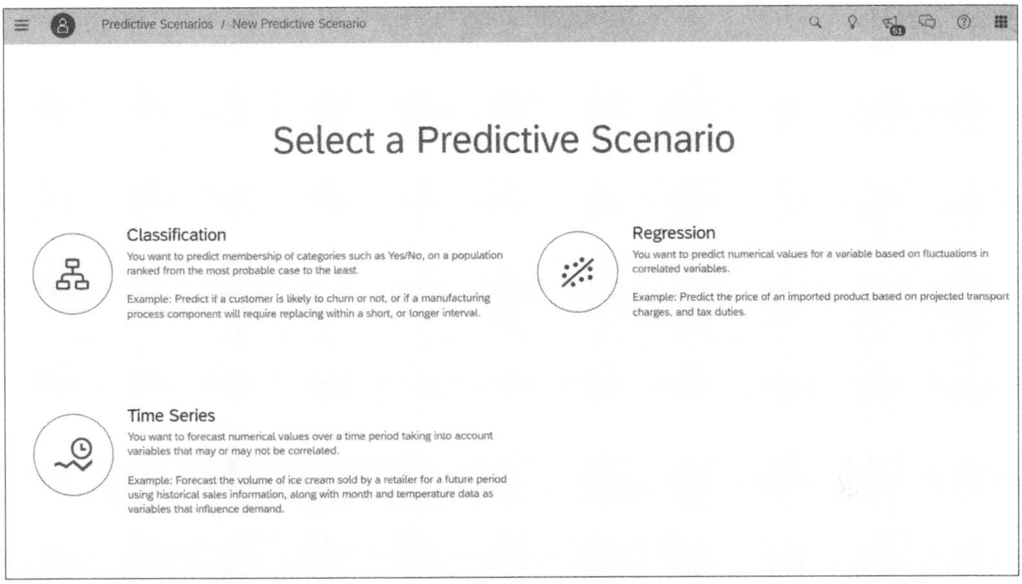

Figure 3.13 Predictive Scenarios in SAP Analytics Cloud Smart Predict

Let's take a high-level look at how to connect SAP S/4HANA Cloud to SAP Analytics Cloud, create a model, and publish it back to SAP S/4HANA.

There are a couple of prerequisite steps needed here:

1. **Preliminary connection configuration**
   First, you must complete the following preliminary steps to configure your connection:
   - Connect your SAP S/4HANA Cloud system to SAP Analytics Cloud to import datasets.
   - Connect SAP Analytics Cloud to your SAP S/4HANA Cloud system so that we can publish new predictive scenarios back to SAP S/4HANA.
2. **Prerequisite roles**
   Next, you must ensure that the relevant roles are assigned based on of your user type, as shown in Table 3.2.

User	User Role	Task
SAP S/4HANA administrator	SAP_BR_ADMINISTRATOR	To configure connection settings
SAP S/4HANA key user	SAP_BR_ANALYTICS_SPECIALIST	To manage predictive scenarios and predictive models
SAP Analytics Cloud administrator	▪ BI admin ▪ Predictive admin	To configure connection settings
SAP Analytics Cloud business user	Predictive content creator	Create predictive models

Table 3.2 Prerequisite Roles

Once the prerequisites are met, the following steps need to be followed:

1. Configure the necessary connections to acquire the data from SAP S/4HANA Cloud import.
2. Configure the ISLM connection by accessing the **Communication Management** tab in the SAP Fiori launchpad.
3. Acquire the data from SAP S/4HANA Cloud in SAP Analytics Cloud as a dataset.
4. Create a predictive model by using the predictive scenario feature in SAP Analytics Cloud smart predict.
5. Train the models in SAP Analytics Cloud smart predict. At this stage, the predictions can be used in the SAP Analytics Cloud user stories. This will help the business user perform an exploration of the different predictive features available.

We'll discuss all implementation steps for side-by-side predictive analytics with SAP Analytics Cloud in detail in Chapter 5, Section 5.4.

Note

Alternatively, you can also publish the models back to the ISLM framework in SAP S/4HANA Cloud. These published models are available as predictive models in SAP S/4HANA Cloud. We are providing this as another option of publishing back into SAP S/4HANA, but the recommended way would be to leverage the new ISLM framework in SAP Data Intelligence to build models and publish them back into SAP S/4HANA.

3.5 SAP Intelligent Robotic Process Automation

In this section, we start with a brief introduction to SAP Intelligent RPA and then discuss the scenarios developed to help SAP S/4HANA business processes and enhance the machine learning functionality.

Robotic process automation (RPA) automatically replicates tedious actions to free-up human workers for high-value tasks. There are two types of automation:

- Unattended: Fully automated process, in which robots work autonomously with human supervision only
- Attended: Partially automated process, in which robots are coworking with humans (also called robotic desktop automation [RDA])

Automation is one of the key drivers of the intelligent enterprise vision. The efforts of the business enterprise world to do research in automation tools facilitated decades of business process automation and drove efficiency in enterprise processes. This helps in reduction of manual and mundane work so that organizations can focus on higher-value activities.

The following are some of the benefits of automation:

- Improve process efficiency, saving time.
- Improve process quality, reducing errors.
- Improve user satisfaction and increase pride by removing boring tasks from the process.
- Reinforce compliance by introducing validations and autoreporting.
- Cut the process execution time, generating immediate savings.
- Allow parallelization and immediate scalability.
- Introduce brand-new technology without any footprint on the legacy system.
- Offer an alternative to APIs, web services, and classical connections.
- Prevent long and expensive development.

SAP is extensively building up RPA bots to integrate into the SAP S/4HANA business processes in order to provide customers with these benefits.

As shown in Figure 3.14, end-to-end automation is helping add value to the intelligent suite by interfacing tasks with conversational AI, performing tasks with intelligent RPA, and learning from expectations with machine learning.

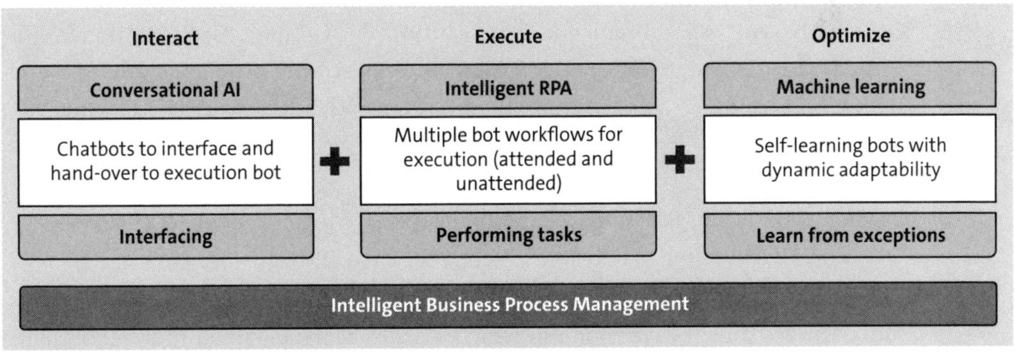

Figure 3.14 Intelligent Robotic Process Automation Overview

This leads us to SAP Intelligent RPA. If RPA is the hands and legs of the robot, machine learning acts as the brain that makes the decisions, and conversational AI plays an interfacing role to receive and disseminate instructions digitally.

Figure 3.15 highlights the four key components that build the pillars of SAP Intelligent RPA:

1. **Design environment**
   A modeler is used to graphically design the bot workflow.
2. **Central repository**
   Used to store the atomic process steps, as well as for reuse during the design phase.
3. **Runtime environment**
   Used to manage bot deployments in various virtual/on-premise/hybrid environments.
4. **Monitoring tool**
   Used to provide an overview while bots are in action, to show the percentage of work completion, progress, service-level agreement (SLA) achievement, risk mitigation, and so on.

	Design Environment	Central Repository	Runtime Environment	Control/ Monitoring Tool
Purpose	Flexibly design robot workflows, record screens	Store atomic steps for full workflow, configuration info	Execution engine for running bot scripts across deployments	Process monitoring, analytics for robot performance/SLAs
Key Capabilities	• End user-friendly workflow designer • Automation methods: – API interfaces – OCR/NLP – Screen scraping – Computer vision • Process mining and discovery • Predefined robot templates	• Multitenancy • Security • Bot version control • Marketplace for third-party prebuilt bot content	• Scaling/load balancing • Public cloud deployment • Workflow management machine/ human • Bot scheduler and time zone manager • Bot failsafe engine • Multiple channels support such as SAP Conversational AI (chatbots) • SAP Data Intelligence	• UX, mobile access • Audit trails • Process insights • Logging and monitoring cockpit

Figure 3.15 Key Components in SAP Intelligent RPA

SAP's aim is to realize the intelligent RPA vision in multiple phases:

- **Phase 1**
  Deliver the core SAP Intelligent RPA solution with four critical components as explained. This will be SAP S/4HANA-focused, out-of-the-box, preconfigured content. It will have the capability to integrate bots with SAP S/4HANA APIs, machine learning services, and SAP Conversational AI.
- **Phase 2**
  Extend to the wider SAP portfolio of applications, such as SAP SuccessFactors, SAP Concur, SAP Ariba, and Qualtrics.
- **Phase 3**
  Move toward a cognitive mode in which we would have gained more experience through multiple bot deployments. This would enable SAP to make the bots more

and more AI-enabled with a prescriptive mode to remodel processes and for self-healing.

In the context of this chapter, we're focusing mostly on phase 1, which takes us now into the next logical step of how SAP Intelligent RPA is used in the context of SAP S/4HANA business processes and how it is enhancing the machine learning functionality for a comprehensive intelligent enterprise. In Chapter 1, we briefly discussed the role played by intelligent RPA in the context of end-to-end scenarios for an intelligent suite. We won't go into detail here as we already talked about the big picture, but we'll take one of those examples and dive in to the details to understand the mechanics behind bot creation and deployment.

Figure 3.16 gives a high-level overview of the architecture involved in the creation of the intelligent RPA bots, and their deployment is shown in Figure 3.17. In Figure 3.16, notice the desktop studio, the desktop agent, the cloud factory, and the best practices to build bots across the different lines of business in SAP S/4HANA. The automation designer uses the desktop studio to build an automation scenario by capturing the screens of the applications. The package containing the scenario is exported from the desktop studio. The administrator imports the package into the cloud factory and configures it by providing credentials, scheduling, and so on. The cloud factory then pushes the scenarios to the available desktop agents for automating the applications. The administrator can typically monitor the performance of the scenarios.

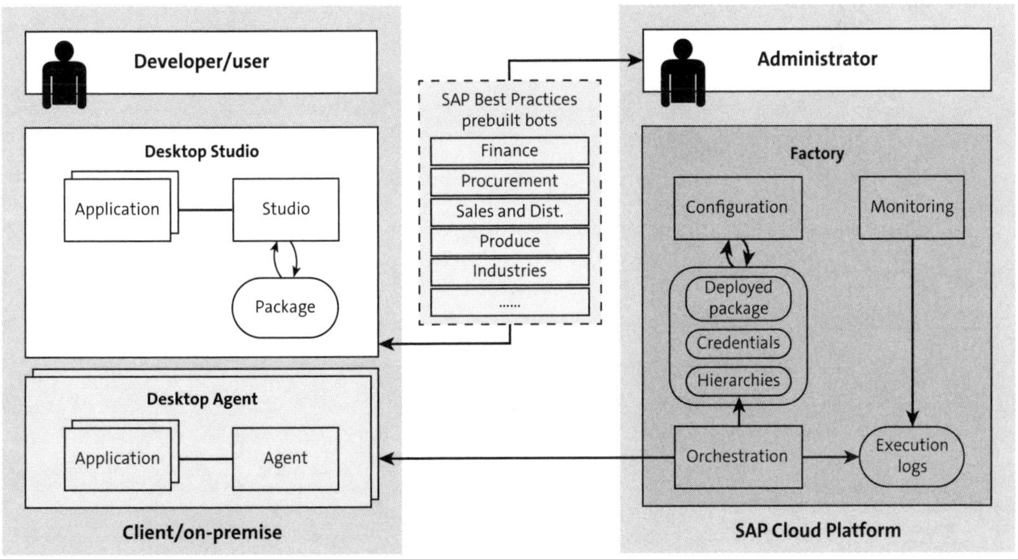

Figure 3.16 Intelligent RPA Architecture

As shown in Figure 3.17, deployment has three main components that we touched on previously. Let's take a closer look at each:

1. The cloud factory is responsible for orchestration and monitoring. It's a cloud-based solution powered by SAP Cloud Platform. No deployment is required—only a web browser.
2. The desktop studio is responsible for project development of process automation. You deploy this only on the process automation developer's workstation.
3. The desktop agent is responsible for executing the process automation:
   - A line of business can run in an attended mode.
   - If some processes can run in such a way that direct human interaction is not required, we are talking about an unattended mode (digital worker mode).
   - Both components will be linked with the cloud securely.

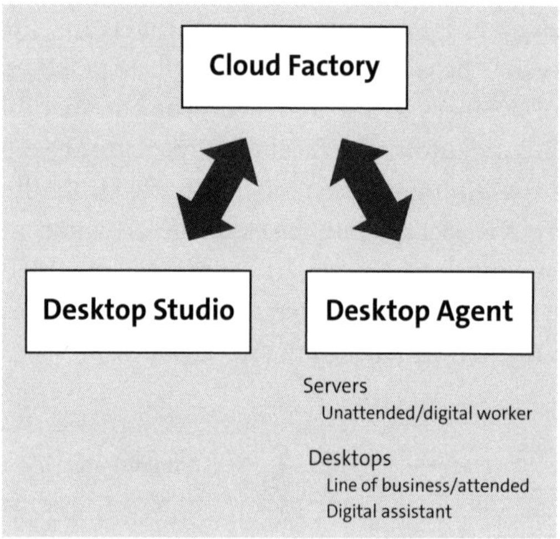

Figure 3.17 Intelligent RPA Deployment

Figure 3.18 shows an overview of the complete process, from bot creation to delivery to a customer.

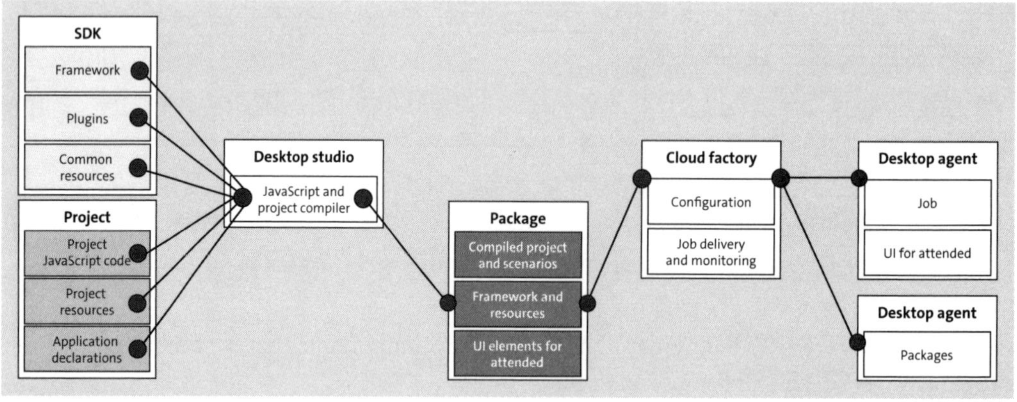

Figure 3.18 Intelligent RPA Delivery

Let's summarize the complete process for bot making:

1. The first step starts with the access and availability to the desktop studio.
2. Next, the software development kit (SDK) includes the framework, plug-ins, and common resources.
3. The project itself includes the code, the resources, and application declarations.
4. These are all merged into one package. The package contains all resources needed, such as the compiled project and scenarios, framework, and so on. There is also a concept of versioning for packages.
5. You import your package within the factory, which will allow you to create configurations and then take care of job delivery and monitoring.
6. Configured packages are then distributed to agents. When a user triggers an agent, it creates a job behind the scenes.
7. With an event, such as a schedule, agents running in unattended mode will process the job.

Now that we've discussed the basic architecture and mechanics behind the creation and deployment of bots, let's briefly look at an example bot.

Figure 3.19 shows the process steps involved in creating a purchase order confirmation received via email automatically in SAP S/4HANA.

Some suppliers do integrate with SAP S/4HANA to send a purchase order confirmation in an integrated way, such as via the Ariba Network or electronic data interchange (EDI), but there are many suppliers who still use email responses to confirm specified quantities of products at specified prices at specified times. When such an email response is received by the buying company, there is still a manual task involved to record the confirmation in the SAP S/4HANA system. But this process can be automated with a bot that extracts the purchase order confirmation email using an optical character recognition (OCR) machine learning service and then triggers an update in the SAP S/4HANA system automatically. There are a lot of business benefits to this:

- Handshake to confirm/update the purchase order product quantity, price, and delivery date
- Removal of a manual step, freeing the buyer from this task
- Allows for better planning because there is better visibility of purchase order status

Because we're discussing the machine learning and predictive analytics aspects of SAP S/4HANA, we wanted to briefly highlight how all these AI technologies and processes are interconnected in the big story of the intelligent enterprise. With this example, you can see how these AI technologies complement each other. While the machine learning service extracts the purchase order confirmation email, the RPA bot automates and updates the purchase order confirmation in the SAP S/4HANA system as explained in the steps shown in Figure 3.19.

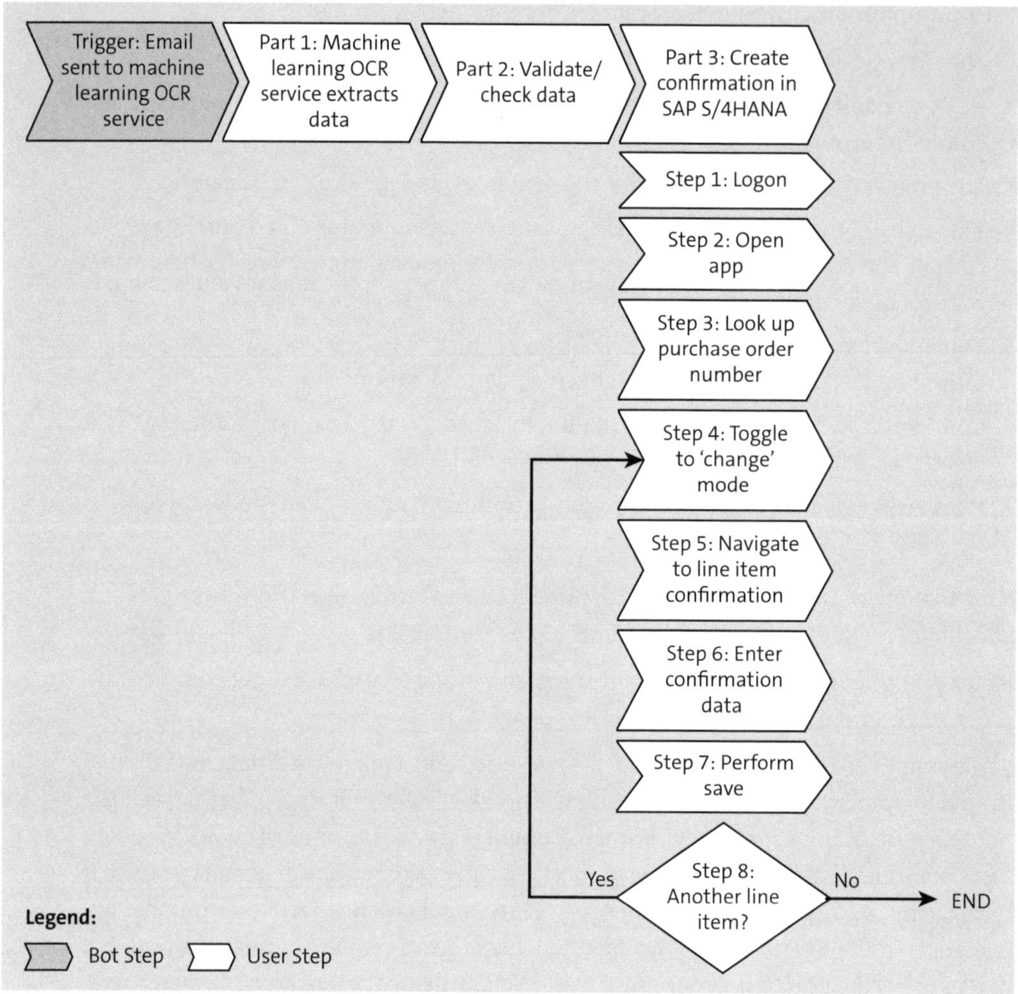

Figure 3.19 Purchase Order Confirmation Bot in SAP S/4HANA

3.6 SAP Internet of Things

In this section, we start with a brief introduction to SAP Internet of Things and then discuss the scenarios developed to help SAP S/4HANA business processes and enhance the machine learning functionality.

Intelligent technologies and services provided by SAP Internet of Things powers the intelligent suite to help customers and partners in the SAP ecosystem, where IoT plays a key role in handling the real-time sensor data and services.

Here are some of the questions we keep hearing from customers:

- How can we support my lines of business with intelligent automations?
- How do we make sense of time series data?

- How can we semantically integrate sensors and devices from different vendors?
- How can we integrate existing IoT data into our business processes?
- How can we use IoT data to develop own IoT-enabled business applications?
- How can we drive my business decisions based on real-time sensor data?
- How can we enhance our LoB processes with operational data from IoT?
- How can we get our business ready for Industry 4.0?

In the end, this boils down to the following: How do we get from sensor values to business insights?

Figure 3.20 gives an overview of how SAP Internet of Things fits into the context of the intelligent enterprise.

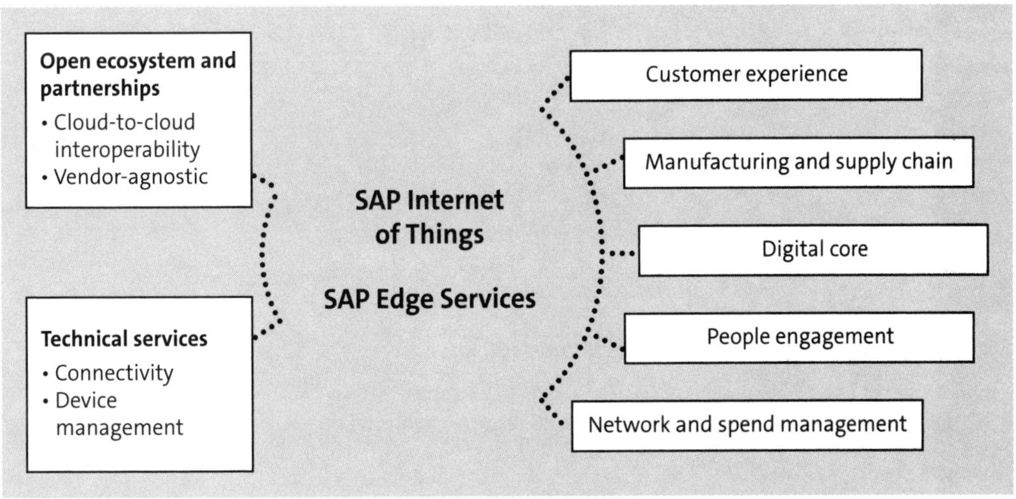

Figure 3.20 SAP Internet of Things: From Internet of Things Data to Business Outcomes

Here are some of the important value propositions from SAP Internet of Things:

- Helps transform your business and step up your productivity with situational awareness based on IoT data
- Increase your business outcome and customer experience with IoT-enabled connected processes
- Extend your business processes from the cloud to the edge
- Contextualize and integrate IoT data with SAP business process information
- Leverage IoT insights with an integrated application developer experience

SAP Internet of Things provides business context to IoT data and enables us to derive meaningful business insights and to tightly embed our custom IoT applications into the SAP S/4HANA business processes.

SAP Internet of Things turns signals into contextual data, eventually converting sensor information into business insights. This enables customers to integrate IoT data with business solutions by transforming raw, technical, signal data into insights that can be embedded in new and existing business applications. This is done by standardizing the signals captured from the sensors into a unified format (the *thing model*) and then assigning relevant context to the data (e.g., business partner, material information, SLA information).

Figure 3.21 explains the process flow for turning signals into contextual data using SAP Internet of Things.

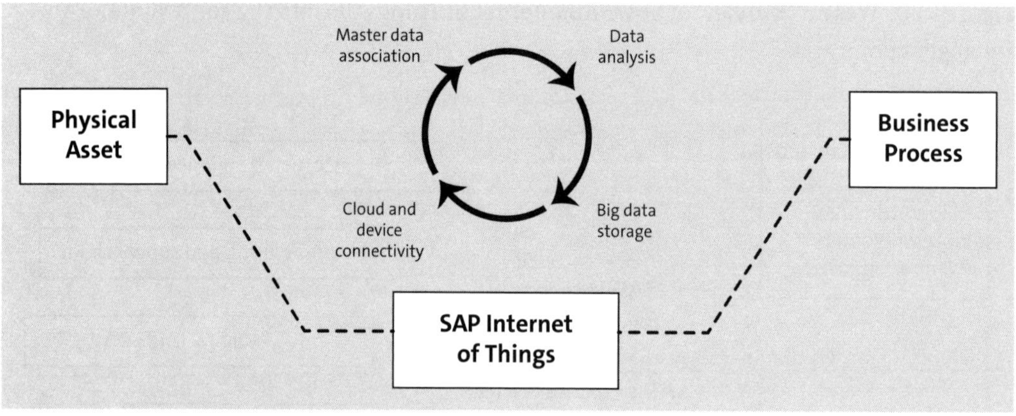

Figure 3.21 SAP Internet of Things Process Flow

Some examples of IoT-enabled use cases are as follows:

- Delivery insights
- Cycle counting
- Replenishment optimization
- Condition monitoring
- Fleet management
- Service management
- Usage insights
- Consumer experience

Let's now dive into the details of one use case, delivery insights, and see how to enhance the SAP S/4HANA business processes with IoT sensor data. Figure 3.22 shows how the IoT sensor data is maintained when material master data is created or triggered.

Next, Figure 3.23 shows how the IoT sensor data is maintained and triggered when a new delivery is created or updated in the SAP S/4HANA sales application. Notifications are automatically sent and the delivery condition is continuously monitored.

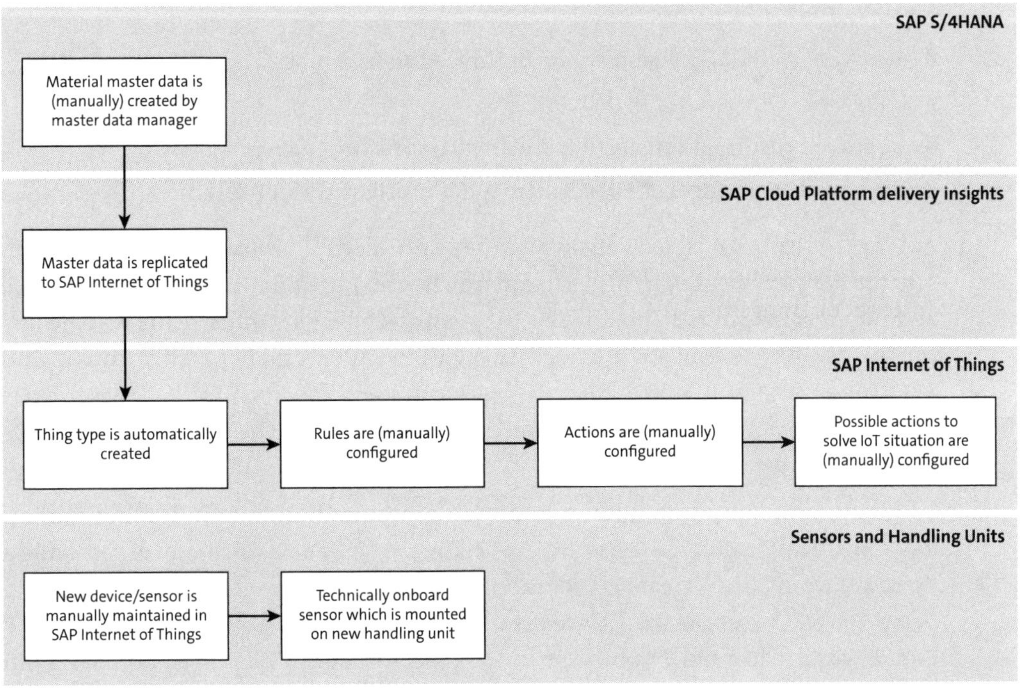

Figure 3.22 Delivery Insights Process Flow: Part I

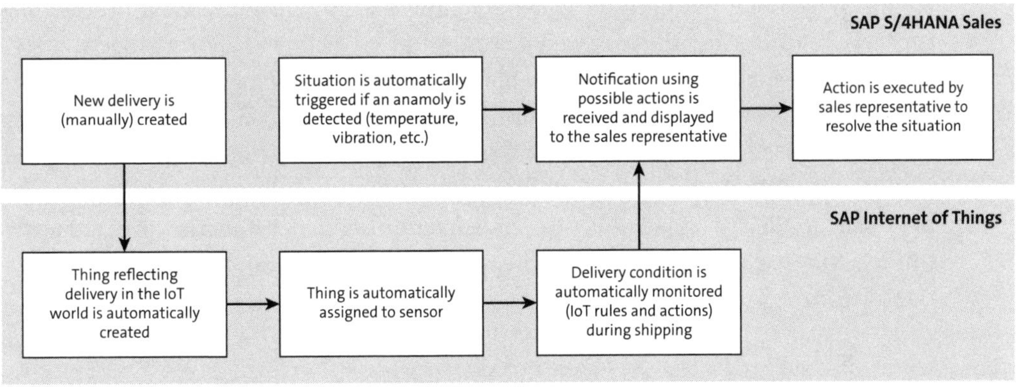

Figure 3.23 Delivery Insights Process Flow: Part II

Delivery insights helps you avoid delays in delivery and to get insights into the status of incoming materials and product data. The system provides notifications and decision support, with recommendation options such as the following:

- Reliable planning delivery of products and materials
- Avoiding production delay and rescheduling
- Decreasing the manual effort to monitor and resolve issues earlier
- Up-to-date lead time of products at the master data level for direct and indirect procurement

The predominant benefits are as follows:

- Seamless material availability and optimized transparency
- Increased transparency in deliveries
- Improved customer satisfaction due to guaranteeing delivery quality
- Improved operational efficiency through proactive decision-making

As you can see, IoT-enabled delivery is based on attaching a sensor to a handling unit. The sensor provides information like temperature, humidity, acceleration, location, and the existence of radiation or specific chemical elements. After being attached by the selling party, the sensor provides its values to SAP Internet of Things, running on SAP Cloud Platform in a tenant that belongs to the selling party. Based on these values and material-specific rules, SAP Internet of Things can trigger the creation of situations in the sales order app in the supplier system.

To resolve a situation, the salesclerk can use a recommendation service that runs on SAP Cloud Platform in the tenant of the selling party. One possible way of resolving a situation would be to create an additional situation in the system of the purchasing party. The purchaser can then make use of a recommendation service that runs on SAP Cloud Platform in a tenant of the purchasing party. Going forward, the purchaser can act accordingly with the sensor data and help address their needs ahead of the time.

The idea here is to explain the relevance of how additional sensor information leveraging SAP Internet of Things complements the SAP S/4HANA business process. Because we're discussing the machine learning and predictive analytics aspects of SAP S/4HANA, we should briefly highlight how all these AI technologies and processes are interconnected in the big story of the intelligent enterprise. As shown in the preceding example focused on delivery insights, the AI technologies complement each other while the machine learning service for supplier delivery predictions alerts you to supplier delivery delays ahead of time. The delivery insights IoT functionality enhances this machine learning service with more real-time sensor data and other recommendations accordingly.

3.7 Summary

In this chapter, we focused mainly on tools, technologies, and services in the context of leveraging predictive analytics and machine learning functionality. We discussed in detail the different approaches of embedding, consuming, and extending machine learning with SAP S/4HANA.

We started with an in-depth understanding of how predictive analytics and machine learning can be embedded into the SAP S/4HANA digital core. Later, we discussed the details of how machine learning services built on SAP Cloud Platform can be consumed by SAP S/4HANA business processes. We also discussed the approaches involving SAP

Data Intelligence. Then we continued into understanding the options for leveraging the predictive analytics services in SAP Analytics Cloud.

Finally, we also briefly discussed the concepts of RPA and IoT technologies in the context of how SAP S/4HANA business processes are leveraging these AI technologies. In these sections, we shared some of the experiences that help explain how machine learning and predictive services are enhanced with RPA and IoT technologies.

In the next chapter, we focus on the core architecture aspects behind the various machine learning and predictive analytics techniques that are implemented with SAP S/4HANA business processes.

Chapter 4
Architecture

Before we dive into the implementation steps, it's important to first get a clear picture of the architecture of machine learning and predictive analytics in SAP S/4HANA. We'll walk through the architecture of the various implementation models in this chapter.

In this chapter, we'll dive into the architecture of SAP S/4HANA with machine learning and predictive technologies. We'll begin with a discussion of the technical challenges involved with implementing machine learning and intelligent business processes. We'll then move on to an overview of the SAP S/4HANA solution architecture before taking a closer look at the technical architecture for embedded machine learning, side-by-side machine learning, and side-by-side predictive analytics. The objective is to provide you with the technical background for using the SAP S/4HANA delivered scenarios and for implementing your own machine learning and predictive analytics use cases.

4.1 Introduction

In this section, we explain the challenges of applying machine learning in the context of SAP S/4HANA. Solving those challenges is the added value that comes from the machine learning approach provided with SAP S/4HANA. We also answer an important, often asked question concerning what intelligence means within the scope of SAP S/4HANA.

4.1.1 Technical Challenges in SAP S/4HANA

Improved processing power, better algorithms, and the availability of big data are facilitating the implementation of machine learning for infusing intelligence into back-office processes and providing an intelligent ERP system. With SAP S/4HANA, SAP delivered the first intelligent ERP system on the market. SAP S/4HANA's underlying in-memory database SAP HANA increases speed, combines analytical and transactional data, and brings innovation with embedded machine learning capabilities. Thus, machine learning is natively integrated into SAP S/4HANA and can be easily used across an entire organization to optimize business operations, improve employee job satisfaction, and create better customer services. With SAP Conversational AI, a natural language experience is provided by SAP S/4HANA, which revisits the way users interact with the system due to enablement of hands-free applications.

However, incorporating machine learning capabilities into ERP solutions is a challenging task. This is due to the high complexity of such systems. SAP S/4HANA, for example, consists of more than 250 million lines of code and 143,000 tables. It supports 25 industry verticals, localizations for 64 countries, and thousands of business processes. Two substantial challenges must be solved:

1. How can we integrate machine learning systematically into business processes for ease of consumption?
2. How can we make machine learning enterprise-ready?

Machine learning models have been created by data scientists for decades. However, often those models resided in special tools and were consumed by experts only, and so they added hardly any value.

The best approach for solution architecture is to systematically build machine learning into business processes so that intelligence is provided to the right person, in the right place, and at the right time. In context of ERP systems, fused machine learning functionality must be enterprise-ready (which we'll discuss at length in Chapter 5, Section 5.5). This covers qualities like compliance, security, scalability, robustness, extensibility, configurability, operations, supportability, globalization or auditing. For example:

- Training and inference processes of machine learning must consider the General Data Protection Regulation (GDPR)
- Machine learning models are new development artifacts that require a solution for lifecycle management, starting from activation and moving through upgrades and operations
- Prediction power of machine learning models decreases with time; degradation strategies must be devised
- Reasoning behind inference results must be explained to end users and recorded for legal auditing

In this chapter, we'll focus on the solution architecture and conceptual foundations that underly enterprise readiness.

4.1.2 How to Operationalize Intelligence

Despite a long history of research and debate, there is still no common definition of intelligence. Scientists have provided various models for mathematical, linguistic, technical, musical, and emotional intelligence, but none of them are generally accepted. But how can we make SAP S/4HANA intelligent without knowing exactly what this means?

The operationalization method is applied to resolve this dilemma and make the term *intelligence* measurable by defining different automation levels, like psychologists are doing with IQ values. Intelligence in the context of SAP S/4HANA is not just an end in itself; rather, it's about increasing automation toward an autonomous ERP system to reduce the total cost of ownership (TCO), such as through faster process runtimes or optimized resources consumption. Thus, the following equation applies: the higher the level of automation of a business process or a system, the higher the level of intelligence is.

In this section, we introduce a methodology for how to measure intelligence of business processes provided by SAP S/4HANA and describe both the business and technology perspectives.

Methodology

SAP S/4HANA operates as a central system for an organization's business processes. Key questions for defining automation levels center on understanding the common structure of all those business processes. As illustrated in Figure 4.1, for a business process, there are four dimensions to be considered for automation. This is an industry standard that's transformed for ERP software.

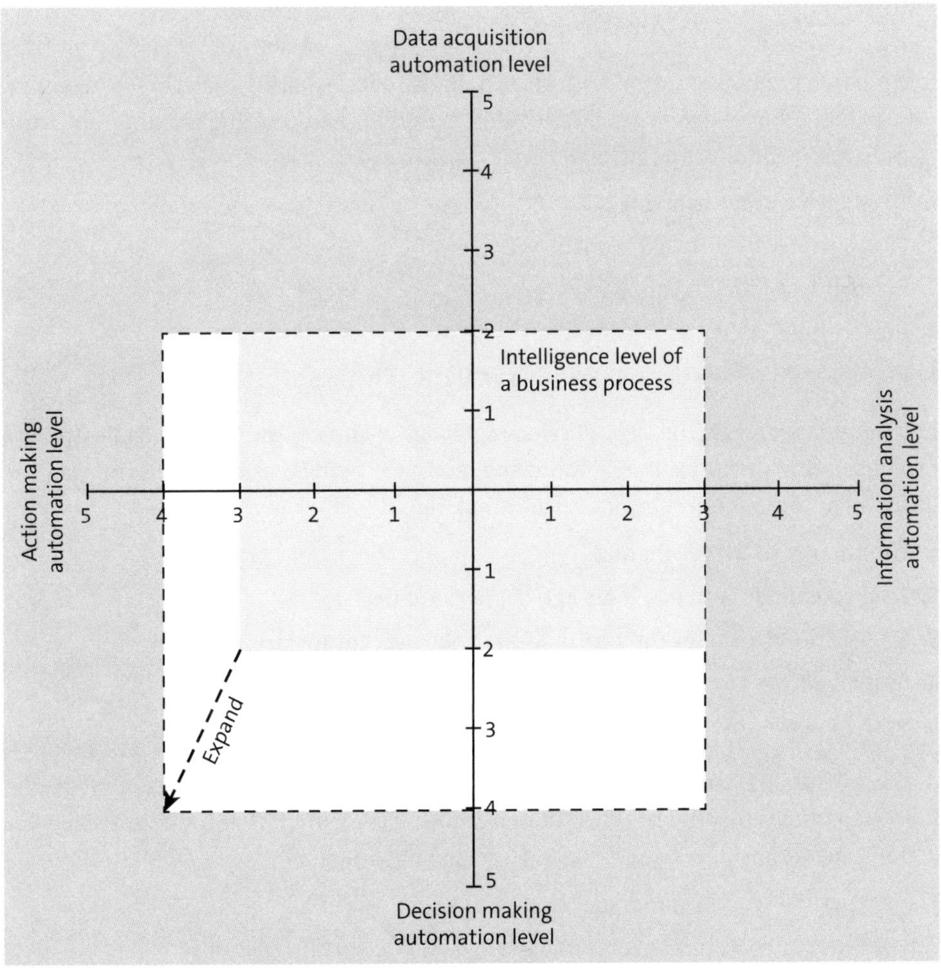

Figure 4.1 Methodology for Operationalizing Intelligence

Each dimension is rated from 1 (low) to 5 (high) for its level of automation. By determining the level of automation for each dimension for a given business process or a system, the overall level of automation can be detected. Thus, the current and target level of intelligence can be determined, and an execution plan can be defined for making the business process more intelligent.

Data acquisition is the process of inputting data into SAP S/4HANA using devices such as a keyboard, scanner, disk, or voice. The following qualities could be used for determining the different automation levels for data acquisition:

1. Manual entry by user
2. Manual entry and data integration
3. Data integration and manual entry (exceptional)
4. Conversational AI and data integration
5. AI-based data extraction and integration (e.g., PDF document is transformed to structured data and entered by robotic bot)

Information analysis is the process of studying and interpreting the data for meaningful findings. The following qualities could be used for determining the different automation levels for information analysis:

1. Descriptive (what happened)
2. Diagnostic (why did it happened)
3. Predictive (what will happen)
4. Prescriptive (what should we do)
5. Cognitive (autonomous self-learning analysis of happenings)

Decision-making is the process of selecting a logical choice from the available options by considering the impacts. The following qualities could be used for determining the different automation levels for decision-making:

1. User makes decisions manually
2. User consumes system events and changes for decisions
3. System provides relevant information to the user for making decisions
4. System actively evaluates and recommends decisions
5. System autonomously makes decisions that are traceable and auditable

Action execution is the process of enforcing instructions to achieve a specific goal. The following qualities could be used for determining the different automation levels for action execution:

1. User performs actions manually
2. User consumes system events and changes for performing actions
3. System provides relevant information to the user for performing actions

4. System actively evaluates and recommends actions
5. System autonomously perform actions that are traceable and auditable

To deepen your understanding, let's apply this methodology to the following sales performance use case: A *sales plan* is a strategy that sets out revenue goals and identifies the steps to meet those targets. Developing such a sales plan is usually a manual process of analyzing historical data in order to forecast the revenue as a key figure. The situation could be improved by applying machine learning to predict the forecast in terms of how sales could develop in the future. Thus, the manual effort for sales planning can be decreased and the actual sales volume can be increased by providing better insights to take action where needed. Figure 4.2 depicts the application of the approach to determine the current and target levels of intelligence for the explained sales performance scenario. As you can see, with predicting the sales revenue based on machine learning, the automation level of the dimensions decision and action making are increased. However, the level of automation for the dimensions data acquisition and information analysis remain unchanged because no additional intelligence is infused into them.

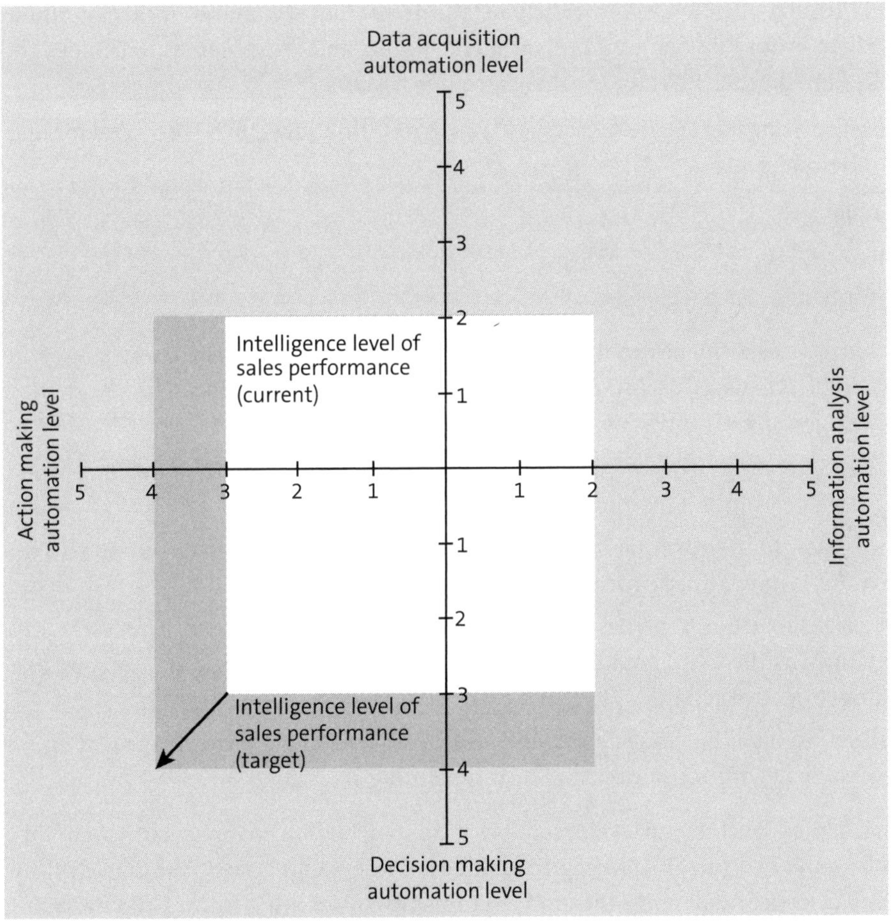

Figure 4.2 Example of Applying Methodology

Business View

Businesses of all sizes aim to increase their top line and bottom line by using a modern ERP system like SAP S/4HANA. Increasing levels of intelligence within business processes support them in achieving these goals by generating new insights from data or by enabling data-driven automation. However, the individual business value of adding intelligence is dependent upon the context or the use case in which it is used. For example, already an incremental improvement through rule-based approaches can be perceived as more intelligent compared to the status quo, thus adding more business value. The proposed framework provides the foundation for an objective discussion with various stakeholders about the current and future state of a business process and the spectrum on which business value can be generated.

Technology View

To make a business process more intelligent, first the current level of automation based on the described methodology should be determined. This is the foundation from which the target level can be derived by solution managers based on business needs. The next question is how to achieve the defined level of automation per dimension. There is not just one tool but various concepts and technologies to achieve the specified intelligence level for the dedicated dimensions.

However, as shown in Figure 4.3, categorization of techniques can be provided for realizing different levels:

- Manual
- Rule-based
- Self-learning

Business processes can be run manually without any kind of automation. To overcome this, various rule-based techniques can be applied to increase the level of automation and intelligence. Let's consider a few rule-based examples:

- An ABAP report performing input or process validations and displaying error messages with resolution instructions to the user
- A workflow for performing individual tasks/decisions moving from one step to the next until a predefined process is complete
- An insight-to-action analytics scenario starting with a key performance indicator (KPI) tile notifying the user about drifting trends, performing root-cause analysis and executing corrective actions
- A situation-handling application informing the user about a problem, providing the necessary data for resolution, and recommending actions

When it comes to achieving level 5, rule-based approaches are no longer sufficient; they must be extended to self-learning techniques. These techniques make sense of raw data and discover hidden insights and relationships by learning from that data, rather than programming explicit rules. Some examples include the following:

- Deep learning for image recognition
- SAP Conversational AI for natural language processing (NLP)
- Bots or intelligent applications based on machine learning models for autonomously making decisions and performing actions

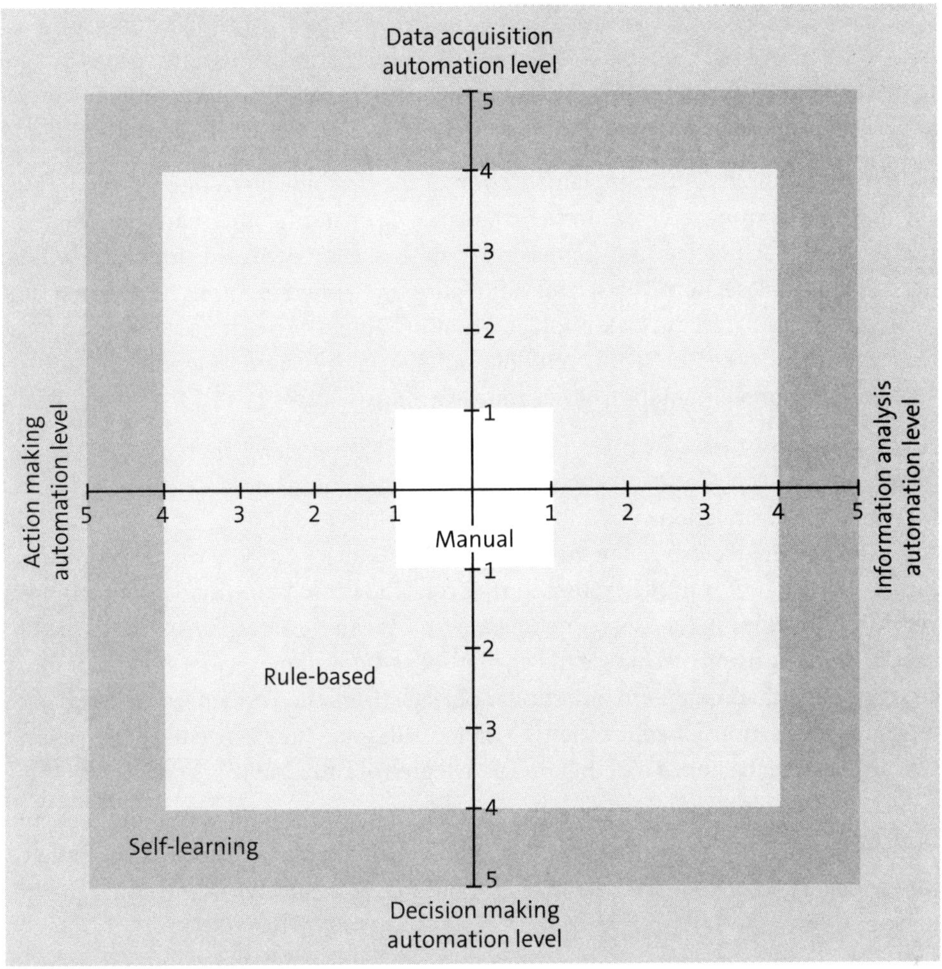

Figure 4.3 Categories for Automation Technologies

In context of SAP S/4HANA, rule-based technologies are well known, so in this book we're focusing on self-learning methods based on machine learning and predictive analytics in order to increase the intelligence level of business processes.

The described approach provides a framework to derive a strategic direction for SAP S/4HANA in terms of making it more intelligent. To infuse SAP S/4HANA with more intelligence, additional business processes must be moved to level 4 and 5, targeting an autonomous ERP system. Thus, the overall intelligence of SAP S/4HANA is reflected by the level of the intelligence of the provided business processes.

4.2 Architecture Overview

In this section, we explain which machine learning application patterns have been identified in the context of SAP S/4HANA. These are basically the underlying use cases that must be solved with the proposed architecture. We also describe the guiding principles behind the solution architecture.

4.2.1 Machine Learning Application Patterns

For defining the solution architecture, it's crucial to understand which technical capabilities are required for the implementation of the machine learning scenarios. Thus, the machine learning use cases were first analyzed concerning this matter. In this context, several machine learning application patterns were identified, and those will be described in this section. The strategic direction is to provide for each pattern a uniform concept and framework for implementation. Thus, the machine learning application patterns can be applied as reusable building blocks by development teams to accelerate the implementation of machine learning use cases.

Matching

Matching assigns relationships and detects similarities and anomalies in a given dataset. As an example scenario, say that as a master data specialist, you want to reduce the number of duplicates in the system during consolidation. While manual matching is very time-consuming for users, intelligent systems can significantly speed up matching decisions by using machine learning methods. The system can present one or more strategies and their quality to link similar objects. Users then only need to approve or reject the suggestions or adjust them to their needs. Matching is needed when at least two artifacts can be linked together due to a degree of similarity.

The process of matching follows a set of rules, which can be dynamically adjusted ("learned") by the system. The learned rules can change over time due to user input or other triggers. Matching can be applied on different content types, including the following:

- Text (e.g., search and replace) or images (e.g., find all dogs in a set of pictures)
- Audio (e.g., natural language processing, when an audio stream is matched to a query)
- Video (e.g., find out which company logos appear and how often during a football game)
- Complex business objects (e.g., matching of invoices to good receipts or finding a duplicate of a customer)

The content to be matched highly influences the type of output and its presentation. One aspect of matching is the *quality* of a match. This means that objects can fully or only partially match together. A *full match* is given when all defined parameters are met, whereas a *partial match* finds only some of the required parameters to fit. The more parameters match, the better the quality of a match is. The following matching types have been identified so far:

- **Relationship matching**
  Logically connects objects of different types—for example, matching multiple invoices to one payment
- **Compatibility matching**
  Fits objects of different types but with shared properties together to form a complete system—for example, assembly of a high/medium/low-end computer (a computer consists of several components, such as a motherboard, a CPU, memory, and a display, which have to fit together—and, for instance, the CPU fits only into motherboards with a specific socket)
- **Similarity matching**
  Merges similar objects of the same type into one—for example, merging multiple similar business partners because they're the same type of object

For developing matching patterns, commonly used algorithms include multiclass classification algorithms like XGBoost/multilayer perceptron, clustering algorithms like K-means, and nonparametric methods like the k-nearest neighbors algorithm.

Recommendations

Recommendations propose datasets or actions based on the current context. For example, as a material requirements planner, you want to see potential solutions for resolving a material shortage issue. Intelligent systems can help users by recommending appropriate content or by suggesting an action or input the user may prefer. In this case, we speak of a *recommendation pattern* and its impact on the UI. We differentiate among three types of recommendations:

1. **Content recommendations**
   The system filters the content down to what may be interesting for the user, based on the user's behavior or the content characteristics. Examples of typical content recommender systems include Amazon and Netflix.
2. **Input assistance**
   The system assists the user by entering data or filtering data. Typical examples might be a search phrase suggestion, an appropriate form template, or a set of suggested default values for certain fields, based on the user input and interaction history.
3. **Solution proposals**
   The system supports users working on complex problems by recommending specific actions or proposed solutions. In some use cases, this might be combined with a simulation of the possible outcome. Typically, solution proposals involve various decision-support systems. Examples use cases include payment and invoice matching, and material shortage scenarios.

As a prerequisite for the implementation of recommendation patterns, you must have historical data concerning actions performed and inputs provided during business processes. For the recommendation type solution proposal, the logging of business processes is mandatory; for content recommendation, the necessary historical data might be derived from application data. For the recommendation type, input assistance in addition to texts/descriptions could be required.

For recommendation patterns, commonly used algorithms include social analysis, multiclass classification algorithms like XGBoost/multilayer perceptron, text analysis/mining, or the recurrent neural network (RNN).

Ranking

Ranking distinguishes between relevant and less relevant datasets of the same type in relation to the current context. For example, as a purchaser, you want to see the top suppliers for a specific product in the context of a given purchasing request. Ranking simplifies complex decisions for business users by showing the best options first. Items in a group are ranked by comparing criteria that are relevant for the user's business context, such as an amount, priority, or score. In a ranked table or list, the results are always sorted to show the most highly ranked items on top. We differentiate between two types of ranking:

1. **Ranking by inherent value**
   Ranking is based on a value that is already available in the existing dataset, such as the price. The value is typically known and understood by the user and requires no further explanation.
2. **Ranking by score**
   Ranking is based on a calculated grade, mark, or score. In this case, users might need to understand the calculation behind the score, especially if machine learning methods are used.

Although you can rank a list of items based on their rating, ranking and rating are two different concepts. A *rating* places a single item on a predetermined scale—for example, rating your service provider on a scale from 1 (very bad) to 5 (very good). *Ranking* is always about comparing a common value across a group of items. In the UI, rankings are usually shown for a list or group.

For developing relevance and ranking patterns, commonly used algorithms include classification algorithms like XGBoost and clustering algorithms like K-means/the gaussian mixture model, as well as nonparametric methods like k-nearest neighbors algorithm.

Predictions

Predictions forecast future data and trends based on patterns identified in past data, taking into account all potentially relevant information. For example, imagine that as a master data manager, you want to estimate the potential number of change requests a team will need to process in the next quarter to balance the workload. Intelligent systems based on predictive models significantly reduce the cost required for companies to forecast business outcomes, environmental factors, competitive intelligence, and market conditions.

There are two main classes of predictive models: parametric and nonparametric. The features of these can be combined to form a third class, semiparametric models. Usually, *parametric models* make specific assumptions with respect to one or more of the population parameters that characterize the underlying distribution. *Nonparametric models* typically include fewer assumptions about structure and distributional arrangement but often encompass strong assumptions about independencies.

Some algorithms for predictive models include ordinary least squares, generalized linear models (GLM), logistic regression, random forests, decision trees, neural networks, and multivariate adaptive regression splines (MARS). Each of those algorithms has a particular use and answers a specific question or uses a certain type of dataset.

Categorization

Categorization assigns datasets to predefined groups (classes). For example, as a service agent, say you want to classify the priority of incoming requests (high/medium/low) based on their content to improve customer service. Categorization can also discover new groups (clusters) in the datasets, such as grouping customers into segments for appropriate product offerings, targeted marketing, or fraud detection.

Categorization is a complex task in which intelligent systems can help increase the level of automation by applying machine learning methods of classification and clustering. Those are used for the categorization of objects into one or more classes and clusters based on their features. Classification and clustering are similar processes, but there is one minor difference. For classification, there are predefined labels assigned to each input instance according to its properties; in clustering, those labels are missing. Because classification uses labels, training and testing datasets are required for verifying models. This is not necessary for clustering. Usually classification is more complex than clustering because there are many levels in classification, whereas only grouping is done in clustering.

For developing categorization patterns, commonly used algorithms include classification methods like XGBoost, neural networks like CNN/RNN/GAN, dimension-reduction algorithms like principal component analysis, and clustering algorithms like K-means/the gaussian mixture model.

Conversational AI

Conversational AI interacts with a system based on natural language conversation and enables a hands-free paradigm. For example, as a purchaser, say you want to create a purchase order by talking with the system. Being able to have a conversation with a digital assistant to complete your business processes is a key part of the user experience for an intelligent application.

SAP Conversational AI understands typical natural language patterns to search for business entities using various parameters, look up a specific business entity by name or ID, retrieve the value of an attribute of a specific business entity, and create simple new entities, including line items. SAP Conversational AI enables a humanized user experience for applications to get tasks done in the context of the business data. Thus, interaction with applications in natural language is supported. Creation of business objects with information prepopulated from a conversational context is ensured. The SAP Conversational AI technology supports sharing of notes, screenshots, and business objects with other users within the context of a conversation. The ability to synthesize business transactions from multiple SAP applications at a single point of interaction is enabled with this technology. Customer skills for using a digital assistant can be created and deployed across applications and channels.

4.2.2 Guiding Principles for Solution Architecture

The defined solution architecture will follow the key guiding principles of SAP S/4HANA, which we'll describe in this section, as depicted in Figure 4.4.

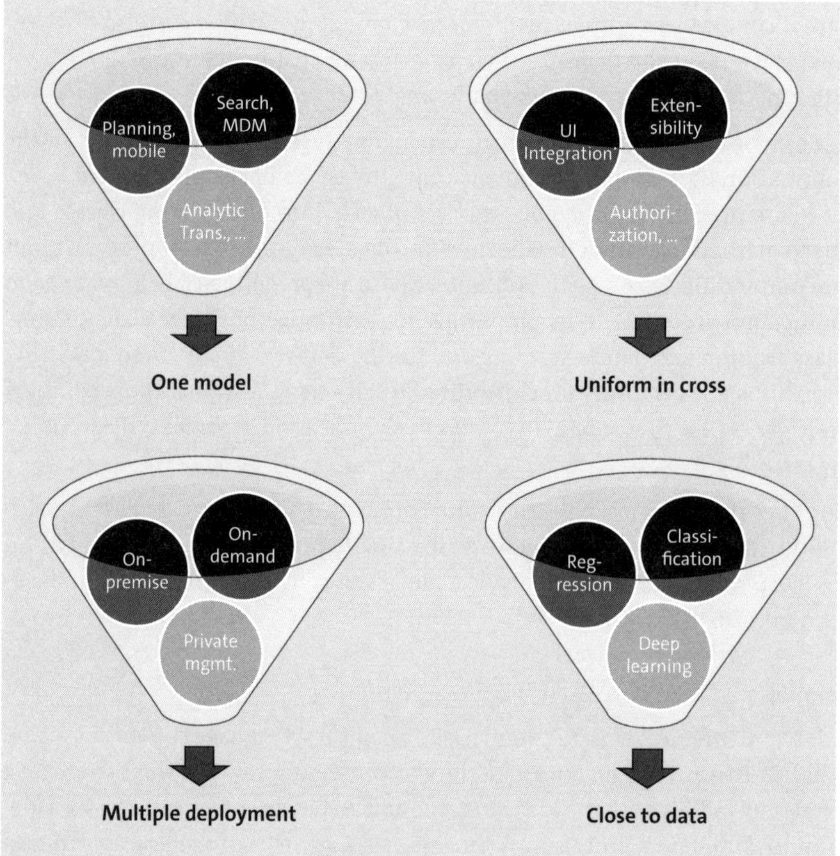

Figure 4.4 Guiding Principles for Solution Architecture

Let's walk through these principles:

- **One model/uniform in cross**
  Often the core data model is defined numerous times for different purposes only because minor metadata must be added. For example, technology solutions for integration, user interface (UI), analytics, or transactions unnecessarily require their own specific business object models. This increases the total cost of development (TCD) as the same content must be provided again and again. Due to incompatible metamodels, cross topics like UI integration, extensibility, and authorization must be solved various times. This results in a high TCO and makes the consumption of solutions difficult for customers. Therefore, one core data model will be defined and be reused in different contexts by domain-specific enhancements.

  The data model of SAP S/4HANA is based on core data services (CDS). Thus, redundant data modeling is avoided and cross topics are solved uniformly. For machine learning, this data model will be reused as much as possible.
- **Multiple deployments**
  SAP S/4HANA supports multiple deployment options. Therefore, the defined solution must be invariant regarding different deployment options and work in on-premise, privately managed, and public cloud environments. This must be ensured also for machine learning applications.
- **Close to data**
  In the context of machine learning, one golden rule is to bring the algorithms to the data and not vice versa. Algorithms consist of minor lines of code and are usually self-contained. In contrast, application data has high volume and holds numerous dependencies. Thus, replication or extraction of application data is usually a very complex and TCO-intensive task (e.g., handling of deltas, meeting performance requirements, interpreting the data semantically) and should be avoided.

4.2.3 Solution Architecture

Improved processing power, better algorithms, and the availability of big data are facilitating the implementation of machine learning functionality in SAP S/4HANA. Furthermore, SAP S/4HANA's underlying in-memory platform, SAP HANA, increases speed, combines analytical and transactional data, and brings innovation with embedded machine learning capabilities. With SAP Data Intelligence, additional features are integrated into SAP S/4HANA and the platform covers the entire spectrum, ranging from simply consuming intelligent services to training and deploying its own machine learning models.

Use cases like ranking, categorization, and prediction can be solved with classic algorithms like classification, clustering, regression, or time series analysis. Usually those algorithms do not allocate much memory and CPU time. Thus, they can be implemented within the SAP S/4HANA stack, where the application data for model training and the machine learning consuming business processes also are located. This embedded machine learning architecture has very low TCO and TCD. As illustrated in Figure 4.5, the embedded machine learning architecture is based on the SAP HANA machine learning provided with the Predictive Analysis Library (PAL) and the Automated Predictive Library (APL) as the necessary algorithms.

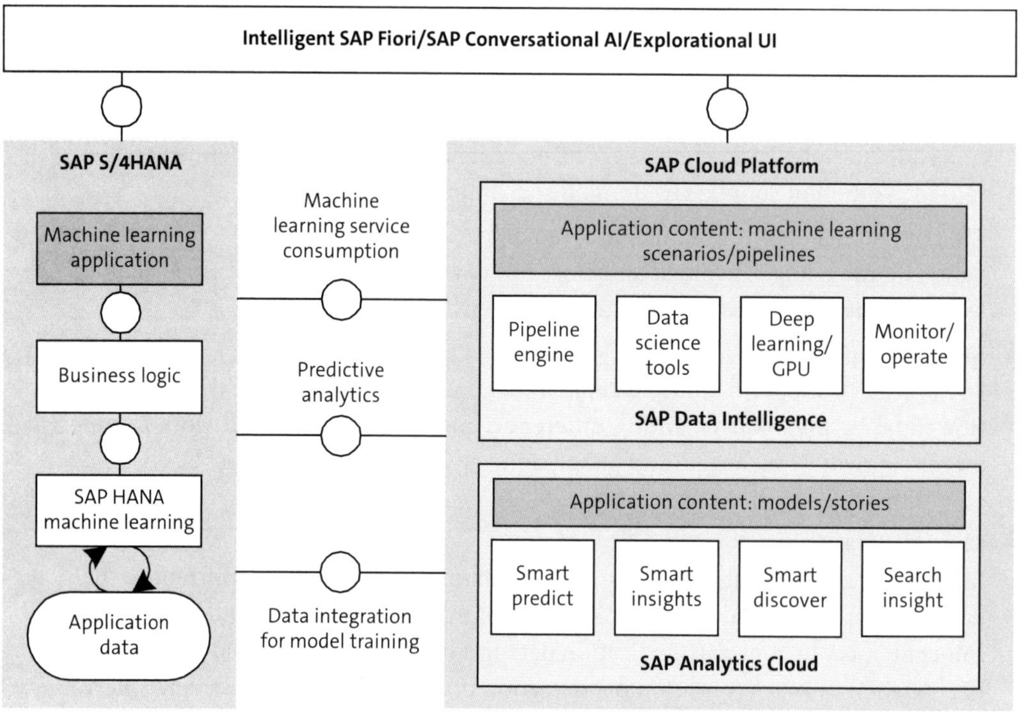

Figure 4.5 Overall Solution Architecture

Use cases like image recognition, sentiment analysis, and NLP require deep learning algorithms based on neuronal networks. For model training, usually these kinds of algorithms demand a huge volume of data and graphics processing unit (GPU) time. Therefore, these kinds of scenarios are scaled out to SAP Data Intelligence to keep the load in the transactional SAP S/4HANA system low. For this side-by-side machine learning approach, the requested data—images, audio, text documents, historical data, application logs—are typically stored not in SAP S/4HANA but rather in the business data lake. In general, the consumption of a trained model is based on remote interfaces. However, for mass processing those interfaces must be bulk-enabled, or local deployment of inference models must be provided.

Predictive analytics uses machine learning to automate data preparation, insight discovery, and insight sharing for business users, operational workers, and data scientists. SAP Analytics Cloud provides predictive analytics features, also known as *smart features*. These features are dedicated to uncovering insights, understanding root causes, and predicting what's likely to happen in the future. This ensures decision-makers have access to the right information at the right time to influence the best possible outcomes. The focus here is on smart predict, which is an explorative approach for unforeseen, exceptional, and irregular needs for machine learning methods.

A conversational UI for natural language interaction with SAP S/4HANA is enabled by SAP Conversational AI. This is a self-learning solution using machine learning functionality to gain knowledge based on historic data and experience. Machine learning requires additional visualization capabilities in the UI, such as for illustrating confidence intervals or forecasting charts. Thus, to embed machine learning capabilities in UIs, intelligent SAP Fiori elements are used.

In the following sections, we'll provide more details about embedded machine learning, side-by-side machine learning, and predictive analytics architecture. The focus here is on the development architecture; the data science tasks are not within the scope of this book.

4.3 Embedded Machine Learning

Embedded machine learning is an appropriate approach for use cases like ranking, categorization, and prediction, in which classic algorithms such as classification, clustering, or regression are sufficient for the implementation. The SAP S/4HANA platform contains with SAP HANA machine learning libraries that can be used for the development of embedded machine learning scenarios without the need to move application data. For a given key question to be solved with machine learning, data scientists perform experiments and explorations to determine the required algorithms and the data attributes for model training. Those are the input for the development of the machine learning use case that is the focus of this section.

As illustrated in Figure 4.6, the solution is based on two main architecture decisions: using CDS views and making use of the machine learning techniques provided by SAP HANA.

The data persistency in SAP S/4HANA is represented by application database tables. On top of these tables, a virtual data model (VDM) is implemented using CDS views. One purpose of these VDM views is to hide the cryptic database model and to provide a reusable semantic layer that can be consumed in different scenarios, such as analytics, planning, search, or transactions. At runtime, the CDS views are consumed via the analytical engine or the Service Adaptation Description Language (SADL), which is part of the ABAP application server. Those frameworks evaluate the metadata of the CDS views in terms of annotations to enable the required functionality for the business process, such as hierarchy handling or analytics capabilities. Out of the ABAP CDS views, the SAP HANA SQL views are generated so that SQL operations can be pushed down to SAP HANA for optimal performance.

The algorithms for embedded machine learning can be performance-intensive as high volumes of application data must be processed. Thus, for performance optimization, the algorithms should be processed close to the application data. SAP HANA provides PAL and APL to offer statistical and data mining algorithms. In addition, specific algorithms can be implemented if required. As shown in Figure 4.6, those algorithms are invoked and orchestrated by the modeling and administration component, which is explained later in this section.

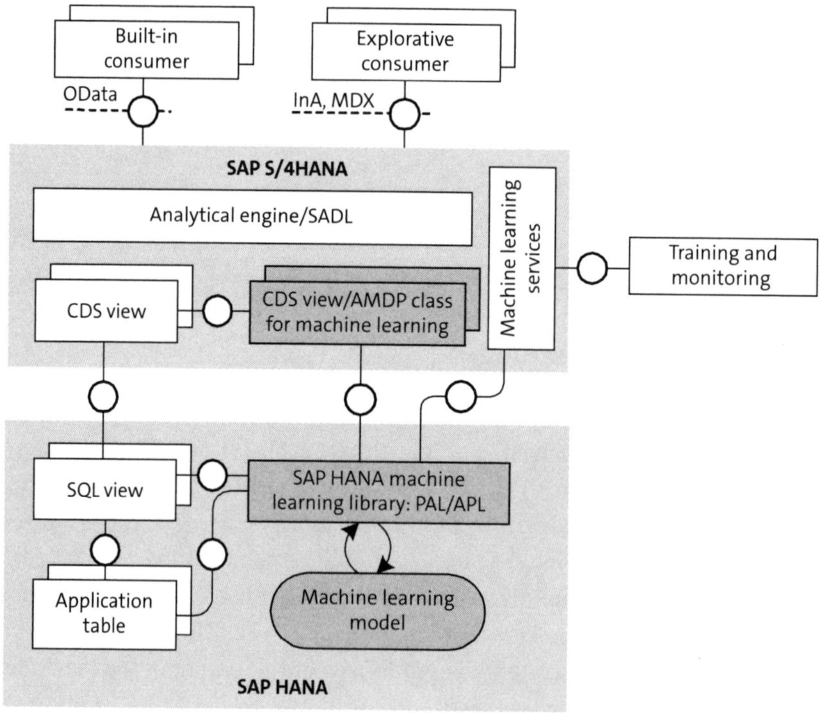

Figure 4.6 Embedded Machine Learning Architecture

The algorithms require application data as input for model training. The VDM, with its SQL views and the application tables, can be reused for this purpose. However, in contrast to the ABAP CDS views, not all metadata is available. This drawback can be eliminated once the ABAP CDS views can be transformed to SAP HANA CDS views. The trained models are exposed to business processes by wrapping them with CDS views. The details of this integration are described in the next section. The CDS views for machine learning can be combined with other VDM CDS views and can then be exposed to the consumers. By consuming machine learning models through CDS views, existing content (e.g., VDM views) and concepts (e.g., authorization, extensibility, UI integration) are reused. This results in a simple and very powerful solution architecture.

The modeling and administration tool illustrated in Figure 4.6 was previously known as the Predictive Analytics integrator (PAi). As of the 2020 SAP S/4HANA release, it is rebranded as the intelligent scenario lifecycle management (ISLM) framework. Its purpose is to provide a common interface for the consumption of machine learning models independent of the underlying predictive engine. The intention is to harmonize the management of machine learning models and to provide a simple common interface to allow applications to interact with different types of supported machine learning libraries without the requirement for applications to develop machine learning engine-specific code. Consumer applications interface only with application programming interfaces (APIs) and do not interact with low-level machine learning libraries.

The ISLM architecture is shown in Figure 4.7. The ISLM framework contains information about the installed SAP HANA libraries. It provides a repository for machine learning models that includes, for example, information concerning model types (e.g., regression, classification, time-series), model data sources (e.g., tables, views), model training data, and model quality figures to allow for assessing metrics and supporting model comparison. The ISLM framework also provides a pluggable infrastructure and adapters for APL and PAL. In addition to hiding the complexity of machine learning libraries from application development, the ISLM framework supports the lifecycle management of the involved artifacts in terms of transport within the system landscape, delivery, and upgrade mechanisms. Furthermore, it provides configuration capabilities for model training based on the customer data in their development landscape. This framework is going to be explained in detail in Chapter 5.

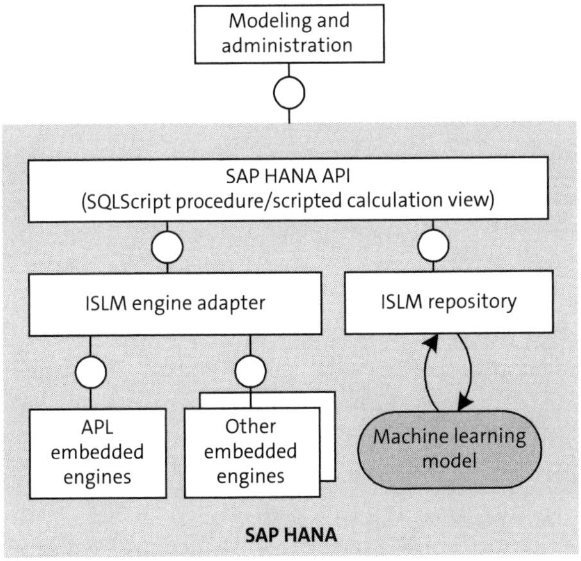

Figure 4.7 ISLM Architecture

Machine learning models are stored in SAP HANA and exposed by generated SQL script procedures or scripted calculation views. But how are those SAP HANA entities consumed from the ABAP application server?

Various techniques are provided by the ABAP application server to access the underlying database system. As already mentioned, existing content and concepts are reused wherever possible. Therefore, CDS views and CDS table functions are the appropriate technologies to consume the machine learning models stored in SAP HANA, as shown in Figure 4.8. SQL script procedures are used to wrap the machine learning models on the SAP HANA level. Those SQL script procedures can't be invoked by standard CDS technology but can be with CDS table functions. These are used for breakout scenarios. They are data dictionary objects and are defined via data definition language (DDL) sources, but in contrast to CDS views, they are not database-independent; they represent native SQL script functions. These functions are implemented using the techniques of ABAP Managed Database Procedures (AMDP)—that is, via ABAP with an implementation in SQL script. On top of the machine learning models exposed via SQL script procedures, corresponding CDS views for machine learning are defined by using CDS table function technology. The CDS table function represents the input and output structure of the SQL script procedure that is called by the AMDP implementation of the CDS table function. The provided CDS view, and respective CDS table function, can be combined with other VDM views and consumed accordingly. For this purpose, OData services can be generated automatically. Existing CDS annotations can be used.

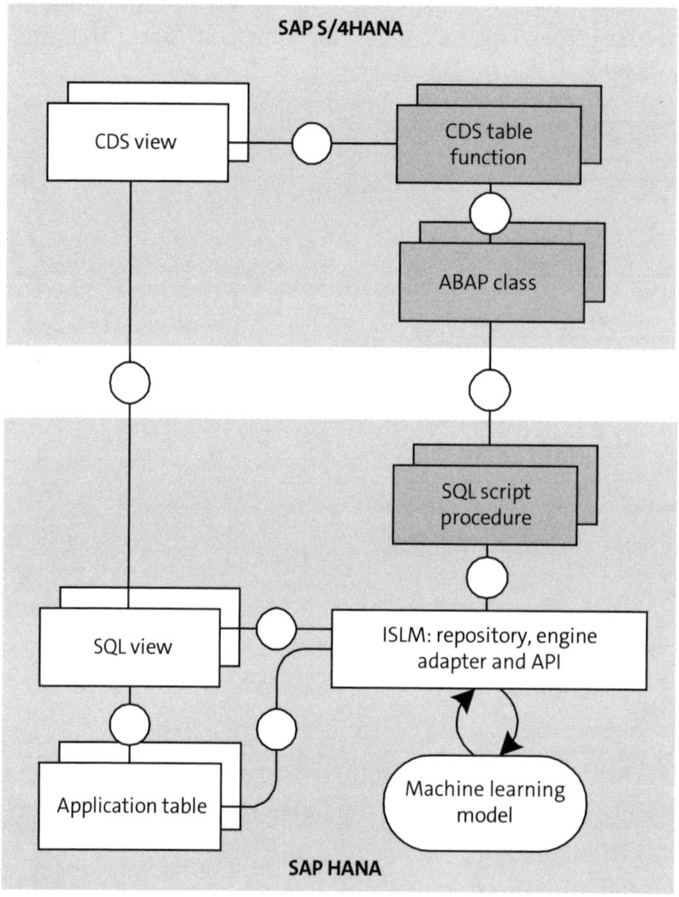

Figure 4.8 ABAP Integration

APL has built-in data science steps like feature engineering and detection of adequate algorithms. Therefore, this library is very easy to use und results in very low TCO and TCD. However, the algorithms are restricted to classification/regression, clustering, time series analysis, and recommendation models. Thus, for breakout scenarios in terms of using other algorithms or fine-tuning the models, PAL is used; it provides more than 100 machine learning algorithms. Figure 4.9 illustrates how this library is integrated for embedded machine learning.

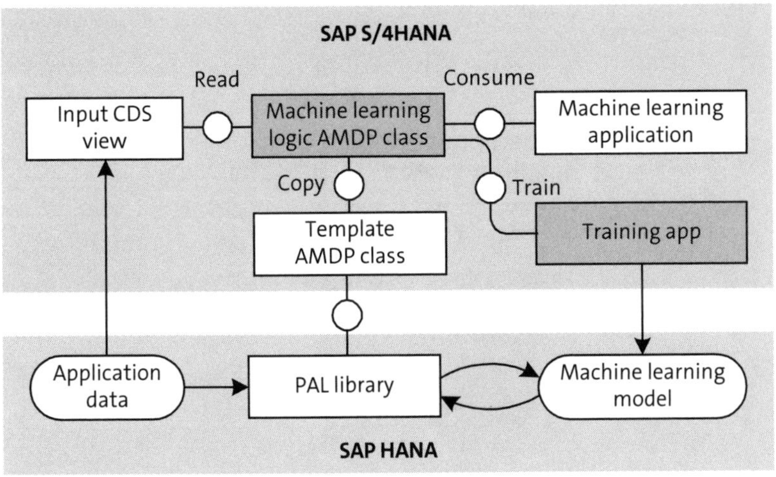

Figure 4.9 PAL Integration

For development efficiency, for each PAL algorithm, a template in the form of an AMDP class is provided. This class implements predefined interfaces and offers standardized methods for training and consumption of models. For extensibility, business add-in (BAdI) enhancement points are provided for those methods. Application developers copy and adopt those templates for specific machine learning scenarios. CDS views as inputs for model training are defined. Consumption and integration of the machine learning models into the business processes is based on the method of the AMDP class. Concrete machine learning model instances are available once the customer triggers training via the training application.

The following qualities are provided by this solution architecture using PAL:

- Seamless integration into the SAP S/4HANA programming model
- Reuse of existing SAP S/4HANA concepts (e.g., authorization, UI integration), CDS view content, and tools
- Powerful extensibility based on enhancement points and BAdIs from start due to reuse of the transactional programming model
- Rich set of PAL algorithms and usage of SAP HANA text analysis
- Supports breakouts per definition, such as orchestration of multiple algorithms or data transformations

- Lifecycle management and operations with SAP S/4HANA ABAP tools and concepts
- GDPR compliance by design; no data transfer
- On-premise and cloud deployment ensured due to one-code-line approach

Lifecycle management and ABAP integration of PAL models are also based on intelligent scenarios, as depicted in Figure 4.10. The details of lifecycle management will be described in Chapter 5, Section 5.5.1.

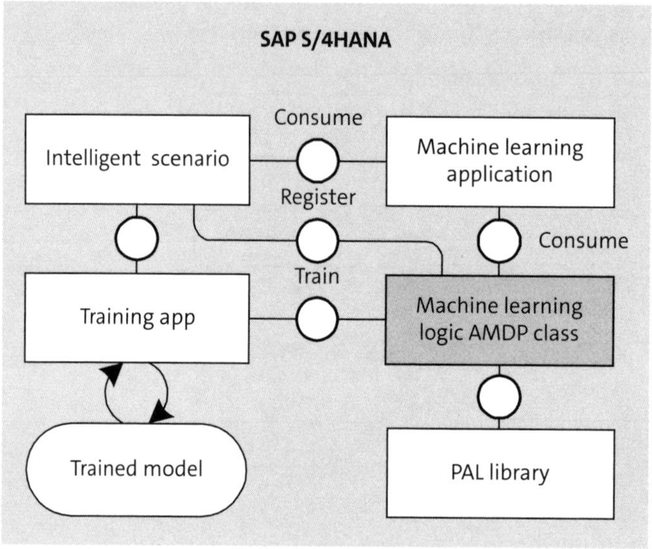

Figure 4.10 PAL Model Training

As already mentioned, PAL algorithms are integrated via AMDP technology from SAP HANA into ABAP. The AMDP class implements an interface with methods for training and consumption. The intelligent scenario is created for the PAL-based use cases, too. An *intelligent scenario* is a business application-specific machine learning design time artifact that includes metadata about an application data source for training (e.g., table or view), an algorithm type (e.g., regression or classification), and a consumption API. The AMDP class for the PAL algorithms is registered in an intelligent scenario. Because the generic training application is based on intelligent scenarios, it triggers the training method of the AMDP class and saves the trained model. Machine learning applications can consume the trained model via the AMDP class.

4.4 Side-by-Side Machine Learning

The journey toward the intelligent enterprise is enabled by the interaction among three main pillars: data, intelligent algorithms, and business processes. Nowadays, enterprises have big and diverse enough data assets to feed intelligent algorithms, which can be used to innovate critical business processes, which will then produce even more fine-grained contextualized data, feeding a virtuous cycle. Embedded machine learning targets scenarios in which the business and machine learning logic reside in the SAP S/4HANA platform, but the term *side-by-side machine learning* is used in the following use cases:

- **SAP S/4HANA machine learning based on SAP Cloud Platform**
  The SAP S/4HANA application and the according business logic are based on SAP Cloud Platform. Such applications should consume the required machine learning services directly from SAP Data Intelligence and SAP HANA machine learning, following the rule of bringing the algorithms to the data.
- **SAP S/4HANA machine learning based on ABAP application server**
  The SAP S/4HANA applications and the related business logic are based on the SAP S/4HANA platform. However, the required machine learning capabilities are not available on SAP S/4HANA—for example, image and language recognition, or sentiment analysis. These features are consumed from SAP Data Intelligence remotely or based on an exchange of trained models.

Side-by-side machine learning fits use cases like image recognition, sentiment analysis, or NLP that require deep learning algorithms based on neural networks. These kinds of algorithms are very demanding regarding consumption of system resources. They usually require huge volumes of data and GPU time for model training. Thus, to keep the load in the transactional SAP S/4HANA system low and ensure acceptable runtime for the business processes, side-by-side machine learning scenarios are scaled out to SAP Data Intelligence. This infrastructure also complements the overall solution in which specific algorithms are not provided on the SAP S/4HANA platform, classic methods (e.g., regression, classification) consume too many resources of the transactional system, or huge volumes of external data (e.g., Facebook, Twitter) are required for model training. In particular, SAP S/4HANA extension applications should consume SAP Data Intelligence capabilities because the application data and business processes are founded on SAP Cloud Platform. Thus, the golden rule of bringing the algorithms to the data is followed.

So because side-by-side machine learning scenarios are based on SAP Data Intelligence, the question arises: How do we integrate this technology into SAP S/4HANA in terms of model training and inference? The answer to this question is given next.

SAP Data Intelligence is an important component of the side-by-side machine learning architecture. It's designed for the cloud and deployable in any cloud, hybrid, or on-premise landscape that addresses the entire end-to-end lifecycle for creating value out of data, combining enterprise AI with intelligent information management, and enabling enterprises to effectively operationalize machine learning and data science on complex and real enterprise landscapes. We introduced SAP Data Intelligence in Chapter 3, Section 3.3.1.

SAP Data Intelligence handles how customers discover, refine, govern, orchestrate, and scale efforts in getting intelligence out of data assets they own or those to which they need to connect. This covers all the data management use cases that deal with different kinds of data (structured, unstructured, streaming), different integration approaches (batch, real-time, near-real-time), or different processing patterns (offline, online, lambda). Key capabilities include the following:

- **Data connectivity and orchestration**
  Leverage central connection management to connect to data wherever it resides—on premise or in the cloud—regardless of the data type—structured, unstructured, streaming—and integrate it with flexible data pipelines. Orchestrate data processing across highly distributed and heterogeneous landscapes, executing any SAP or non-SAP processing engines close to the data sources to minimize the amount of data to be moved or replicated.
- **Data governance and cataloging**
  Access an advanced metadata management system/catalog, enabling data lineage, data quality, profiling, data discovery, and searching of datasets to ensure auditability and governance. This gives IT team members the flexibility and control they require to ensure trusted and accurate data is easily discoverable by the teams that need it, all integrated within a single solution.
- **End-to-end lifecycle management of machine learning models**
  Streamline data science and machine learning projects, from modeling and development to operations, across all enterprise data assets to manage the end-to-end lifecycle. A central repository enables versioning and a tailored lifecycle management process for machine learning projects.

  Support data discovery, access and preparation, and experimentation in Jupyter Notebook; leverage a library of pretrained models for the most common functional services; and support deployment, (re)training, serving, and monitoring of all models. Access ready-to use, adaptable business content in terms of operators and templates. Finally, SAP Data Intelligence provides an environment for model deployment and operation, a means to integrate results back into an application or employ delayed consumption, and continuous testing and maintenance of all models in production.
- **One integrated solution**
  SAP Data Intelligence includes data pipelining, orchestration, machine learning, and metadata cataloging in a single solution. This is very valuable: all hyperscalers have different services for these functionalities, which have to be integrated, while the main pure players and niche players focus only on a subset of these areas.
- **Hybrid and multicloud deployment**
  SAP Data Intelligence is available both as a service in SAP Cloud Platform and as a bring-your-own-license product. It has been natively built on Kubernetes since its initial design, allowing it to deploy the very same solution in any hyperscaler, private cloud, or on-premise data center.

- **Native integration capabilities**
  Besides reusing all relevant open-source technologies and open standards, SAP Data Intelligence is also capable of natively integrating and reusing SAP data sources and engines. SAP Data Intelligence pipelines can natively integrate into ABAP applications; orchestrate SAP Business Warehouse (SAP BW) process chains, SAP Data Services jobs, and SAP HANA flowgraphs; execute SAP HANA streaming analytics scenarios; and integrate into all SAP Cloud Platform applications.

Many of these capabilities may exist today in a customer's information management landscape, but they are typically offered in a myriad of different ways across several disparate toolsets that require different skills and different frameworks, whereas SAP Data Intelligence provides a single, comprehensive way to manage all data types cohesively and intelligently.

SAP Data Intelligence is designed to be used by different user profiles throughout the enterprise. From a business user with technical affinity to developers and data scientists, there are modules, services, and tools for all levels. The typical lifecycle SAP Data Intelligence supports comprises the following phases:

1. **Data management**
   With a comprehensive set of intelligent information management capabilities, it's possible to manage the data that drives the entire AI process, making it possible for IT enterprise architects, data engineers, and data management professionals to eliminate data silos and ensure that the data needed for data science teams is made available in a governed manner. With tools to help profile, prepare, and merge data, data science teams can rapidly get to the datasets they need to move on to the next phase.
2. **Experimentation**
   For data scientists developing models, SAP Data Intelligence provides the tools they are experienced with, like Jupyter Notebook, and the frameworks they require, like R, Python, or TensorFlow. Once provisioned, data science teams can work in a Jupyter Notebook environment to develop models using open-source frameworks and SAP frameworks (like APL and PAL), pretrained services, and a visual pipeline GUI to orchestrate data ingestion, training steps, or multiple models. These experiments for a specific business problem are put under version control so teams can explore various potential solutions and models and easily test and iterate until they land on a model they would like to pursue in production.
3. **Productization**
   Models can rapidly be delivered into production by packaging the required assets (like pipelines and model assets) into a machine learning scenario. In production, a team responsible for machine learning operations can easily take over and connect production data to the new scenario, (re)train, deploy models to model servers, integrate with business applications, and start generating insights. Once in production, models can be centrally managed from a single cockpit in which the ongoing testing, retraining, and quality of those models can be monitored.

Models and pipelines can be reused, recombined, and traced through the entire process and to jumpstart new scenarios. This includes reuse of datasets as well. The machine learning functionality of SAP Data Intelligence is used directly by data scientists through a set of tools run as web applications either inside SAP Data Intelligence or indirectly by applications that call representational state transfer (REST) APIs of SAP Data Intelligence. Both tools and backends providing REST APIs are based on the system application server, which helps to delegate aspects like user authentication and authorization to the platform. Tools include design-time tools for data scientists, like the open-source JupyterLab component and the SAP Data Intelligence pipeline modeler. Moreover, there are tools for managing the lifecycle of a machine learning project and the lifecycle of datasets used in a machine learning context.

As illustrated in Figure 4.11, SAP Data Intelligence provides a data lake for business data. Thus, application data can be extracted from SAP S/4HANA for training machine learning models.

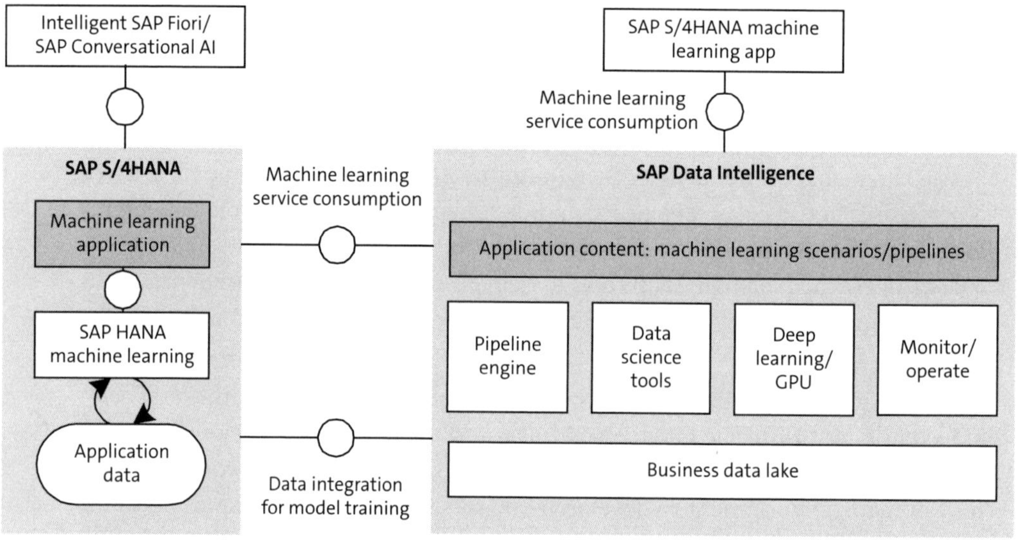

Figure 4.11 Side-by-Side Machine Learning Architecture

For pre- and postprocessing of the application data, the pipeline engine offers a graphical programming model to create pipelines. A *pipeline* orchestrates data ingestion, training, and inference activities. It consists of operators and data flows connecting those operators. Operators can be predefined connectors to integrate into data sources leveraging proven SAP technologies from SAP Data Services, SAP HANA smart data integration, SAP Landscape Transformation Replication Server (via an ABAP agent), freely programmable options (like a Python operator), or serving operators for exposing a REST endpoint. For machine learning cases, SAP Data Intelligence provides specific machine learning operators that call functional services (like for image classification) or core services (for model serving and training) provided by SAP Data Intelligence or by hyperscaler services. With the operator concept, SAP Data Intelligence is completely open to third-party frameworks a data scientist would like to work with. There is also specific support for SAP HANA-embedded machine learning; that is, a connection to any SAP HANA, PAL, or APL data stored in SAP HANA can be used from an SAP Data Intelligence context.

The pipeline engine orchestrates complex data flow pipelines, is based on scalable infrastructure provided by SAP Cloud Platform/Kubernetes, supports heterogenous execution runtimes (e.g., R, Python, Spark machine learning), and enables connectivity to SAP S/4HANA. Based on the application data, exploration and feature engineering is performed by data scientists to define machine learning models. For this, common data science tools like Jupyter Notebook and Python are supported. For deep learning scenarios, a framework is provided that enables training on a GPU infrastructure.

To implement machine learning use cases, applications must define machine learning scenarios and model pipelines. SAP Data Intelligence organizes each machine learning use case by the artifact machine learning scenario. This bundles all design-time entities to solve a machine learning business question. In addition, it is the bracket keeping track of consumed and produced artifacts, like datasets and machine learning models, as well as the metrics reported by these, and is thus the basis for machine learning scenario lifecycle management. Thus, the machine learning scenario artifact contains all development entities that are required for the implementation of a specific machine learning use case.

Inference and training processes are developed as pipelines comprising sequential and parallel tasks. In particular, for each machine learning scenario, a training pipeline is provided, which receives the training data from SAP S/4HANA and processes the data to train the algorithms for a specific use case. Structured data is handled by a CDS operator and stored, such as in SAP HANA, whereas unstructured data is managed by an object store and big data storage. Often, saved application data is deleted after a training run. However, there are use cases with high retraining frequency, in which, after an initial upload, the deltas periodically must be received and stored for the next training run. The application data can be persisted. Alternatively, streaming based on continuous data transfer without persistency can be applied. The training and inference pipelines are exposed by REST services. Those are invoked by SAP S/4HANA applications remotely and integrated into the business processes and UIs. Thus, machine learning capabilities are provided as built-in functionality to the right person, in the right place, and at the right time.

The operation and monitoring of machine learning models are managed with different administration applications:

- The machine learning operations cockpit provides a view of currently deployed models and their runtime KPIs and produced artifacts. It allows for manual activation of machine learning scenarios and calling the associated pipelines. It also supports landscape management, model configuration, and provisioning of machine learning scenarios to other tenants.

- Scenario scheduling allows automated calls to the pipeline API.
- The debrief cockpit provides data scientists with KPIs for the created inference pipelines/models during the exploration and retraining phases.

Figure 4.12 drills into the details of the integration between SAP Data Intelligence and SAP S/4HANA. On the SAP S/4HANA side, an intelligent scenario is the artifact corresponding to the machine learning scenario on the SAP Data Intelligence side. An intelligent scenario is also a design-time artifact that represents a machine learning use case and contains metadata like the name and description of the use case. In particular, it encompasses the ABAP class, which implements the consumption API of the machine learning model.

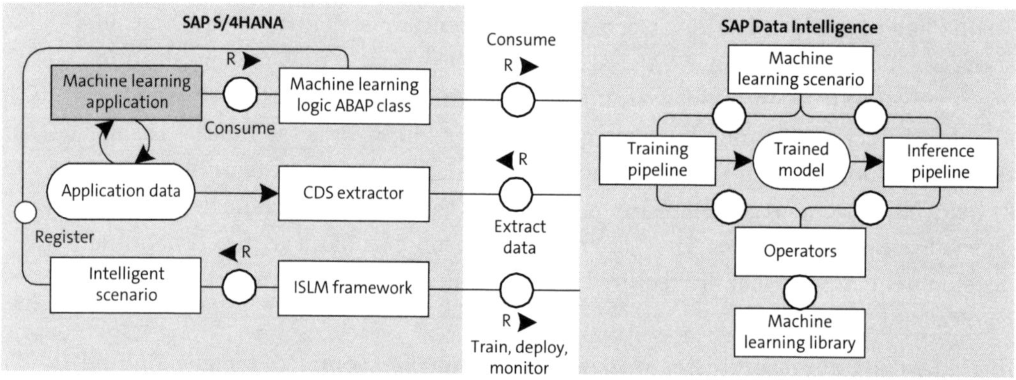

Figure 4.12 SAP Data Intelligence Integration into ABAP

As already mentioned, in SAP Data Intelligence, a pipeline for model training and consumption is provided for each machine learning use case. The training pipeline reads the necessary application data based on the CDS extraction view, which is provided by the SAP S/4HANA application. The pipelines are exposed via REST services and can be invoked from SAP S/4HANA. The training REST service has a standardized signature and can be called generically the ISLM framework. This component is in charge of the lifecycle management of the machine learning models and provides capabilities like scheduling of training jobs or monitoring. The signature of the REST services for model consumption is scenario-specific (e.g., forecast revenue for sales orders or predicting debt default risk) and is handled by the machine learning logic ABAP class shown in Figure 4.12. This ABAP class basically wraps the REST service into an ABAP API that can be consumed by machine learning applications in order to integrate inference results into business processes and UIs. Optionally, inference results can be cached for scenarios in which performance optimization is required.

Figure 4.13 illustrates the details of the integration with a machine learning logic class `CL_ML_LOGIC`.

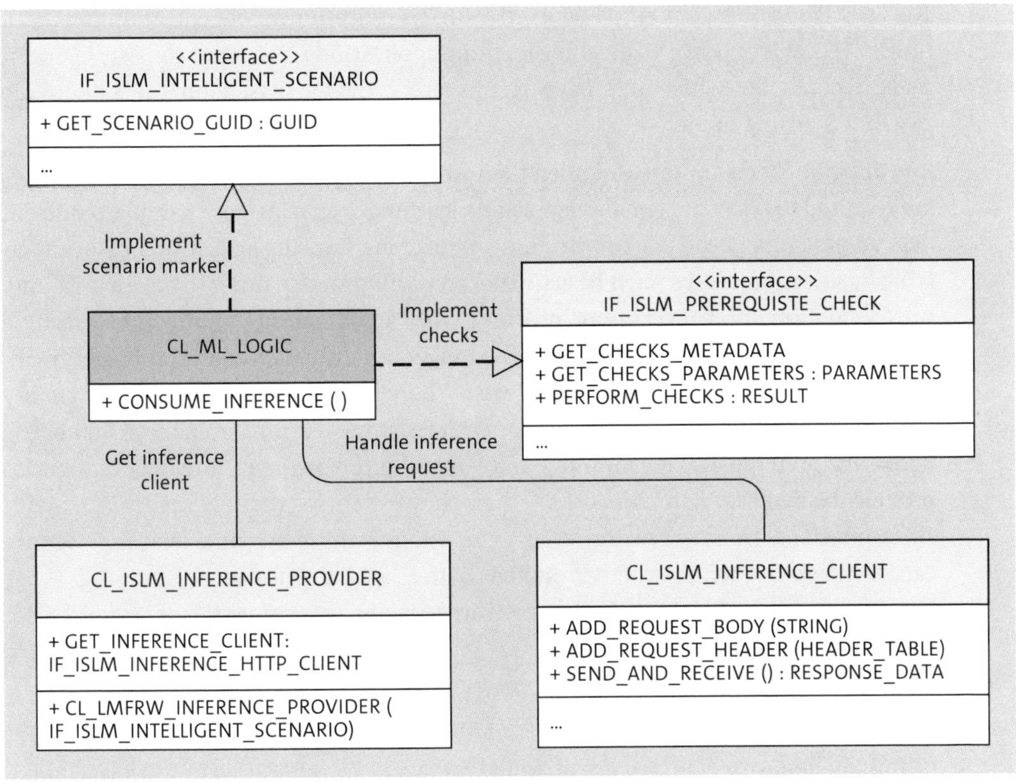

Figure 4.13 Design of ABAP Machine Learning Logic Class

As described, the machine learning logic ABAP class is registered to an intelligent scenario on the SAP S/4HANA side. To have a harmonized programming model across all machine learning use cases, the ABAP class is standardized by implementing interfaces. Side-by-side machine learning scenarios have to register the machine learning logic ABAP class in the ISLM framework via an intelligent scenario. During development, changes to the intelligent scenario are expected. Therefore, the artifact is saved as a draft initially. The draft status controls the transportation of the scenario registration content in the ISLM framework.

Once the intelligent scenario is finalized, it has to be published. The intelligent scenarios need to implement the IF_ISLM_INTELLIGENT_SCENARIO marker interface to be identified as an intelligent scenario in the ISLM framework. This is a mandatory procedure for side-by-side machine learning scenarios for implementation of the GET_SCENARIO_GUID method. This method returns the corresponding machine learning scenario of SAP Data Intelligence. Thus, the development artifacts between SAP S/4HANA and SAP Data Intelligence are connected. The IF_ISLM_INTELLIGENT_SCENARIO interface defines the required behaviors of an intelligent scenario in the ISLM framework. This enables the operational part of the intelligent scenario to be more concrete and stable. This implementation will be validated during the registration process.

The CONSUME_INFERENCE method wraps the REST service for model consumption as an ABAP API that is exposed to machine learning applications. The input consists of one or more records for which inference results are provided based on the underlying machine learning model.

It is difficult for customers to understand which technical and business prerequisites are required to train and consume machine learning scenarios. For example, sufficient data volume must be in place for training of machine learning algorithms. Also, underlying business processes must be activated and configured in order to have a meaningful foundation for the training process. With an increasing number of machine learning scenarios, such evaluation can no longer be performed by the customers manually due to high TCO and huge complexity. Therefore, a prerequisite check capability is required to automatically validate for each machine learning scenario whether the necessary prerequisites for training and consumption are met. Intelligent scenarios provide the necessary prerequisite checks by implementing the IF_ISLM_PREREQUISITE_CHECK interface. Thus, the readiness and consistency of the machine learning use case can be checked, such as whether enough data is available for model training. Those checks are performed by the ISLM framework to evaluate whether the prerequisites for model training are fulfilled.

To enable the ABAP class for customer extension, enhancement spots are provided. Thus, customers can enhance the consumption logic or add specific transformations as examples. Reusable routines are provided by the CL_ISLM_INFERENCE_PROVIDER and CL_ISLM_INFERENCE_CLIENT utility ABAP classes. These classes support methods like getting the currently activated model for inference, receiving the URL endpoint of a REST service provided by SAP Data Intelligence, creating an ABAP REST client object based on the REST service URL, triggering training for a machine learning scenario, or reading monitoring data for a trained model. The interactions among those classes in the context of training and inference processes are shown in Figure 4.14.

The training process is triggered by the ISLM framework for a selected intelligent scenario. Because an ABAP class is registered for each intelligent scenario, the ABAP implementation for the selected intelligent scenario can be determined. Thus, the corresponding machine learning scenario on the SAP Data Intelligence side can be read. With the ID of the machine learning scenario, the URL for the training REST service can be determined by invoking the necessary reuse routine of the machine learning utility class. Once the URL for the training REST service is available, an ABAP REST client object can be created based on the URL. This proxy object is used to call the training method from ABAP and run the necessary job. The status of the training—whether the job was successful or failed—is displayed by the ISLM framework for the end user.

The inference process is usually triggered by the machine learning application, as shown in Figure 4.15. The API for the inference calls is provided by the machine learning logic ABAP class, which is known for the machine learning application during design time. Like the training process, the URL for the inference REST service is determined by reusing the CL_ISLM_INFERENCE_PROVIDER machine learning utility class. Based on the URL, a REST client object is created in ABAP. This proxy object is used to call the REST service for inference. The input parameters for the inference call are passed by the machine learning application. The calculated results are also processed by the machine learning application—for example, integrated into business processes or illustrated in the UI.

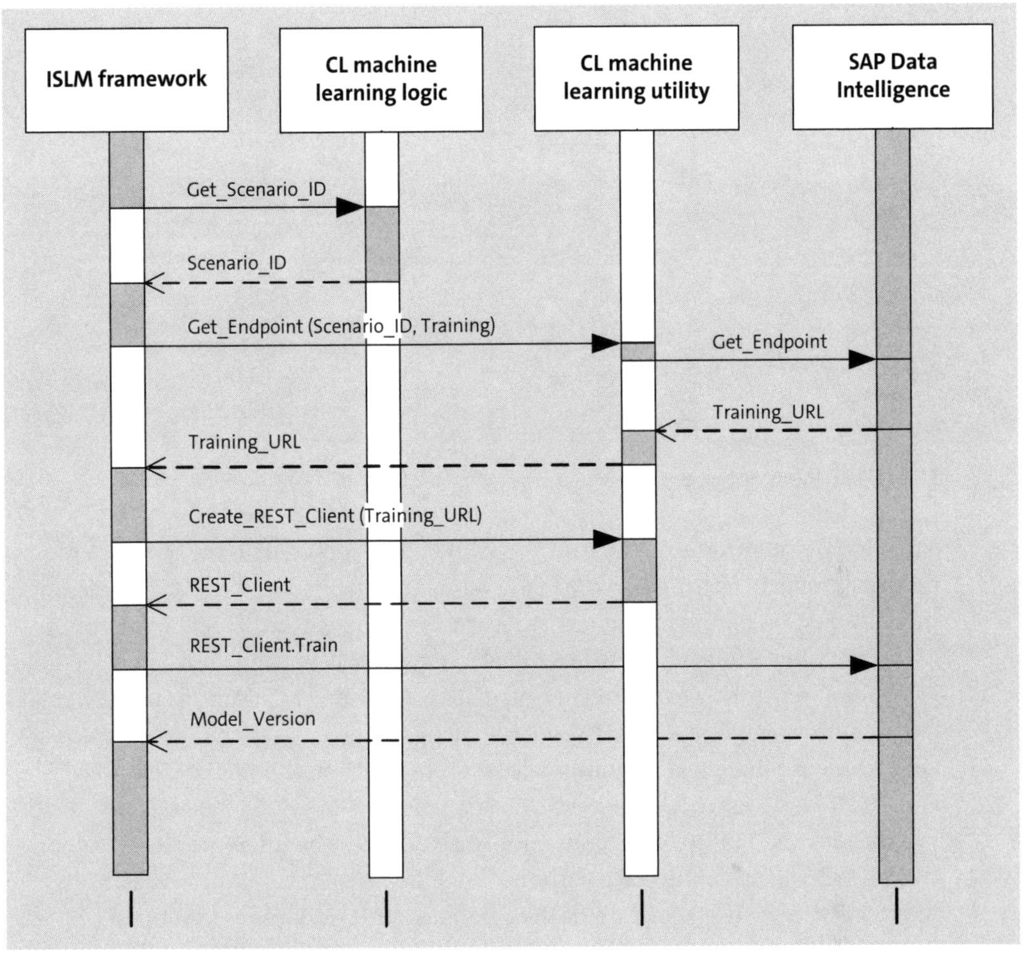

Figure 4.14 Training Sequence

The REST inference call is handled by the CL_ISLM_INFERENCE_CLIENT machine learning utility class. As a first step, an HTTP client is created based on the URL of the REST service. This proxy object is used to send the request and receive the according response. The request message must be compiled by setting header and body attributes. The response message must be transformed into ABAP structures for consumption by business processes.

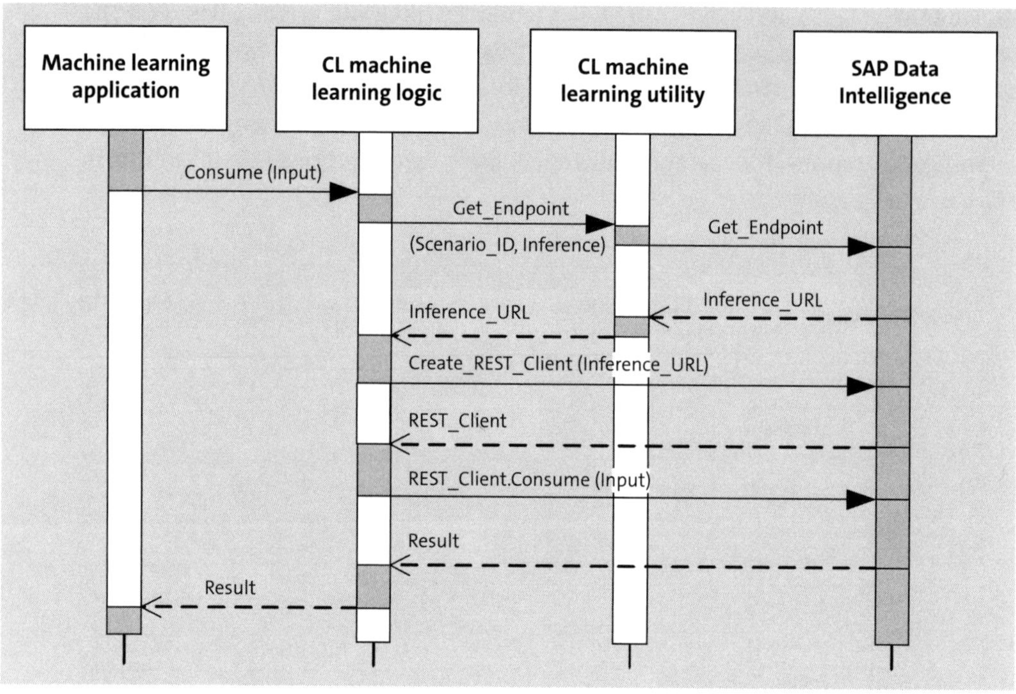

Figure 4.15 Inference Sequence

Now that we've introduced the basic structure of the REST client, the question of how to manage dynamic changes to the REST endpoints can be addressed. Machine learning models must be retrained regularly to learn from changes to the data environment. Each retraining results in a new machine learning model version with a new REST service for inference. However, consuming machine learning applications usually breaks if the underlying inference REST service is changed. The mitigation of this issue is covered by the machine learning utility class. From a technical point of view, new inference REST services result in new URL endpoints. The `CL_ISLM_INFERENCE_PROVIDER` machine learning utility class dynamically determines the inference REST endpoints and provides a stable interface with the machine learning applications. For this, the request and response structure of the inference REST service must remain stable or be enhanced compatibly.

As mentioned earlier, on the SAP Data Intelligence side, the machine learning scenario and pipeline artifacts must be defined. The training and inference pipelines cover the necessary machine learning logic. Pipelines are modeled graphically based on operators for transformation, validation, or incorporation of algorithms. The machine learning scenario constitutes a link among all development artifacts in SAP Data Intelligence to resolve lifecycle management. The training and inference pipelines are exposed by REST services to the SAP S/4HANA platform. The machine learning logic ABAP class is required for wrapping those REST services and making them consumable by ABAP methods. Figure 4.16 illustrates the basic steps for the training and inference pipelines, which can vary per use case.

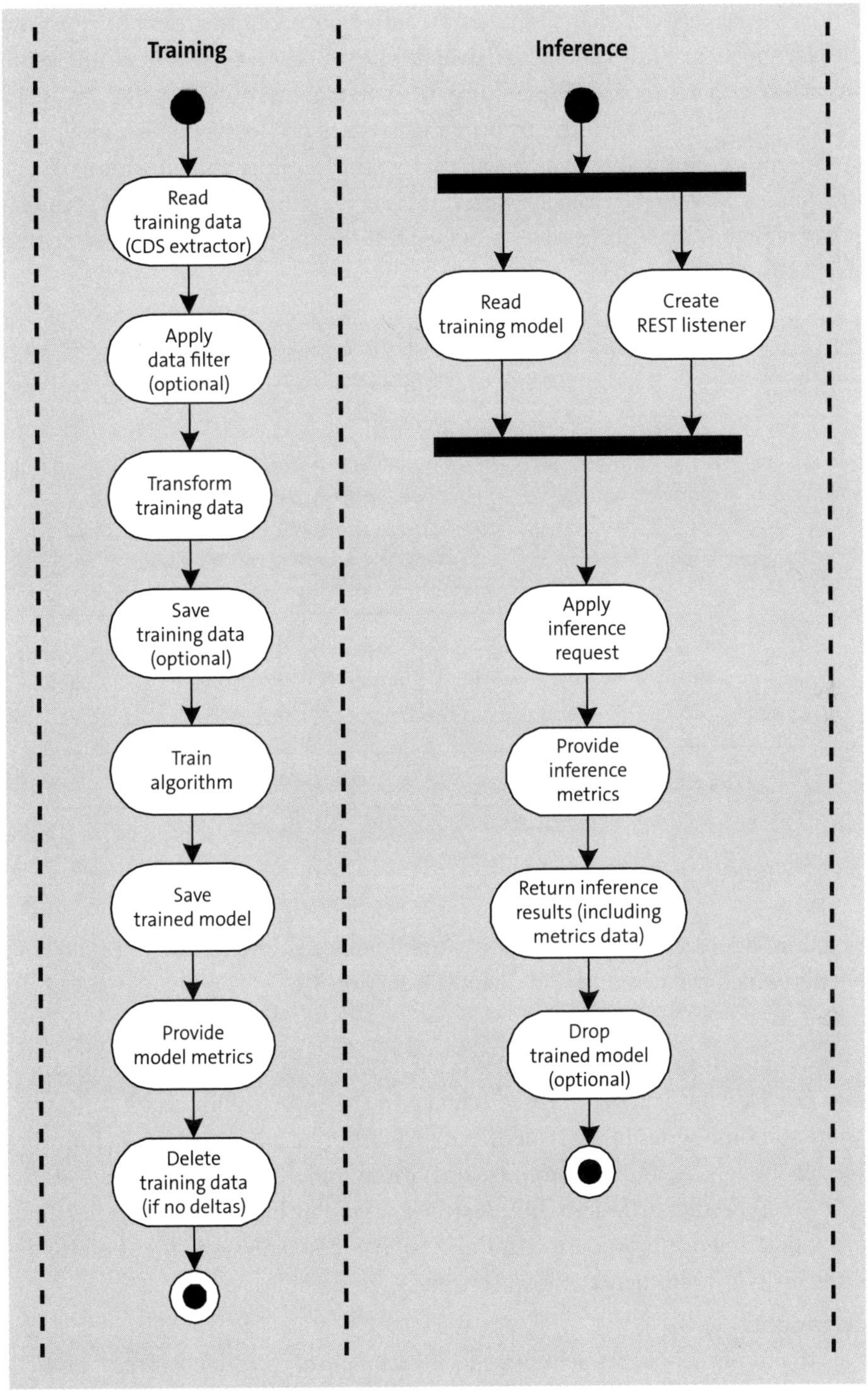

Figure 4.16 Design of Training and Inference Pipelines

4.5 Side-by-Side Predictive Analytics

In a side-by-side setup with SAP Analytics Cloud, smart assist expands the conversation beyond just visualizations, charts, and data connectivity, offering easy-to-consume machine learning capabilities embedded deeply within SAP Analytics Cloud, alongside and integrated with analytics and planning. The smart features within SAP Analytics Cloud make insight discovery faster by bringing data science to the masses. Intelligent machine learning technology drives the analytics process while reducing human bias, ensuring that decision makers can act with confidence. Figure 4.17 depicts the architecture of the predictive analytics services provided by SAP Analytics Cloud, including several smart assist features.

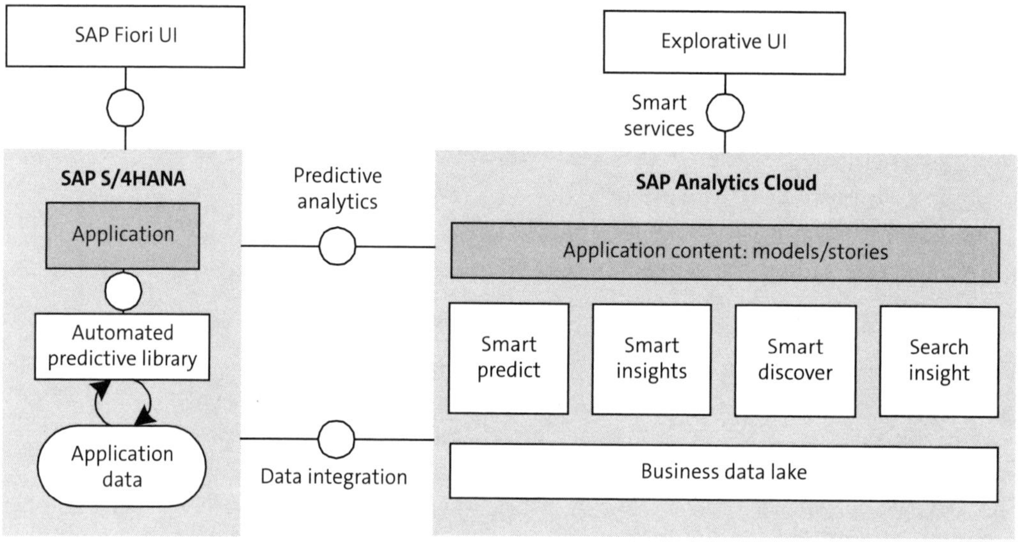

Figure 4.17 Predictive Analytics Architecture

As depicted in Figure 4.17, the smart assist functionality is covered by the following components (which we introduced in Chapter 3, Section 3.4.2):

- **Smart predict**
  Augments existing business intelligence capabilities by learning from your historical data to predict what is most likely to happen in the future. With classification, regression, and time series forecast algorithms, smart predict creates high-performing and stable models to help you optimize operations and drive strategic decisions for growth. Smart predict makes machine learning accessible by automatically handling complex data science procedures. This allows business analysts to focus on implementing insights to influence existing business processes.
- **Smart insights**
  Reveals the top contributors behind a specific data point or variance. This feature is particularly useful when we are looking at aggregated data, such as revenue. Smart

insights helps us to understand revenue contribution trends at a more granular level, such as by region or product. With the click of a button, machine learning algorithms run in the background to analyze all the data relevant to the selected information. First, suggested visualizations ranked in order of relevance are presented. Then, these visualizations can be added to an SAP Analytics Cloud story along with a text explanation thanks to the natural language generation engine.

- **Smart discovery**
  Acts as a digital business analyst. It automates the data exploration process to reveal information that is statistically relevant. Trained machine learning models that are specific to customer data generate an overview of significant patterns, outliers, key drivers, and influencers. Then, an interactive what-if simulation is created based on the model, allowing you to explore the possible results of changing factors and variables. With the autogenerated story from smart discovery, the data speaks for itself. Intelligent algorithms remove the need for analytics to be fully human-led, reducing bias to ensure that important trends are not missed.
- **Search to insight**
  Makes content creation easier with the power of natural language query. To use search to insight, simply ask a question about the data, and SAP Analytics Cloud will intelligently create relevant visualizations that can be added to stories.

A *dataset* is an SAP Analytics Cloud entity representing a collection of homogenous data (basically a database table) that can be produced and consumed in different SAP Analytics Cloud workflows. Typical workflows are data acquisition, consisting of creating a dataset from data located in an external system, and story building, consisting of creating a dashboard using a dataset. The physical location of data held by the dataset is complemented by a dataset description that comprises information for each column like the data type, length, format, and semantic information (e.g., latitude, longitude, country), plus more predictive-specific information like the category (e.g., nominal, ordinal). The data integration infrastructure of the SAP Analytics Cloud platform provides two data access modes for the smart assist features:

1. **Data-acquisition mode**
   In this mode, data is either transferred from external systems like SAP S/4HANA or uploaded from the end user environment to the SAP Analytics Cloud platform and processed. This offline approach safeguards the resource consumption of the external systems and allows a scale-out of predictive processing on SAP Analytics Cloud. Its drawback is that the offline data can become outdated with time and must be updated on a regular basis.
2. **Live-connection mode**
   In this mode, data stay on external systems like SAP S/4HANA and are processed there as well. In the context of smart assist, usually the SAP HANA database with APL installed is required. With the live connection, real-time access to data is provided so

that the latest state of the application data is always considered. However, for processing APL directly on the online data, greater CPU time and RAM in the external system are consumed. Currently, live-connection mode is implemented for access to application data on the database system but not on the application server level.

Datasets can have one of two origins. They can be either the result of a data acquisition or can be the result of a predictive model application. This relates to the role a dataset can have:

- **Training dataset**
  A dataset can be used as the training dataset for a predictive model. In this context, the dataset contains valid values for the model's explanatory variables and target variables. The data will be used to optimize the model fit.
- **Apply-in dataset**
  A dataset can be used as the input for the application of a predictive model. In this context, the dataset contains values for the model's explanatory variables but not for the target variables, the goal being precisely to evaluate the value of that target variable using a trained model.
- **Apply-out dataset**
  A dataset can be used as the output for the application of predictive model. In this context, at the end of the application phase, the dataset will contain the estimated value of the target variable for all the records of the apply-in dataset.

Smart assist can be consumed in SAP Analytics Cloud stories and from there integrated into SAP S/4HANA UIs. A *story* is a presentation-style document that uses charts, visualizations, text, images, and pictograms to describe data. Once a story is created or opened, pages, sections, and elements can be added and edited. The story toolbar is divided into different categories such as **File**, **Insert**, **Data**, and **Tools** to help users find options and perform tasks more efficiently. Owners of stories can share them with other users and grant permissions for these stories. After stories are shared, users with view permissions can analyze the data by navigating within the stories.

We'll focus mainly on the smart predict architecture, which enables the implementation of predictive analytics generically. The remaining smart assist features are provided as built-in functionality in the SAP Analytics Cloud infrastructure for direct usage. Another reason for concentrating on smart predict is that most of the smart assist functions are based on this service. Figure 4.18 illustrates the detailed architecture of smart predict.

In SAP Analytics Cloud, predictive features essentially fall into smart predict—which provides for managing model authoring and storage and for training models and applying them—and smart assist—which is an umbrella for a number of predictive features built on top of the predictive cloud services, like smart discovery and smart insight. Smart predict bundles two different authoring experiences, a purely form-based and easy-to-consume automated UI geared toward the nontechnical user and, in the future, it is planned to provide a pipeline-based editor targeting the maverick data scientist. In addition to providing support for managing model specifications, smart predict core services implement the concept of a task, typically encapsulating model training or applying work and a mechanism for executing such tasks in a regulated fashion, preventing the overload of underlying automated engines. Tasks pertaining to models authored through the automated UI and handled by the automated engine are exposed as a Cloud Foundry application. The associated composer models are processed by an embedded composer engine by delegating the execution to the APL in SAP HANA. Execution of automated tasks involves interaction with the automated engine through a synchronous and stateful KxCommonInterf/Thrift protocol.

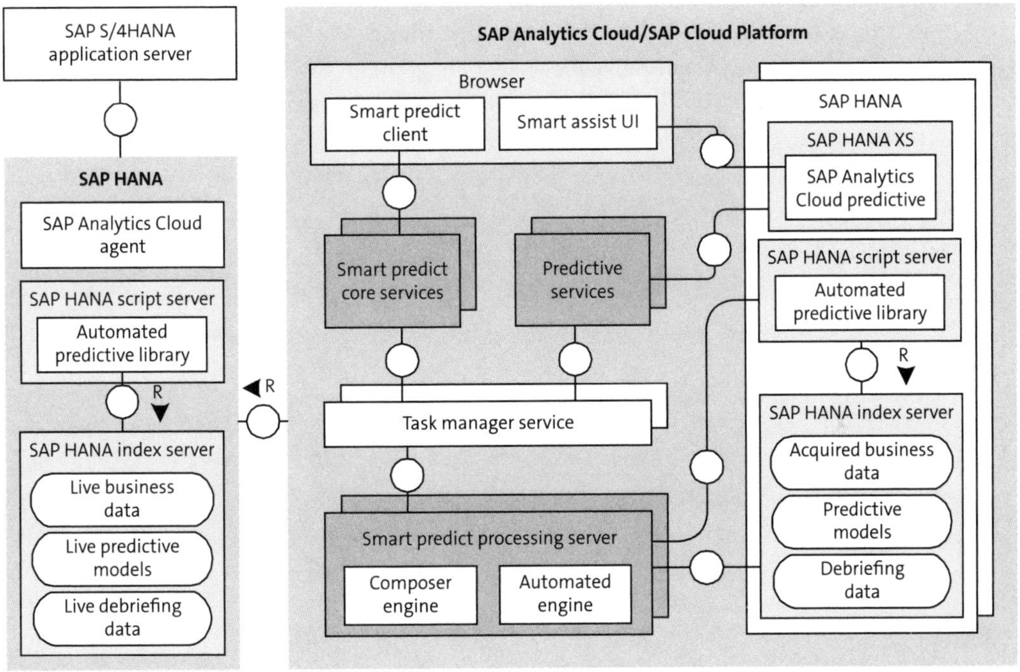

Figure 4.18 Smart Predict Architecture

The smart predict core services provide for automated and composer models for authoring, debriefing, entity management, and task management services. The automated engine supports a native mode in which the engine itself executes the predictive algorithms based on the data it fetches over open database connectivity (ODBC) from the SAP HANA instance. In a delegated mode, the automated engine hands over the processing of the predictive algorithms to APL running in the SAP HANA instances, preventing the data from being moved from the database to the engine. The SAP HANA instance of SAP Analytics Cloud stores the predictive models (physical model settings, scoring equations, and model raw statistics), the predictive model debriefing information (model statistics), and the physical data that are accessed and produced during the execution of the predictive workflows.

4.6 Summary

In this chapter, we explained what makes SAP S/4HANA intelligent and how this is related to automation. We gave a brief overview of technical challenges that must be solved in the context of SAP S/4HANA regarding machine learning and predictive analytics. For understanding the structure of the already delivered scenarios, we shared the application patterns that have been identified so far. To enable you to use the existing SAP S/4HANA machine learning and predictive scenarios, we explained the solution architecture behind them.

This knowledge is the foundation for the next chapter, in which we will describe how to implement your own machine learning and predictive analytics use cases step by step according to the SAP S/4HANA programming model. We will also share with you how to resolve advanced aspects like lifecycle management, GDPR, and model degradation, which are crucial in the context of enterprise applications such as SAP S/4HANA.

Chapter 5
Technical Implementation

In this chapter, you'll learn about the technical implementation of machine learning and predictive analytics for SAP S/4HANA. You'll learn to implement your own machine learning and predictive analytics application based on SAP S/4HANA technology, and you'll get the necessary technical background for using the SAP S/4HANA-delivered scenarios.

Now that we've covered the technical architecture underpinning machine learning and predictive analytics in SAP S/4HANA for both embedded and side-by-side models, it's time to dive into the technical implementation. In this chapter, we'll begin with a comparison of the three models (embedded, side-by-side machine learning, and side-by-side predictive analytics), then discuss how to develop and train each model. Finally, we'll discuss the key processes behind managing your machine learning/predictive analytics applications.

5.1 Approach Comparison

Because this chapter deals with the implementation steps for embedded machine learning, side-by-side machine learning, and predictive analytics, first we'll summarize the necessary development artifacts for each of those approaches. We also provide decision criteria for when to use which technique.

Figure 5.1 illustrates which objects embedded machine learning, side-by-side machine learning, and predictive analytics have in common and which artifacts have to be implemented specifically by developers. Independent of the underlying machine learning technology, the artifacts in the middle of Figure 5.1—namely, the machine learning application, intelligent scenario, and CDS view—must be provided. The machine learning application basically integrates the machine learning capability into the business processes and the related user interfaces (UIs). The intelligent scenario constitutes the link among all development artifacts that are required for the machine learning application. This is especially used to handle lifecycle management aspects. The core data services (CDS) view is defined to read the application data for model training. For this ABAP artifact, a SQL view is generated on top of the application tables during the activation phase. All SQL requests are pushed down from ABAP to the SQL view to achieve optimal performance.

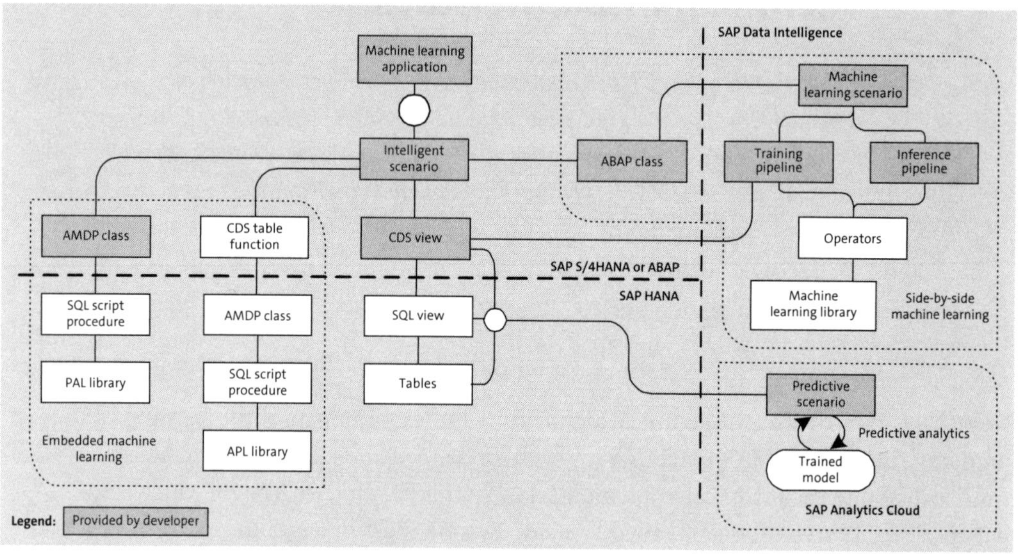

Figure 5.1 Development Artifacts

For embedded machine learning scenarios that are based on the Automated Predictive Library (APL), no development is required because the necessary artifacts are generated automatically. This is possible because the APL incorporates machine learning logic for training and inference (inference meaning the prediction-making process) in the library implementation itself. Thus, if the supported APL algorithms are sufficient for a use case, this is the approach of choice from a development point of view. For embedded machine learning scenarios that are based on the Predictive Analysis Library (PAL), an ABAP Managed Database Procedure (AMDP) class must be provided. This class contains training and inference methods implemented in SQL script to cover the necessary machine learning logic. PAL is used for breakout scenarios in which specific data validations, transformations, or feature reductions are required. The AMDP class has to implement a predefined interface and is owned by the developer so that complex logic can be developed.

For side-by-side machine learning use cases based on SAP Data Intelligence, training and inference pipelines must be provided to cover the necessary machine learning logic. Pipelines are modeled graphically based on operators for transformation, validation, or incorporation of algorithms. A machine learning scenario constitutes the link among all development artifacts of SAP Data Intelligence to resolve lifecycle management. The training and inference pipelines are exposed by representational state transfer (REST) services to the SAP S/4HANA platform. The ABAP class is required for wrapping those REST services and making them consumable by ABAP methods.

For predictive analytics use cases based on SAP Analytics Cloud, predictive scenarios must be created with the smart predict tool, which reads data from SAP S/4HANA to train the necessary algorithm. Smart predict accesses CDS views, SQL views, or database tables of SAP S/4HANA for exchanging the required training data. Smart predict also uses APL for embedded machine learning. While embedded machine learning is for implementing regularly used machine learning applications, smart predict focuses on explorative use cases for unforeseen, exceptional, and irregular needs for machine learning methods. The trained models are stored in SAP Analytics Cloud so that predictions can be applied and illustrated accordingly.

Table 5.1 summarizes the technical capabilities of the different approaches and provides decision criteria for when to use which approach. In the next section, we explain how to implement embedded machine learning, side-by-side machine learning, and predictive analytics based on the SAP S/4HANA infrastructure and programming model.

Embedded Machine Learning	Side-by-Side Machine Learning	Predictive Analytics
Capabilities		
▪ Data remains in the SAP HANA database ▪ Real-time inference ▪ Large library of machine learning algorithms (100+) ▪ Accessible via standard development tools ▪ Complexity can be hidden in view and table functions ▪ Integrated with ABAP management and transport system ▪ Inherited privacy and security features ▪ Model management ▪ Model performance monitoring and automated retraining	▪ SAP Data Intelligence provides cloud services for scalable and distributed training and inference ▪ Support for open-source frameworks and algorithms (TensorFlow, Scikit-Learn, R) ▪ Pretrained models for certain business use cases ▪ Access via standard REST application programming interface (API) ▪ Algorithms and pre-trained models for working with unstructured data/fuzzy data (image, audio, video, text, etc.) ▪ Deep learning algorithms ▪ GPU support	▪ Easy-to-consume machine learning functionality embedded within SAP Analytics Cloud ▪ Bringing data science to the masses ▪ Supports APL algorithms like classification, regression, and time series forecast ▪ Enables automatically handling complex data science procedures ▪ Ensures data access to SAP S/4HANA based on CDS views, SQL views, and database tables ▪ Allows graphical illustration and sharing of results based on SAP Analytics Cloud stories and wizards

Table 5.1 Decision Criteria for Different Approaches

Embedded Machine Learning	Side-by-Side Machine Learning	Predictive Analytics
Decision Criteria		
■ Business and machine learning logic reside in SAP S/4HANA ■ Simple use cases like forecasting or trending in which algorithms have low demand for data, RAM, and CPU time ■ Data located in SAP S/4HANA is sufficient for model training; no need for huge external data source for training ■ Required algorithms are provided by SAP HANA machine learning (PAL, APL, text analysis)	■ Machine learning logic resides on SAP Data Intelligence while the business logic can be based on SAP S/4HANA or SAP Cloud Platform ■ Complex use cases like image recognition or natural language processing (NLP) that, among other needs, require neural networks with high demand for data, RAM and CPU/GPU time ■ Huge volume of external data is required for model training; main focus is on processing unstructured data ■ Third-party machine learning libraries are required	■ Explorative use cases for unforeseen, exceptional, and irregular needs for machine learning methods ■ Allows business analysts to focus on implementing the insights to influence existing business processes, disregarding the data science tasks ■ Prediction and classification use cases are required to be defined

Table 5.1 Decision Criteria for Different Approaches (Cont.)

Note

Where there are gaps in infrastructure, we'll describe the required concepts and UI mockups so that workarounds can be implemented, or the necessary frameworks can be developed on your own. The implementation steps throughout this chapter reflect the SAP S/4HANA 2020 and SAP Data Intelligence 2001.1.10 releases. The focus of the selected use cases is on demonstrating the technical implementation.

5.2 Implementing Embedded Machine Learning Applications

As illustrated previously in Figure 5.1, there are two flavors of embedded machine learning depending on the underlying machine learning library. APL has a built-in implementation of data science steps like data preprocessing or transformations. Thus, developers don't need to explicitly implement those aspects but get all the artifacts generated by the intelligent scenario lifecycle management (ISLM) framework. However, due to this automation, the number of APL algorithms and the usage scope are limited. Therefore, PAL is used for breakout scenarios in which APL is not sufficient—for example, if the required algorithms for the use case are not supported by APL or the prediction accuracy of the generated model isn't satisfactory. PAL provides more algorithms than APL and allows optimized implementation of a machine learning use case based on the exploration of data scientists. However, the drawback is that the development effort is higher compared to APL cases as an AMDP class must be implemented.

In this section, we'll explain the necessary implementation steps for both flavors of embedded machine learning. The implementation is based on the ISLM framework, which consists of the two SAP Fiori applications illustrated in Figure 5.2: the Intelligent Scenarios app and the Intelligent Scenario Management app. You can find these applications in the SAP Fiori launchpad under **Intelligent Scenarios** by assigning the `SAP_BR_ANALYTICS_SPECIALIST` role to the user.

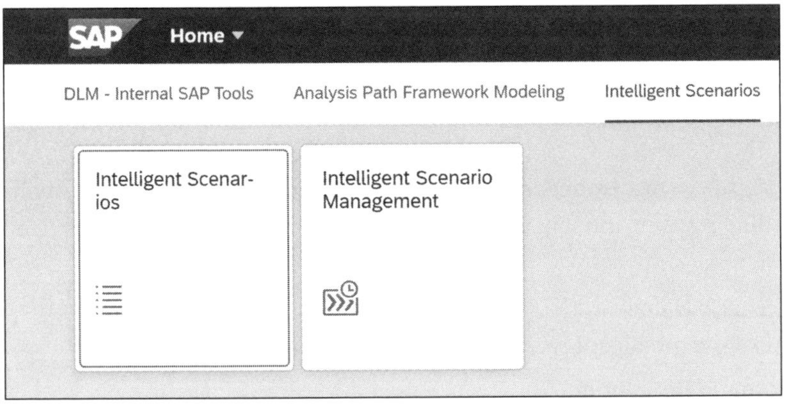

Figure 5.2 ISLM Design-Time and Configuration-Time SAP Fiori Applications

The Intelligent Scenarios app is a design-time application for developers to define the intelligent scenario artifact that incorporates all development objects for a use case, independent of whether embedded machine learning or side-by-side machine learning is applied. The Intelligent Scenario Management app is a configuration-time application in which customers can handle the full lifecycle management of machine learning models, including model training and activation.

Let's start with the implementation steps for embedded machine learning based on APL.

5.2.1 Generated Approach Based on the SAP HANA Automated Predictive Library

Because our objective is to illustrate the technical implementation steps, we'll select a simple use case in which the tax amount will be predicted for sales orders. The implementation of all machine learning use cases starts with the creation of the *intelligent scenario* design-time artifact. For this, you have to launch the Intelligent Scenarios SAP Fiori application (refer back to Figure 5.2). In this application, the **Create** button offers the option to create an intelligent scenario for **Embedded** machine learning or **Side-by-Side** machine learning, as shown in Figure 5.3.

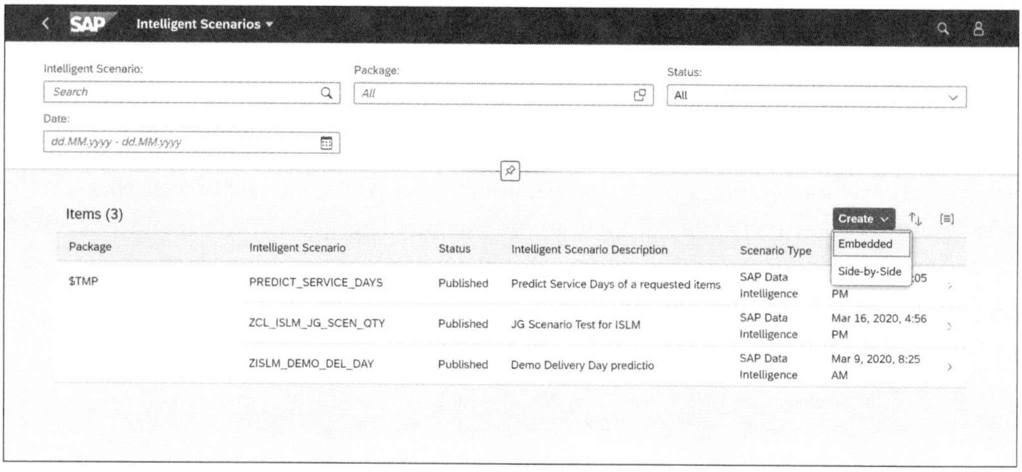

Figure 5.3 Create Intelligent Scenarios

For our use case, select the **Embedded** option from the dropdown and create your embedded machine learning intelligent scenario with the following entries as depicted in Figure 5.4:

❶ **Intelligent Scenario Name**
Unique name of the intelligent scenario for identification.

❷ **Intelligent Scenario Description**
Text to describe the machine learning key question to be solved by the defined intelligent scenario. We've entered "Predict Tax Amount" based off of our example.

❸ **Intelligent Scenario Type**
Type of the intelligent scenario which represents the type of the underlying algorithms, e.g. regression or classification.

❹ **Package**
Package to store the development artifacts.

❺ **Extensible checkbox**
Activation flag to allow extensibility of the intelligent scenario.

❻ **Add Model dropdown**
SAP HANA machine learning library to be used for the intelligent scenario (APL, PAL/HEMI).

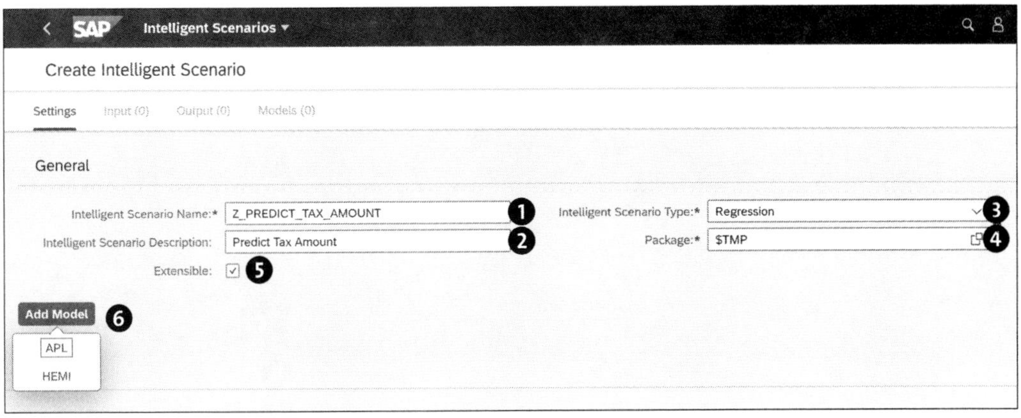

Figure 5.4 Maintain Intelligent Scenario and Select APL Model

Because our sample use case will be based on code generation, APL is used as the machine learning library. Once **APL** is chosen from the **Add Model** dropdown, the following additional information must be maintained, as shown in Figure 5.5:

❶ **General**
Includes the **Name**, **Description**, and **Package** of the intelligent scenario.

❷ **Training Dataset**
CDS view to read and train the machine learning model. Data scientists recommend which application fields are required so that the application developer can define the CDS view accordingly.

❸ **Apply Dataset**
CDS view to define the consumption signature for the machine learning model.

❹ **Target**
Field of the training CDS view to be predicted.

❺ **Keys**
Key fields of the CDS view, which are determined automatically.

❻ **Apply Output Configuration**
The checkboxes here control the prediction output fields to be generated as part of the inference API. **Score** is selected by default and represents the regression prediction. **Confidence** stands for the error base value on the prediction. **Percentile** indicates the value below which a given percentage of observations in a group falls.

Figure 5.5 Maintain Additional Information for APL-Based Intelligent Scenario

By completing the entry dialog in Figure 5.5 and clicking the **Add** button, the details of the intelligent scenario are displayed. In Figure 5.6, you can see the **Input** tab with the following values:

- ❶ **Input**
  Displays the input for the intelligent scenario, which is basically the CDS view for model training.
- ❷ **Key role**
  Key fields of the CDS training view are displayed.
- ❸ **Input role**
  Input fields of the CDS training view with which the underlying APL algorithm is trained are displayed.
- ❹ **Target role**
  The target field of the CDS training view is displayed, which will be predicted based on training using the input fields.
- ❺ **Publish/Save Draft**
  When you click one of these buttons, the target field of the CDS training view is displayed, which will be predicted based on training using the input fields.

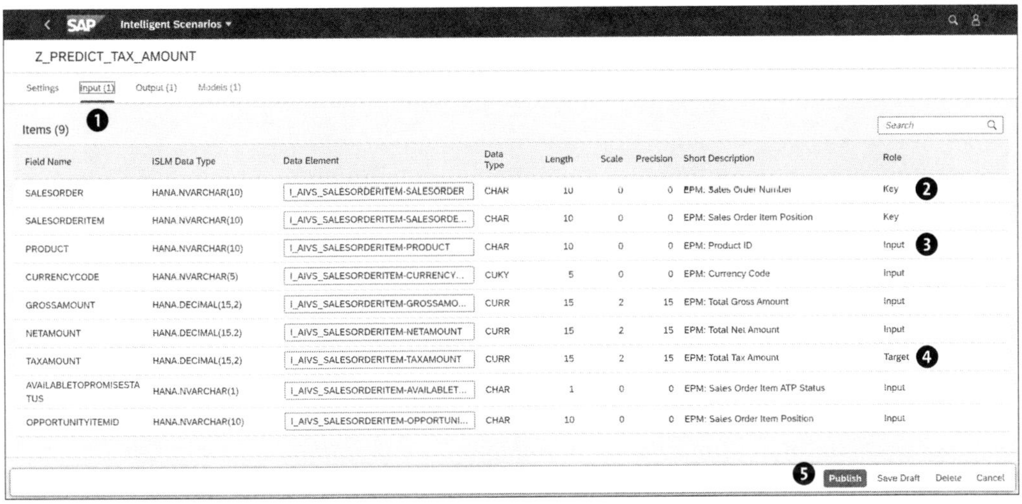

Figure 5.6 Display Input Section of Defined Intelligent Scenario

You can also navigate to the **Output** tab, as depicted in Figure 5.7, which contains the following values:

❶ **Output**
Displays the output for the intelligent scenario, which is basically the consumption API of the trained model.

❷ **Key role**
Key fields of the consumption API are displayed.

❸ **Prediction role**
Prediction fields of the consumption API with which the predicted results are exposed are displayed.

❹ **abap.applyclass**
When you click the **API Details** button, you can view the generated consumption API as an ABAP method that can be integrated into business processes and UIs.

❺ **abap.applyview**
When you click the **API Details** button, you can also view the generated consumption API as a CDS table function that can be integrated into business processes and UIs.

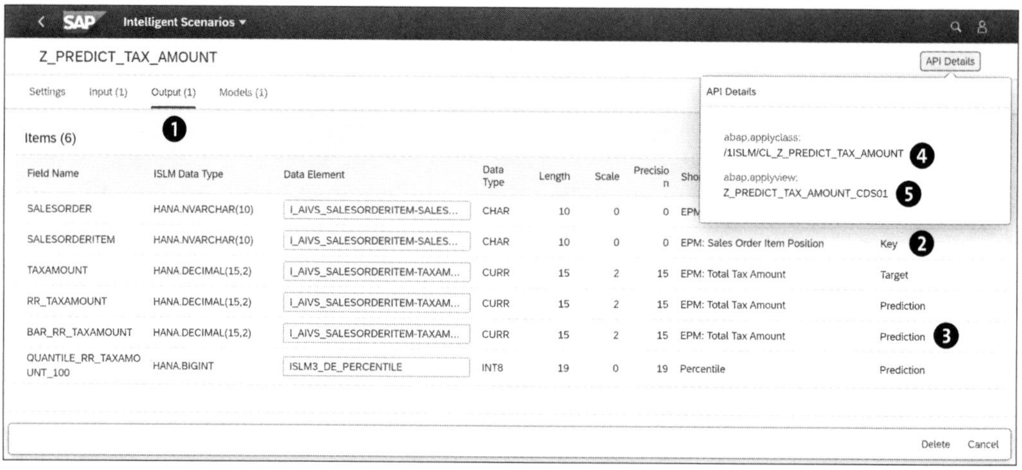

Figure 5.7 Display Output Section of Defined Intelligent Scenario

As already mentioned, using APL-based intelligent scenarios is the simplest approach for implementation of machine learning use cases. Thus, with the creation of the intelligent scenario, the development task is already complete. All of the other necessary development artifacts are generated automatically. For example, an ABAP method and a CDS table function are generated for the consumption of the trained model, as illustrated in Figure 5.7. Those APIs can be used to integrate machine learning results into business processes and UIs during design time. Meaningful results are provided by the consumption APIs when the model is trained on productive data.

For this, you have to launch the Intelligent Scenario Management app (refer back to Figure 5.2), select the defined intelligent scenario for the use case, and click the **Train** button, as depicted in Figure 5.8.

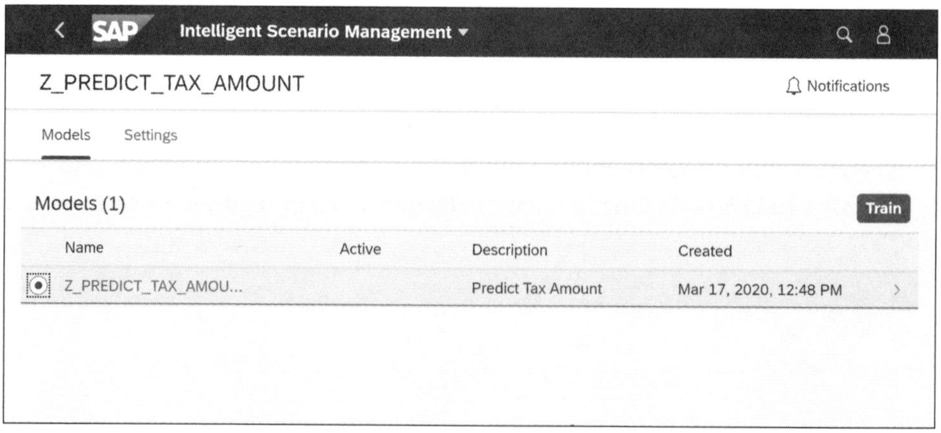

Figure 5.8 Trigger Training of Defined Intelligent Scenario

To reduce the volume of the training data, filter criteria can be defined in the subsequent screen, as shown in Figure 5.9 under the **Training Filters** section. In our example, the sales order number is restricted and only the products **HT-100** and **HT-800** are considered. The available filter conditions are derived from the structure of the underlying CDS view for model training and vary therefore for each use case.

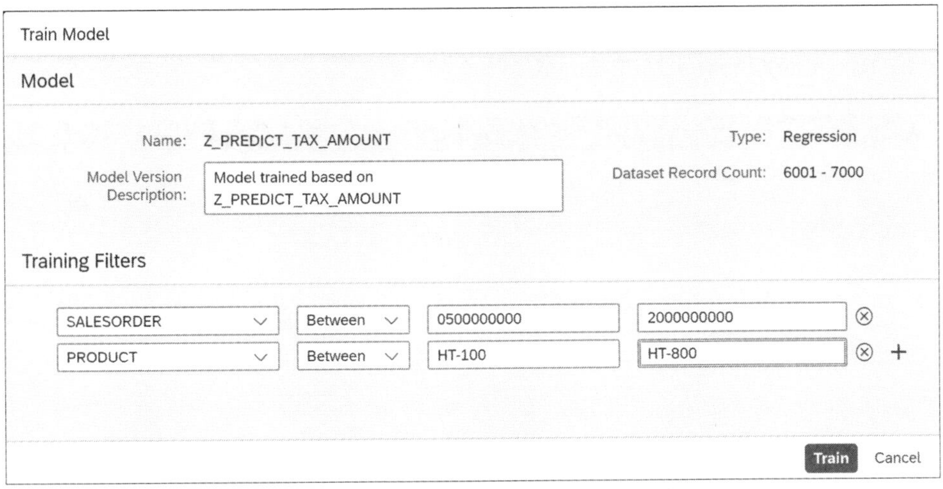

Figure 5.9 Define Filters to Reduce Volume of Training Data

Once you've set your filters, click the **Train** button. However, filtering is optional; you could also click the **Train** button without specifying filter conditions. As illustrated in Figure 5.10, each training run results in a new version of the model. However, only one model can be active for productive use. The following values are important:

❶ **Model Versions**
Displays the different model versions that resulted from training runs.

❷ **Activate**, **Deactivate**, **Retrain**, and **Delete**
Models can be activated for productive usage, deactivated, retrained if the data environment changed, or deleted by clicking these buttons.

❸ **Status**
This attribute indicates whether the model is activated for productive usage (**Active** status).

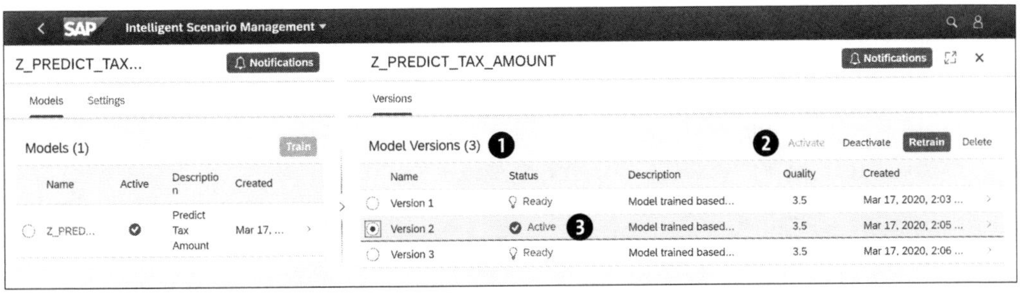

Figure 5.10 Train Multiple Model Versions, Activate One

To decide which model version will be activated for productive usage, you need to drill down to the model debriefing information by clicking on one of the listed model versions, thus arriving at the screen shown in Figure 5.11. Here, the **Predictive Power** and **Prediction Confidence** KPIs are displayed. The closer those numbers are to 1, the better the model quality is. Key influencers are also provided to show which input parameters most contributed to the trained model.

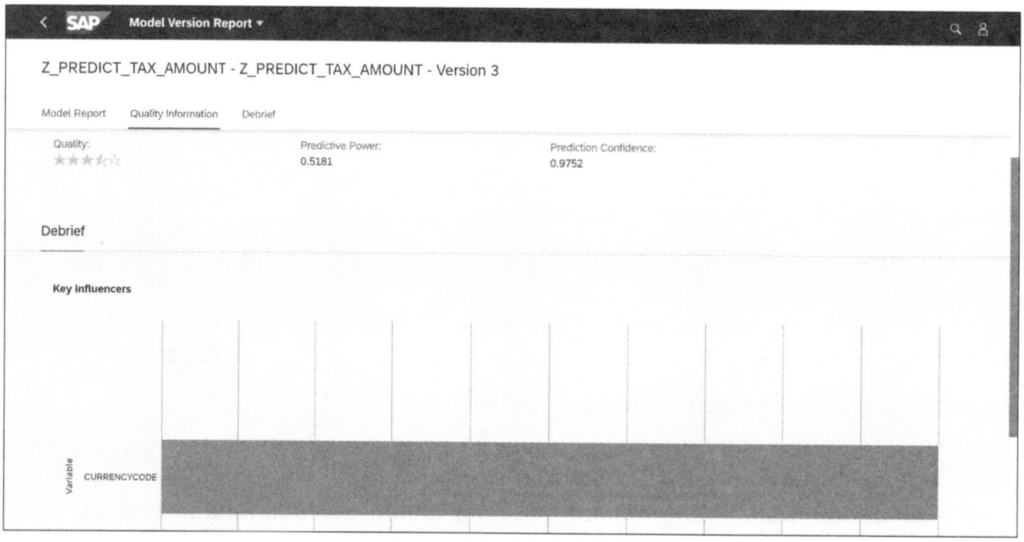

Figure 5.11 Display Model Accuracy KPIs

Listing 5.1 shows the generated CDS table function from Figure 5.7, which you access via the **API Details** button. You can use standard development tools, such as ABAP in Eclipse, to display the CDS view. With this CDS view, the trained model can be consumed and the inference results integrated into business processes or UIs. As already mentioned, an ABAP method for consumption is provided as well.

```
@EndUserText.label: 'ISLM AutoGenerated View   Z_PREDICT_TAX_AMOUNT_CDS01 '
@AbapCatalog.sqlViewName: 'Z_D7A03819460EAE' --ISLM Random generated SQL name
@AbapCatalog.dataMaintenance: #DISPLAY_ONLY
@ClientHandling.type: #CLIENT_DEPENDENT
@ClientHandling.algorithm: #SESSION_VARIABLE
@VDM.private: false
@VDM.viewType: #COMPOSITE
define view Z_PREDICT_TAX_AMOUNT_CDS01
 as select from Z_PREDICT_TAX_AMOUNT_TF01
       (  p_clnt: $session.client  )
    as a
 association [1..1] to I_AIVS_SALESORDERITEM as _CKE_toBase on a.SalesOrder =
_CKE_toBase.SalesOrder and
```

```
a.SalesOrderItem = _CKE_toBase.SalesOrderItem
{key a.SalesOrder as SalesOrder,
key a.SalesOrderItem as SalesOrderItem,
    a.TaxAmount as TaxAmount,
    a.Rr_TaxAmount as Rr_TaxAmount,
    a.Bar_Rr_TaxAmount as Bar_Rr_TaxAmount,
    a.Quantile_Rr_TaxAmount_100 as Quantile_Rr_TaxAmount_100,
 /* Associations */
    @ObjectModel.text.element: null
    @Consumption.hidden: true
    _CKE_toBase as _CKE_toBase
}
```

Listing 5.1 Generated CDS View for Model Consumption

The CDS view can be, for example, executed in ABAP in Eclipse. There, the **Data Preview** tool can be used to show the results of the CDS view, as depicted in Figure 5.12. The **Rr_TaxAmount** column contains the predicted tax amount for dedicated sales orders.

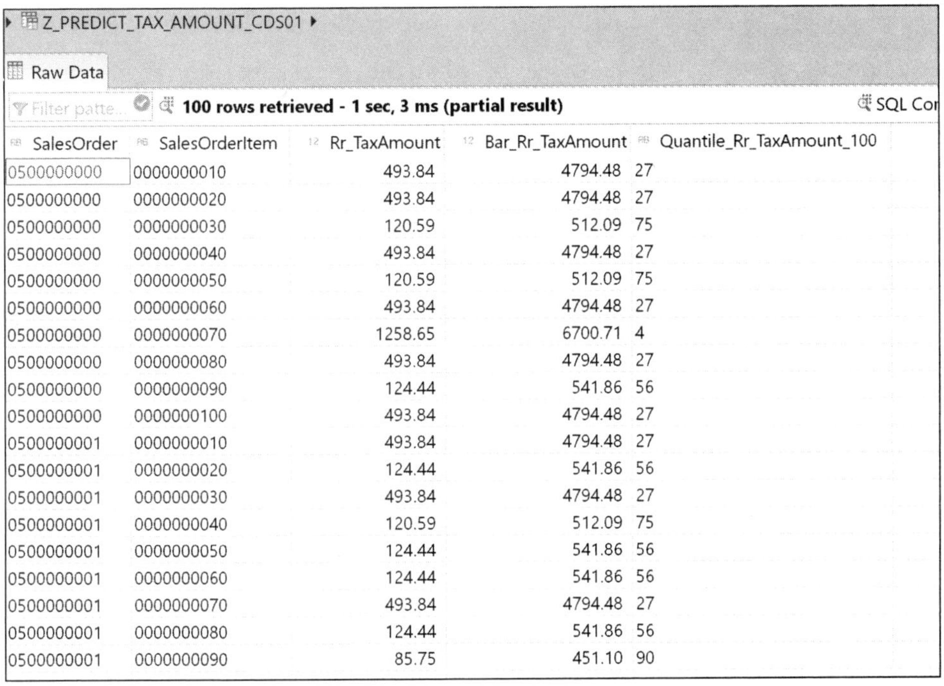

Z_PREDICT_TAX_AMOUNT_CDS01

Raw Data

Filter patte... 100 rows retrieved - 1 sec, 3 ms (partial result) SQL Co

SalesOrder	SalesOrderItem	Rr_TaxAmount	Bar_Rr_TaxAmount	Quantile_Rr_TaxAmount_100
0500000000	0000000010	493.84	4794.48	27
0500000000	0000000020	493.84	4794.48	27
0500000000	0000000030	120.59	512.09	75
0500000000	0000000040	493.84	4794.48	27
0500000000	0000000050	120.59	512.09	75
0500000000	0000000060	493.84	4794.48	27
0500000000	0000000070	1258.65	6700.71	4
0500000000	0000000080	493.84	4794.48	27
0500000000	0000000090	124.44	541.86	56
0500000000	0000000100	493.84	4794.48	27
0500000001	0000000010	493.84	4794.48	27
0500000001	0000000020	124.44	541.86	56
0500000001	0000000030	493.84	4794.48	27
0500000001	0000000040	120.59	512.09	75
0500000001	0000000050	124.44	541.86	56
0500000001	0000000060	124.44	541.86	56
0500000001	0000000070	493.84	4794.48	27
0500000001	0000000080	124.44	541.86	56
0500000001	0000000090	85.75	451.10	90

Figure 5.12 Display Inference Results in ABAP in Eclipse

5.2.2 Coded Approach Based on the SAP HANA Predictive Analysis Library

PAL provides over 100 machine learning algorithms and is used when APL isn't sufficient for the embedded machine learning scenarios. This can be the case, for example, if the required algorithm isn't supported by APL or the automated data science procedures of APL don't result in adequate prediction accuracy. PAL is used for these kinds of breakout scenarios.

For PAL, the machine learning logic is implemented by the developer, not just generated by the framework as was the case with APL. The advantage is that the developer has full control and can optimize the machine learning logic implementation for the specific machine learning use case according to the recommendation of data scientists. The developer has to implement an AMDP class, which implements predefined interfaces and provides methods like training and consumption. Those methods essentially represent the machine learning logic and cover, for example, preprocessing steps, data transformations, launching the necessary PAL algorithms, and providing an API for consumption of the trained model.

In this section, to keep the focus on the technical implementation, we'll again review a simple use case. Imagine that as a cash specialist, you categorize cash flows by different liquidity items, such as operations and investments. By leveraging machine learning capabilities, you can gain predictive insights into liquidity items by comparing actual and predicted results. The actual liquidity items that are assigned can then be replaced by the predicted liquidity items that are proposed by the machine learning service.

The implementation of this use case is based on the PAL classification algorithm random decision tree and starts with launching the Intelligent Scenarios application (refer back to Figure 5.2). Next, you'll create your intelligent scenario as depicted in Figure 5.13. This time, we'll select **Classification** as the **Intelligent Scenario Type** based off of our use case. Under **Add Model**, you must select **HEMI**, for the SAP HANA embedded machine learning interface, which represents the PAL library.

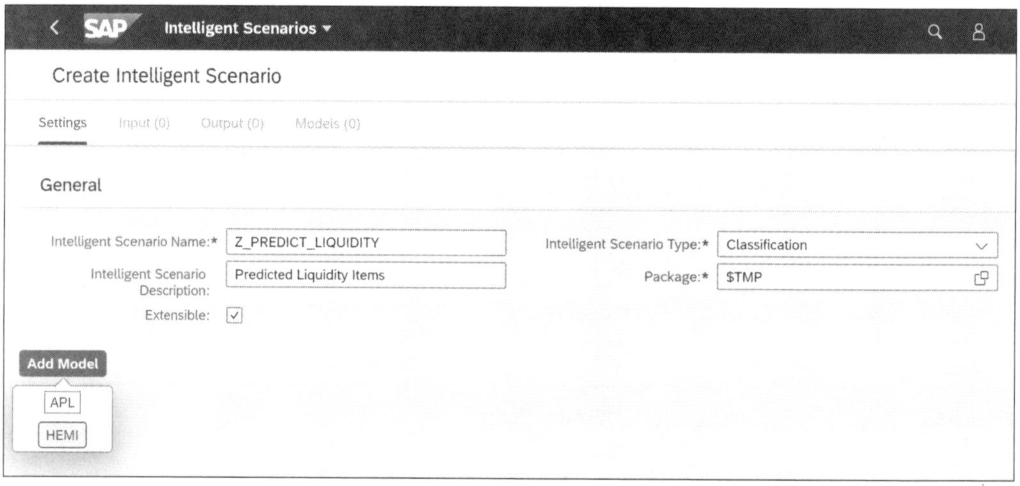

Figure 5.13 Maintain Intelligent Scenario and Select HEMI/PAL

On the next screen, you must register the CL_FQM_ML_HEMI_RDT AMDP class by entering it in the **HEMI Class Name** field, which implements the already mentioned machine learning logic (see Figure 5.14). Click the **Add** button to move forward.

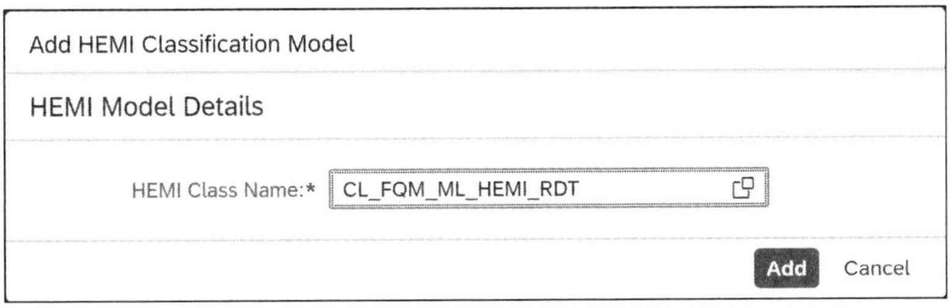

Figure 5.14 Register AMDP Class for Implementing Machine Learning Logic

The ABAP class implements methods for providing metadata like the underlying CDS view as input for training or the API signature for consuming the trained model. These kinds of metadata are processed by the ISLM framework and are also illustrated in the UI, as shown in Figure 5.15.

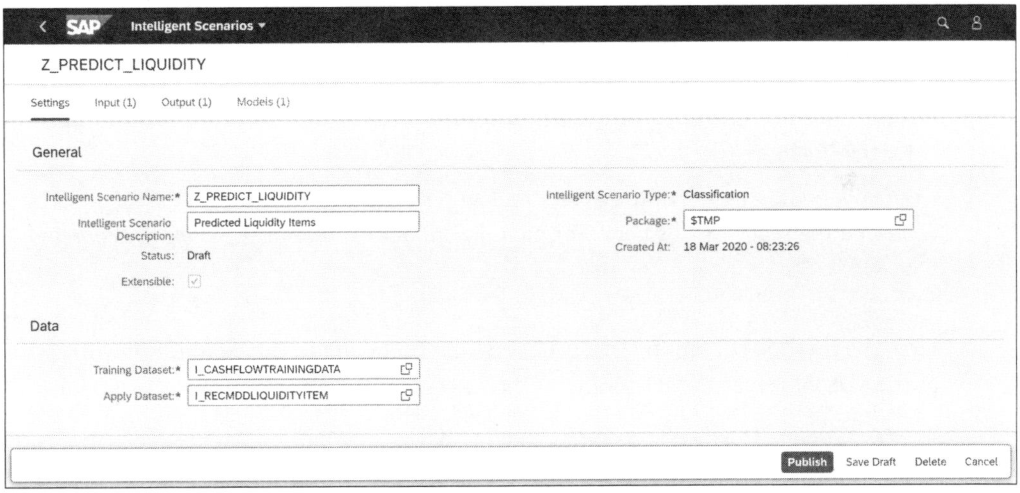

Figure 5.15 Intelligent Scenario for Embedded Machine Learning Case Based on PAL

Other elements shown in Figure 5.15, like the **Input** tab, **Output** tab, **Publish** button, **Save Draft** button, and the semantics of the fields, remain the same as for APL in the previous section. Therefore, we can focus now on the implementation of the CL_FQM_ML_HEMI_RDT AMDP class in the ABAP Class Builder, as shown in Figure 5.16.

The ABAP class must implement different interfaces, so the structure and behavior of those classes are standardized, which is a prerequisite for uniform handling by the ISLM framework.

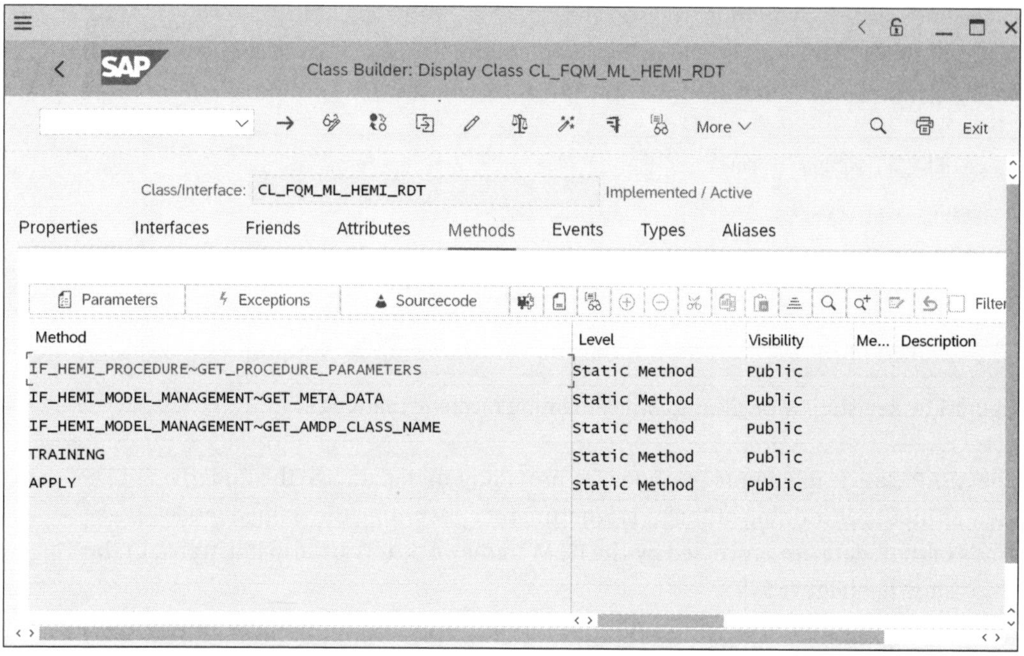

Figure 5.16 Methods of ABAP Class for Machine Learning Logic

Further Resources

An in-depth discussion of the ABAP development steps using the ABAP Class Builder is beyond the scope of this book. For more information, go to *https://help.sap.com* and search for "Class Builder".

First, the method shown in Listing 5.2 provides metadata to the ISLM framework, such as the CDS view for model training or the API signature for consuming the model.

```
METHOD IF_HEMI_MODEL_MANAGEMENT~GET_META_DATA.
    DATA LR_BADI TYPE REF TO CL_FQM_ML_DETECT2_BADI.
    TRY.
      GET BADI LR_BADI.
      CATCH CX_BADI_CONTEXT_ERROR.
      CATCH CX_BADI_NOT_IMPLEMENTED.
    ENDTRY.
    CALL BADI LR_BADI->GET_META_DATA
      IMPORTING
        ES_META_DATA = ES_META_DATA.
    ES_META_DATA-TRAINING_DATA_SET = 'I_CashFlowTrainingData'.
    ES_META_DATA-APPLY_DATA_SET     = 'I_RecmddLiquidityItem'.
    ES_META_DATA-FIELD_DESCRIPTIONS = VALUE #( ( NAME = 'LiquidityItem' ROLE =
CL_HEMI_CONSTANTS=>CS_FIELD_ROLE-TARGET )
```

```
    ( name = 'OriginSystem' role = CL_HEMI_CONSTANTS=>CS_FIELD_ROLE-key )
    ( name = 'OriginApplication' role = CL_HEMI_CONSTANTS=>CS_FIELD_ROLE-key )
    ( name = 'OriginDocument' role = CL_HEMI_CONSTANTS=>CS_FIELD_ROLE-key )
    ( name = 'OriginTransaction' role = CL_HEMI_CONSTANTS=>CS_FIELD_ROLE-key )
    ( name = 'OriginTransactionQualifier' role = CL_HEMI_CONSTANTS=>CS_FIELD_
ROLE-key )
    ( name = 'CashFlow' role = CL_HEMI_CONSTANTS=>CS_FIELD_ROLE-key )
    ).
  ENDMETHOD.
```

Listing 5.2 Coding for Handling ISLM Metadata

Next, the method shown in Listing 5.3 is encapsulated as business add-in (BAdI) for extensibility. It implements the training logic in terms of reading the data based on the CDS view and calling the related PAL random decision tree algorithm.

```
METHOD TRAINING
    BY DATABASE PROCEDURE FOR HDB LANGUAGE SQLSCRIPT OPTIONS SUPPRESS SYNTAX
ERRORS READ-ONLY
    USING CL_FQM_ML_DETECT_BADI=>TRAINING.
    CALL "CL_FQM_ML_DETECT_BADI=>TRAINING"(
      it_data                  => :it_data,
      it_param                 => :it_param,
      et_model                 => :et_model,
      et_variable_importance => :et_variable_importance,
      et_error_rate            => :et_error_rate,
      et_confusion_matrix      => :et_confusion_matrix
    );
  ENDMETHOD.
METHOD IF_FQM_ML_DETECT_BADI_INTF~TRAINING
    BY DATABASE PROCEDURE FOR HDB LANGUAGE SQLSCRIPT OPTIONS SUPPRESS SYNTAX
ERRORS.
    lt_data = select CompanyCode, CertaintyLevel, Currency, PlanningLevel,
CashPlanningGroup, GLAccount, BusinessPartner, Customer, Supplier,
PartnerCompany, CostCenter, ProfitCenter,
    Segment, BusinessArea, Material, WBSElement, PaymentMethod, FinancialAccount
Type, FinancialTransactionType, BankAccountInternalID, LiquidityItem
    from :it_data;
   /* Call PAL random decision trees algorithm */
CALL _sys_afl.pal_random_decision_trees(:lt_data, :it_param, :et_model, :et_
variable_importance, :et_error_rate, :et_confusion_matrix);
  ENDMETHOD.
```

Listing 5.3 Coding for Model Training

Finally, the method shown in Listing 5.4 is encapsulated as a BAdI for extensibility. It implements the consumption logic in terms of applying requests to the trained model and providing inference results.

```
METHOD APPLY
    BY DATABASE PROCEDURE FOR HDB LANGUAGE SQLSCRIPT OPTIONS SUPPRESS SYNTAX
ERRORS READ-ONLY
    USING CL_FQM_ML_DETECT_BADI=>PREDICT_WITH_MODEL_VERSION.
  CALL "CL_FQM_ML_DETECT_BADI=>PREDICT_WITH_MODEL_VERSION"(
    it_data   => :it_data,
    it_model  => :it_model,
    it_param  => :it_param,
    et_result => :et_result
  );
  ENDMETHOD.
METHOD IF_FQM_ML_DETECT_BADI_INTF~predict_with_model_version
    BY DATABASE PROCEDURE FOR HDB LANGUAGE SQLSCRIPT OPTIONS READ-ONLY.
    lt_data_uuid = select to_nchar(sysuuid) as id, OriginSystem,
OriginApplication, OriginDocument, OriginTransaction,
OriginTransactionQualifier, CashFlow,CompanyCode, CertaintyLevel,
Currency, PlanningLevel, CashPlanningGroup, GLAccount, BusinessPartner,
Customer, Supplier, PartnerCompany, CostCenter, ProfitCenter, Segment,
BusinessArea, Material, WBSElement, PaymentMethod, FinancialAccountType,
FinancialTransactionType, BankAccountInternalID
    FROM :it_data;
lt_data_predict = select id, CompanyCode, CertaintyLevel, Currency,
PlanningLevel, CashPlanningGroup, GLAccount, BusinessPartner, Customer,
Supplier, PartnerCompany, CostCenter, ProfitCenter,Segment, BusinessArea,
Material, WBSElement, PaymentMethod, FinancialAccountType,
FinancialTransactionType, BankAccountInternalID
    FROM :lt_data_uuid;

    /* Execute  prediction */
    CALL _sys_afl.pal_random_decision_trees_predict(:lt_data_predict, :it_
model, :it_param, lt_result);

    /* Prediction results are mapped to the composite keys */
    et_result =
 select OriginSystem, OriginApplication, OriginDocument, OriginTransaction,
OriginTransactionQualifier, CashFlow, r.score as LiquidityItem, confidence FROM
:lt_data_uuid as u INNER JOIN :lt_result as r ON u.id = r.id;
    ENDMETHOD.
```

Listing 5.4 Coding for Model Inference

Model training, activation, and debriefing for PAL are handled in an analogous way as for APL, as illustrated previously in Figure 5.8, Figure 5.10, and Figure 5.11. Inference results are exposed by the APPLY method shown in Listing 5.4 and integrated into business processes.

Note

Data scientists work usually in the Python environment for exploration, while SAP S/4HANA developers code in ABAP/SQL script. To improve collaboration, a go-between SAP HANA Python Client API is provided (see *http://s-prs.co/v515801*). Thus, PAL and APL algorithms can be incorporated with Python during exploration and be transformed to ABAP/SQL script for application development in SAP S/4HANA.

5.3 Implementing Side-by-Side Machine Learning Applications

The side-by-side approach is used when embedded machine learning does not meet the requirements of the machine learning scenario. This can be the case if the appropriate algorithm is not provided by SAP HANA machine learning, external data is required for model training, or the algorithm has high hardware requirements (e.g., neuronal networks need a GPU and demand a lot of memory). For side-by-side machine learning, an ABAP class, CDS view, and intelligent scenario must be developed on the SAP S/4HANA ABAP platform, whereas training and inference pipelines are implemented in SAP Data Intelligence.

In this section, our use case for illustrating the technical implementation steps is about predicting the delivery delay of products. We first explain the necessary developments in SAP Data Intelligence, then continue with the ABAP side.

5.3.1 Required Development in SAP Data Intelligence

As illustrated in Figure 5.17, the launchpad for SAP Data Intelligence provides access to various SAP Fiori applications, of which the following three are key for machine learning implementation:

1. **Connection Management**
   For the development of the introduced use case, the Connection Management app is required for setting up the connectivity to the SAP S/4HANA system in order to read the training data remotely.
2. **ML Scenario Manager**
   The ML Scenario Manager app is used to define all the required implementation artifacts.
3. **Modeler**
   The Modeler app is used to create the necessary pipelines.

Note

Setting up the connection to SAP S/4HANA is not machine learning-specific and won't be elaborated further here. For more information, refer to *http://s-prs.co/v515802*.

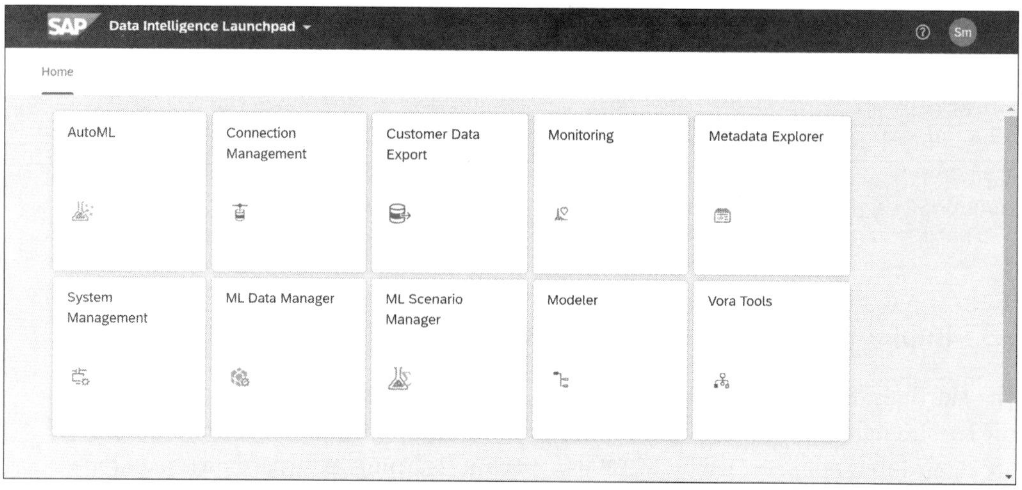

Figure 5.17 SAP Data Intelligence Launchpad

We'll begin with the ML Scenario Manager application. With this app, the machine learning scenario central artifact is created in SAP Data Intelligence as shown in Figure 5.18. For this, a unique **Name** must be provided ("Delivery Delay Prediction" in our example) and a description must be maintained in the **Business Question** field.

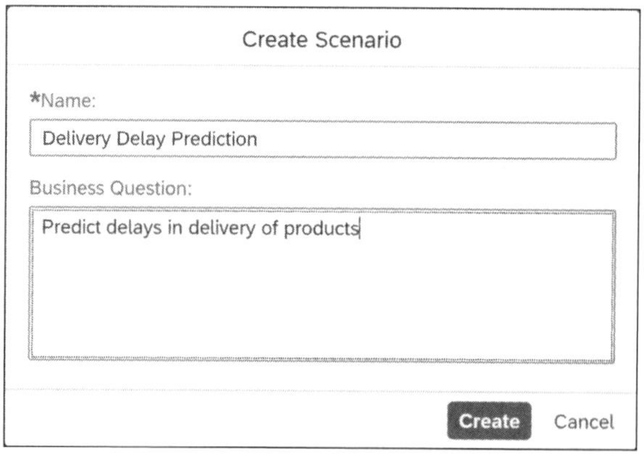

Figure 5.18 Create Machine Learning Scenario

Click the **Create** button. The details of the machine learning scenario are depicted in Figure 5.19. As already mentioned, the artifact encompasses all development and operation artifacts for the use case, such as pipelines, Python coding, and deployments of the model, as follows:

❶ **Datasets**
Datasets that are stored in SAP Data Intelligence and are used for the implementation of the use case, such as during data science exploration.

❷ **Notebooks**
SAP Data Intelligence integrates Jupyter Notebook to provide an exploration environment for data scientists. Thus, exploration and development of the use case are combined to increase efficiency.

❸ **Pipelines**
A pipeline is a graphical way of programming training and inference logic for a use case.

❹ **URL**
Each machine learning scenario has as a unique identifier, which in the current version of SAP Data Intelligence is part of the URL. We'll use this identifier later to bind the machine learning scenario with the ABAP content.

Further Resources

We'll be going through the steps specific to machine learning functionality within SAP Data Intelligence in this section. For more information on creating entities in SAP Data Intelligence, refer to *http://s-prs.co/v515803*.

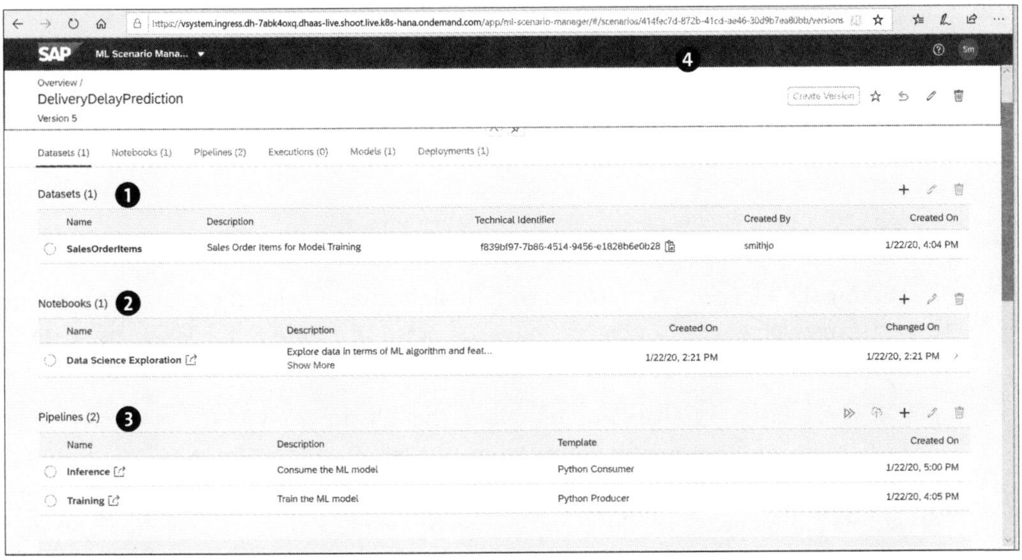

Figure 5.19 Machine Learning Scenario Details

The Jupyter Notebook integration is shown in Figure 5.20. In this environment, data scientists perform exploration for a use case. This includes analyzing the data with graphical illustrations, validating different algorithms, or performing transformations. As depicted in Figure 5.20, the Scikit-Learn machine learning library is used for the implementation of our machine learning use case to apply the linear regression algorithm.

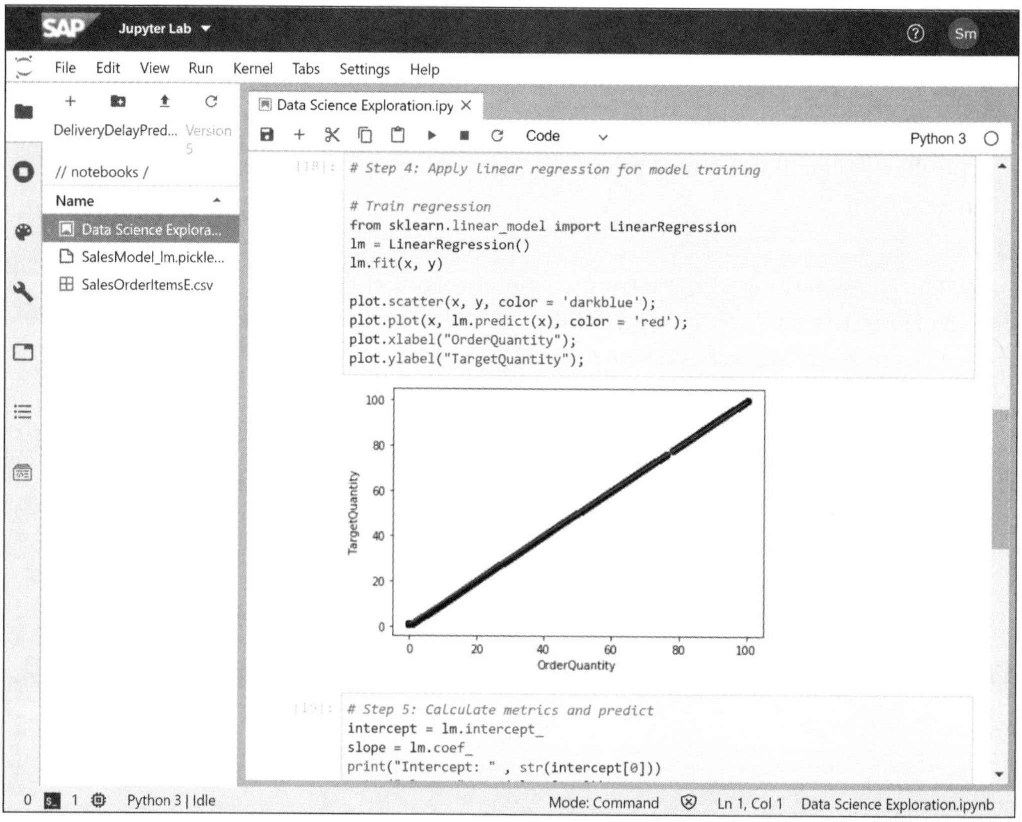

Figure 5.20 Jupyter Notebook for Data Science Exploration

By clicking the **+** icon in the **Pipelines** section of the machine learning scenario (see Figure 5.19), new pipelines can be created as depicted in Figure 5.21. For our use case, a training and an inference pipeline are required.

Pipelines can be created based on templates that already contain the main steps. For our use case, select **Python Producer** from the **Template** dropdown for the training pipeline and the **Python Consumer** template for the inference pipeline. As of the time of writing (summer 2020), the ISLM framework expects only one training and one inference pipeline for a side-by-side machine learning use case. To identity those pipelines, ISLM assumes the **[ISLM_TRAINING_PIPELINE]** string to be included as part of the description. This might be replaced in future versions with an explicit flag.

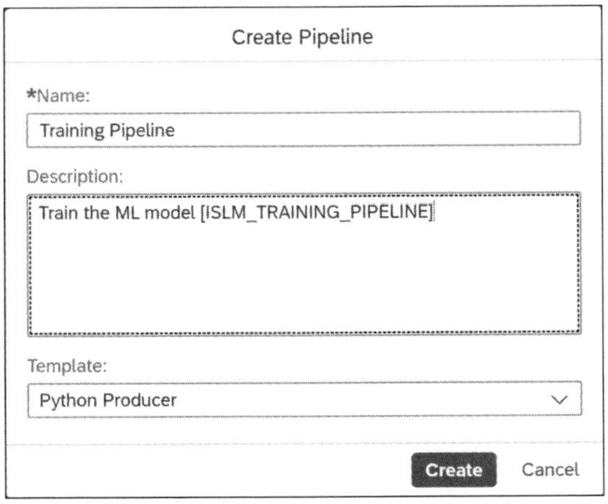

Figure 5.21 Create Training Pipeline

Figure 5.22 shows the training pipeline modeled for our machine learning use case (which you can navigate to via the previous screen, shown in Figure 5.19). Basically, the pipeline reads the application data online, transfers it in comma-separated value (CSV) format, trains the algorithm, stores the trained model, and calculates accuracy metrics, as follows:

❶ **Workflow Trigger**
The operator triggers the execution of the pipeline.

❷ **Cloud Data Integration**
The operator extracts application data from SAP S/4HANA based on the CDS extraction view for training data. The properties of this operator are detailed in Figure 5.23.

❸ **Flowagent CSV Prod...**
The operator transforms the application into a CSV file.

❹ **Algorithm**
The operator trains the machine learning algorithm based on the application data. This operator is coded in Python.

❺ **Artifact Producer**
The operator stores the trained model.

❻ **Submit Metrics**
The operator processes metrics for model accuracy.

❼ **Graph Terminator**
The operator terminates the pipeline and releases the used system resources.

Listing 5.5 shows the coding for operator ❹, **Algorithm**, shown in Figure 5.22. The coding implemented in Jupyter Notebook (refer back to Figure 5.20) is transferred to the structure, which is required by the operator. The input port of the operator receives the application data with which the Scikit-Learn linear regression algorithm is trained.

Model metrics and the blob ID of the trained model are exposed by output ports of the operator.

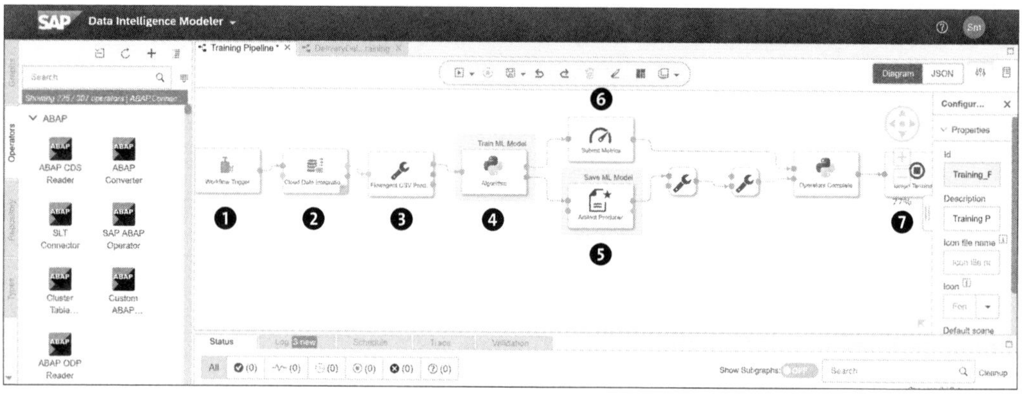

Figure 5.22 Training Pipeline

```
# Python script to perform training on input data & generate Metrics &
 Model Blob
def on_input(data):

# Step 1: Import libraries
import pandas as pd
import io
import numpy as np
import logging

logging.basicConfig(level=logging.DEBUG, format='%(asctime)s - %(levelname)s -
%(message)s')
api.logger.info("Your message")

# Step 2: Read data and display it
dataset = pd.read_csv(io.StringIO(data), sep=",")
x = dataset['OrderQuantity'].values.reshape(-1,1)
y = dataset['TargetQuantity'].values.reshape(-1,1)

# Step 3: Apply linear regression for model training
from sklearn.linear_model import LinearRegression
lm = LinearRegression()
lm.fit(x, y)

# Step 4: Calculate metrics and predict
intercept = lm.intercept_
slope = lm.coef_
metrics_dict = {"Intercept: " , str(intercept[0]), "Slope: ", str(slope[0,0])}
api.send("metrics", api.Message(metrics_dict))
```

```
# Step 5: Store the model for future inference
import pickle
model_blob = pickle.dumps(lm)
api.send("modelBlob", model_blob)

api.set_port_callback("input", on_input)
```


Listing 5.5 Coding for Operator Responsible for Model Training

By clicking on operator ❷, **Cloud Data Integration**, in Figure 5.22, its properties can be maintained, as shown in Figure 5.23. According to the SAP S/4HANA machine learning programming model, training data is accessed via CDS extraction views. Thus, capabilities like initial load, delta handling, and extensibility are enabled. Let's walk through the operator parameters:

❶ **Connection ID**
ID to connect to SAP S/4HANA system.

❷ **Source**
CDS view to extract data for model training.

❸ **Extraction Mode**
Extraction mode to transmit all data (**Full**) or only changes (**Delta**).

❹ **Columns**
Columns of the CDS view extractor that will be considered (filtering of columns).

❺ **Data Preview button**
Preview the extracted data.

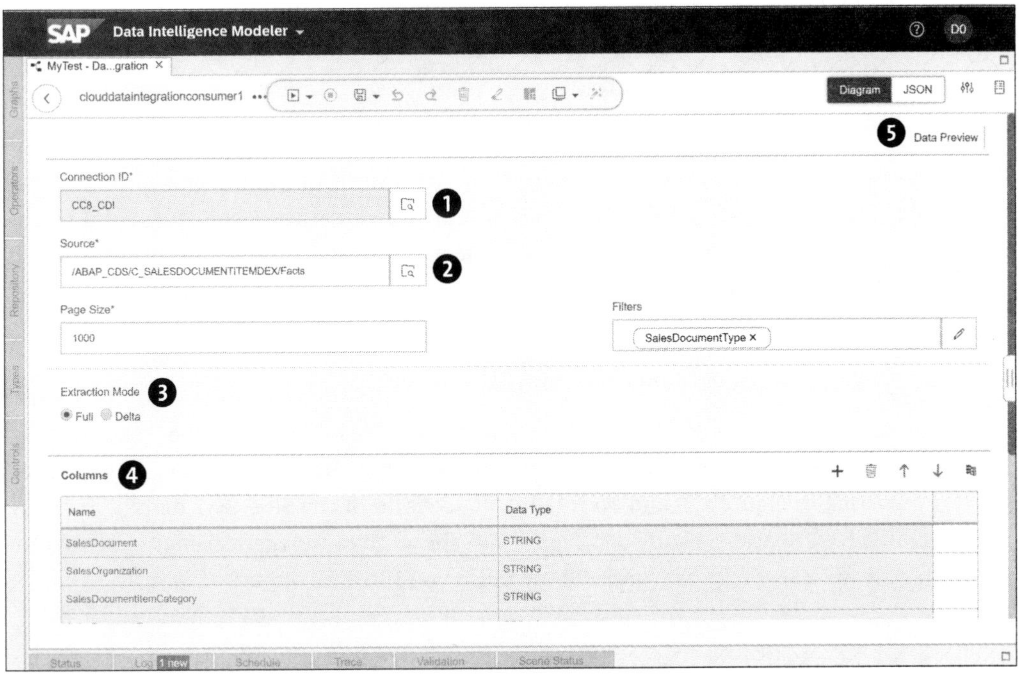

Figure 5.23 Cloud Data Integration Operator for Reading CDS View Data

Next, let's review the inference pipeline for our machine learning use case, which is illustrated in Figure 5.24. In this pipeline, the trained model is loaded into RAM memory and a listener for REST inference calls is instantiated. The details are as follows:

❶ **Submit Artifact Name**
This operator allows for maintaining the ID of the trained model.

❷ **Artifact Consumer**
This operator loads the trained model based on its ID.

❸ **OpenAPI Servlow**
This operator listens for inference requests and sends inference responses back. The underlying communication protocol is REST.

❹ **Properties**
Displays the properties of the selected operator.

❺ **Model Inference**
This operator performs inference requests and returns the inference results.

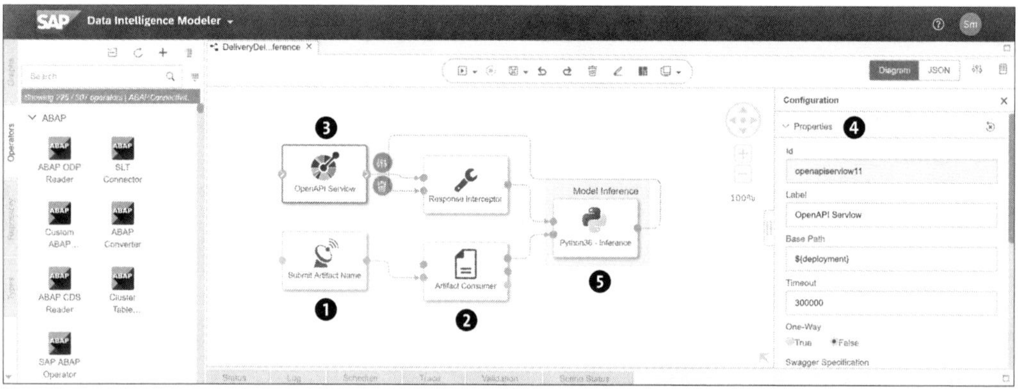

Figure 5.24 Inference Pipeline

The last step of the content development for SAP Data Intelligence is to flag the machine learning scenario as released; only released scenarios are considered by the ISLM framework in ABAP. For this, the star icon must be clicked in the scenario details screen (refer back to Figure 5.19).

5.3.2 Required Development in ABAP

The ABAP development of our machine learning use case starts with the creation of the intelligent scenario. For this, the Intelligent Scenarios application must be launched (refer back to Figure 5.2) and an intelligent scenario of the **Side-By-Side** type must be created (refer back to Figure 5.3). Figure 5.25 shows the intelligent scenario created for our machine learning use case, and the fields you must fill out are as follows:

❶ **Intelligent Scenario Name**
Unique name of the intelligent scenario for identification.

❷ **Intelligent Scenario Description**
Text to describe the machine learning key question to be solved by the defined intelligent scenario.

❸ **Intelligent Scenario Type**
Type of the intelligent scenario, which for our side-by-side machine learning use case is SAP Data Intelligence.

❹ **Package**
Package in which to store the development artifacts.

❺ **Prediction Class**
The ABAP class that implements methods for metadata provisioning and inference calls. This class must be implemented specifically for the scenario using ABAP Class Builder by applying the defined interface as shown in Figure 5.26.

❻ **External ID**
ID of the corresponding machine learning scenario artifact that was created for our machine learning use case in SAP Data Intelligence (refer to Figure 5.19 ❹). Based on this ID, the ABAP content and SAP Data Intelligence content are bounded.

❼ An intelligent scenario can be saved as draft if additional changes are expected; otherwise, it's published for use.

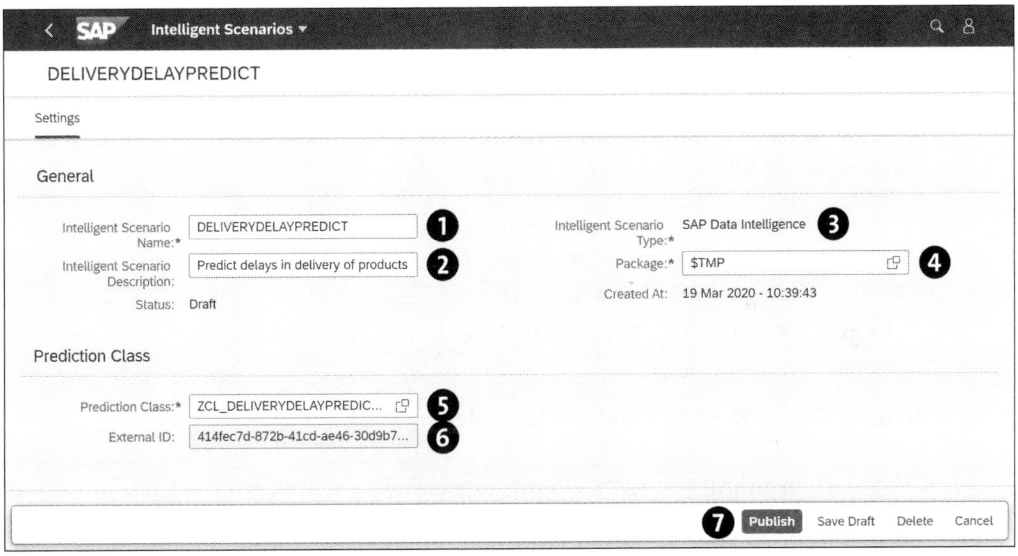

Figure 5.25 Create Intelligent Scenario for Side-by-Side Machine Learning

The ABAP prediction class registered in the intelligent scenario (see Figure 5.25) implements metadata methods and the consumption API of the trained model, as shown in

Figure 5.26. For the implementation, the ABAP Class Builder can be used, as illustrated in Figure 5.26. To standardize the structure and behavior of the class, it must implement the IF_ISLM_INTELLIGENT_SCENARIO interface.

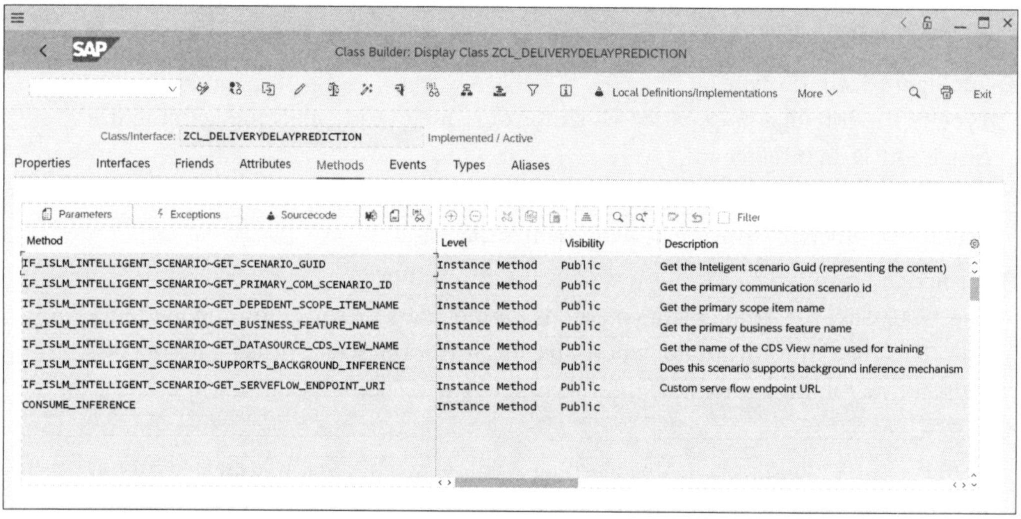

Figure 5.26 Methods and Interfaces of Registered ABAP Class

Metadata provisioning is implemented via methods instead of via dedicated input fields in the UI. Thus, the UI is kept simple, but additional metadata can be considered easily. Examples of such metadata methods include the name of the CDS view for data extraction and the ID of the business feature for scope item activation. The ID of the machine learning scenario artifact that was created in SAP Data Intelligence (refer to Figure 5.19 ❹) is maintained as an attribute of the class and is exposed with the method shown in Listing 5.6.

```
METHOD IF_ISLM_INTELLIGENT_SCENARIO~GET_SCENARIO_GUID.
    "ISLM Specific: Retuen the ML scenario GUID via IF_ISLM_INTELLIGENT_
SCENARIO interface implementation
    rv_scenario_guid = me->mv_scenario_guid.
ENDMETHOD.
```

Listing 5.6 Coding for GET_SCENARIO_GUID Method

With the ID, the ABAP and SAP Data Intelligence content are correlated and processed accordingly by the ISLM framework. The method shown in Listing 5.7 implements the API for consuming the inference results using the REST protocol.

```
METHOD CONSUME_INFERENCE.
DATA:
  lv_json_request_data  TYPE string,
  lr_islm_inference_provider TYPE REF TO cl_islm_inference_provider,
```

```
  lr_islm_inference_client TYPE REF TO cl_islm_inference_http_client.

* Initialise inference provider from ISLM
  lr_islm_inference_provider = new cl_islm_inference_provider( me ).
  lr_islm_inference_provider->get_inference_client(
  IMPORTING eo_http_client = lr_islm_inference_client
                      es_bapireturn  = DATA(ls_bapiret) ).

* Check inference client is avlaible    IF lr_islm_inference_client IS NOT INITIAL.
    lv_json_request_data ='{ "quanity":12 , "product_code":125}'.
    lr_islm_inference_client->add_request_body(
    EXPORTING  iv_header_payload = lv_json_request_data).
    lr_islm_inference_client->send_and_receive(
        IMPORTING
        ev_http_return_reason = DATA(lv_http_return_reason)
        ev_response_data      = DATA(lv_json_response_data)
        es_bapireturn         = DATA(ls_bapireturn)
        ev_http_return_code   = DATA(lv_http_return_code)
       ).

    lr_islm_inference_client->close( ).
  ENDIF.
ENDMETHOD.
```

Listing 5.7 Coding for CONSUME_INFERENCE Method

With this, the implementation of our side-by-side machine learning case is completed. The consumption API of the ABAP class provides the inference results, which can be integrated into business processes and UIs.

However, to provide meaningful values, first the intelligent scenario must be trained. For this, the Intelligent Scenario Management app can be used (refer back to Figure 5.2). The necessary steps are already described in the context of the embedded machine learning implementation (refer back to Figure 5.8 and Figure 5.10). The only difference is that there are **Execute** and **Deploy** buttons provided for side-by-side machine learning. The execution step runs the training pipeline, whereas the deployment step performs the inference pipeline and provides an inference serving container.

To connect the ISLM framework to SAP Data Intelligence, the administrator must map the intelligent scenario to the HTTP destination of SAP Data Intelligence as shown in Figure 5.27. The UI and the menu path for accessing the transaction can vary depending on the SAP S/4HANA release, so be sure to consult the release-specific documentation for ISLM at the SAP Help Portal (*https://help.sap.com*). Basically, the administrator has to access the Customizing transaction and maintain the HTTP destination for the intelligent scenario.

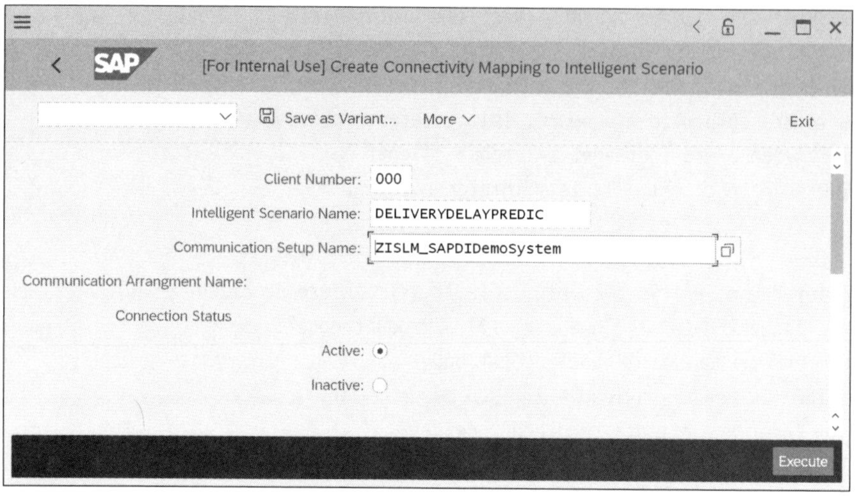

Figure 5.27 Configure Connectivity between ISLM and SAP Data Intelligence

5.4 Implementing Side-by-Side Predictive Analytics Applications

The side-by-side predictive analytics approach based on SAP Analytics Cloud is used for explorative scenarios in which ad hoc analysis for business insights is required. The results are shared among business analysists but usually not permanently stored. The underlying algorithms are provided by APL. Thus, if the explorative scenario will be used on a regular basis as part of the transactional processes, then it's implemented as an embedded machine learning use case founded on APL (refer back to Section 5.2.1).

To illustrate the implementation steps, in this section we'll examine a simple product forecasting scenario in which sales for different product categories are predicted based on the time series algorithm in order to consider seasonal trends. After you log into SAP Analytics Cloud and upload the example dataset, you can create a predictive scenario by selecting **Create • Predictive Scenario** in the menu, as shown in Figure 5.28.

A predictive scenario is a set of use cases with common characteristics. As depicted in Figure 5.29, there are different types of predictive scenarios:

- *Classification scenarios* predict the value of a target variable that can only have two possible values, like yes or no, and 0 or 1. Examples of classification scenarios include customer churn, with the target variable predicting whether a customer will leave or not; propensity to buy, with the target variable predicting whether a customer will buy a product offered to him or not; and fraud, with the target variable indicating whether a transaction or claim was fraudulent or not.
- *Regression scenarios* predict the numerical value of a target variable depending on the variables describing it. Examples of regression scenarios include predictions of the number of customers visiting a shop during lunch time, the revenue of a customer in the next quarter, and the sales price of used cars.

- *Time series scenarios* predict the value of a variable over time, taking into account further descriptive variables. Examples of time series scenarios include the prediction of revenue for a product line over the next few quarters, the number of bicycles hired in a city over the next few days, and travel expenses in the next few months.

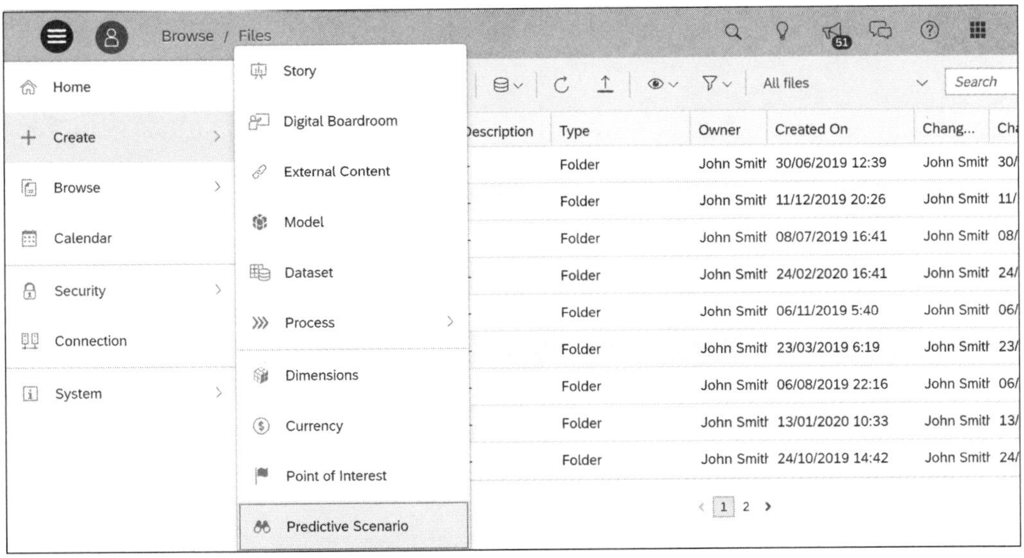

Figure 5.28 Menu Path to Create Predictive Scenario

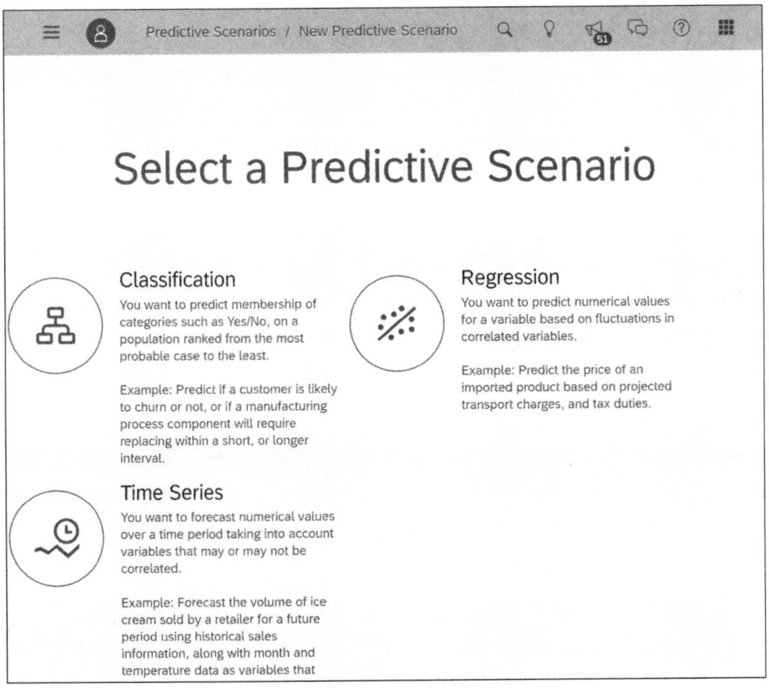

Figure 5.29 Types of Predictive Scenarios

For the given use case, a time series predictive scenario addresses our needs. Thus, create a scenario by clicking **Time Series** on the screen shown in Figure 5.29. On the following **New Predictive Scenario** pop-up, shown in Figure 5.30, enter a **Name** and a **Description** for your predictive scenario.

Figure 5.30 Create Time Series Predictive Scenario

After the values in the pop-up are maintained, you can click on the **Save** button so that the dialog in Figure 5.31 is shown, in which properties of the model must be entered as follows:

❶ **Name**
The dataset to use as input to train the model. This can be a local source or a connection to a remote system.

❷ **Signal Variable**
The target variable, which will be predicted based on historical data.

❸ **Date Variable**
An input variable used to train the time series algorithm. The date variable contains the time dimension.

❹ **Segmented By**
This variable allows you to automatically create a one-time series per segment. Here, a one-time series per product category is created, and the product category is contained in the product category column of the dataset. Therefore, the column is used as the variable.

❺ Once all properties are maintained, training can be launched by clicking the **Train** button.

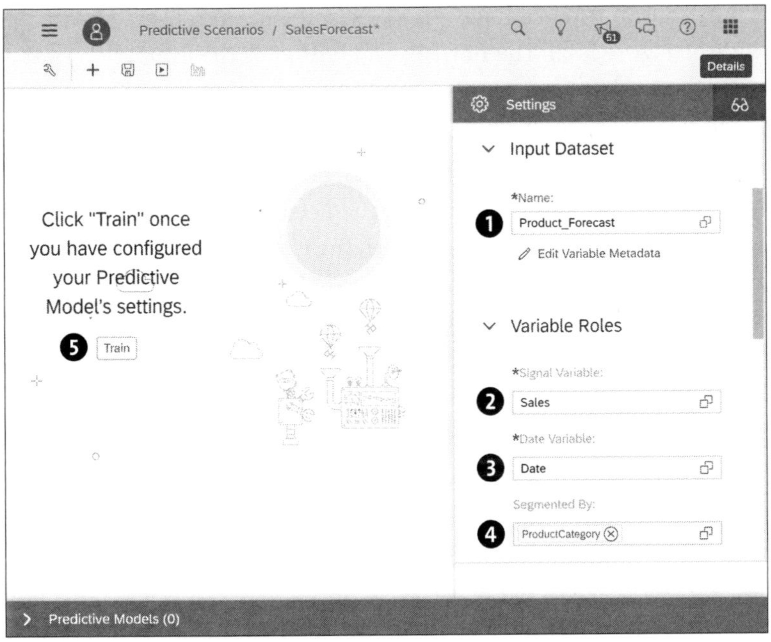

Figure 5.31 Create Predictive Model

Upon completion of the training process, the model debriefing screen shown in Figure 5.32 is displayed.

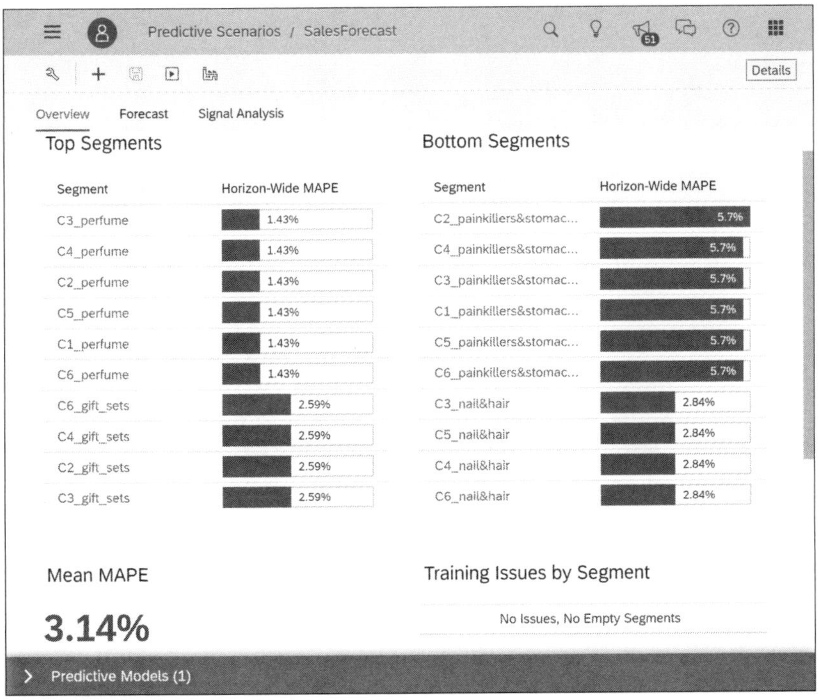

Figure 5.32 Model Debriefing: Overview

The horizon-wide mean absolute percentage error (horizon-wide MAPE) for each of the segments is shown (see Figure 5.32). The horizon-wide MAPE is the evaluation of the error rate when using the model to estimate the future values of the signal. The average quality of the segmented time series is 3.14%.

Click on the **Forecast** tab in the top menu to see the detailed time series chart and all the forecasts, as shown in Figure 5.33. Outliers have been detected by the tool in this example. These are values that could not be explained with the given data. These are indicators that we might want to include more explanatory variables, like information about special events or marketing campaigns. Outliers are points at which the predictive curve is very distant from the real curve. They're represented by a red circle on the plot. An outlier is detected when the absolute value of the residuals is over twice the standard deviation computed on the estimation dataset.

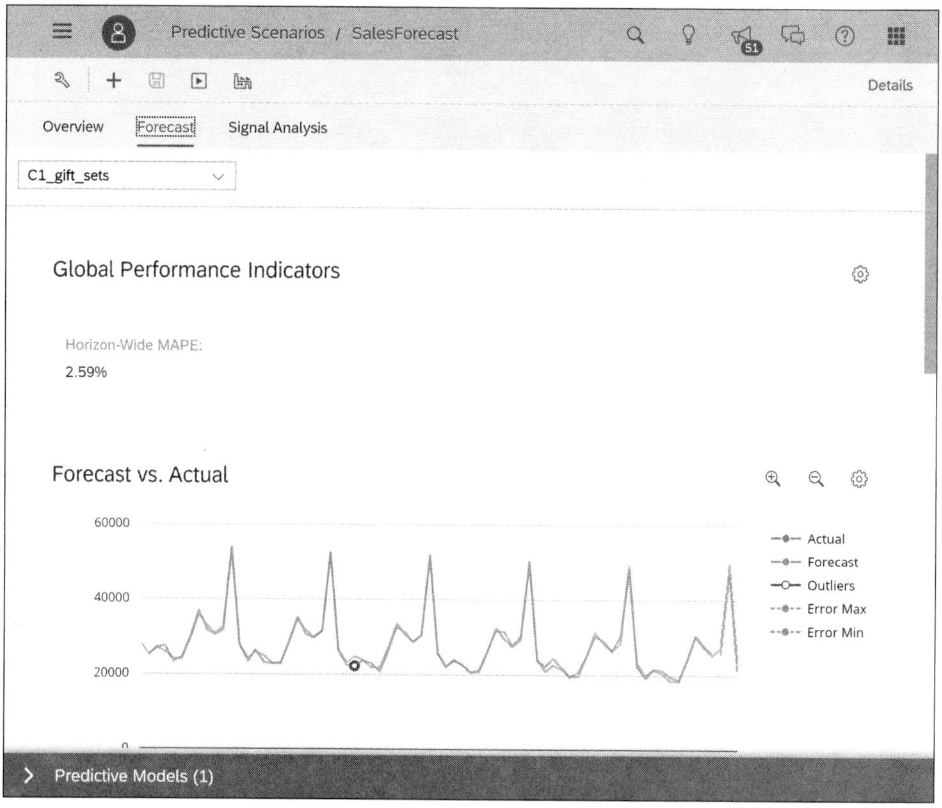

Figure 5.33 Model Debriefing: Forecast

Click on the **Signal Analysis** tab in the top menu to see the details of the signal analysis, as shown in Figure 5.34. Signal analysis displays the descriptive and statistical information about the model and shows the detected signals, like cycles, trends, and fluctuations. The trend is the general orientation of the signal. The cycles are periodic elements that can be found at least twice in the training dataset. The fluctuation is what is left when the trend and the cycles have been extracted.

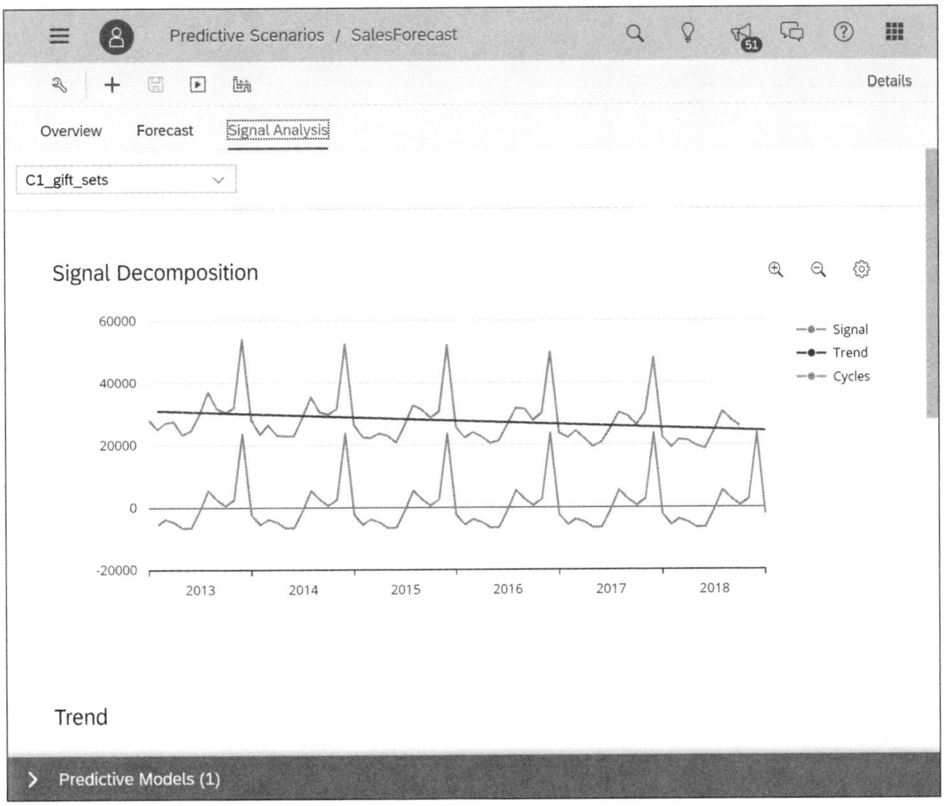

Figure 5.34 Model Debriefing: Signal Analysis

After you take a deeper look at the forecast and are convinced of its quality, you can apply it. To do this, click on the **Factory** icon shown in Figure 5.34; choose the **Output Dataset** for saving the predictions, as illustrated in Figure 5.35; and click the **OK** button.

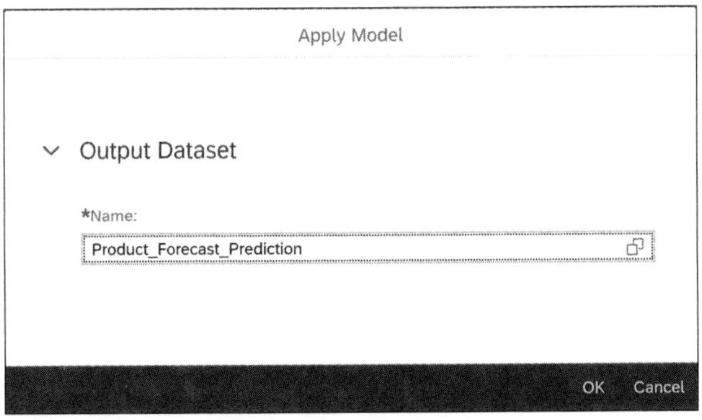

Figure 5.35 Dataset for Storing Prediction Results

As illustrated in Figure 5.36, at the bottom you can see that the status of the model is **Applied**. When the run is finished, the result can be displayed.

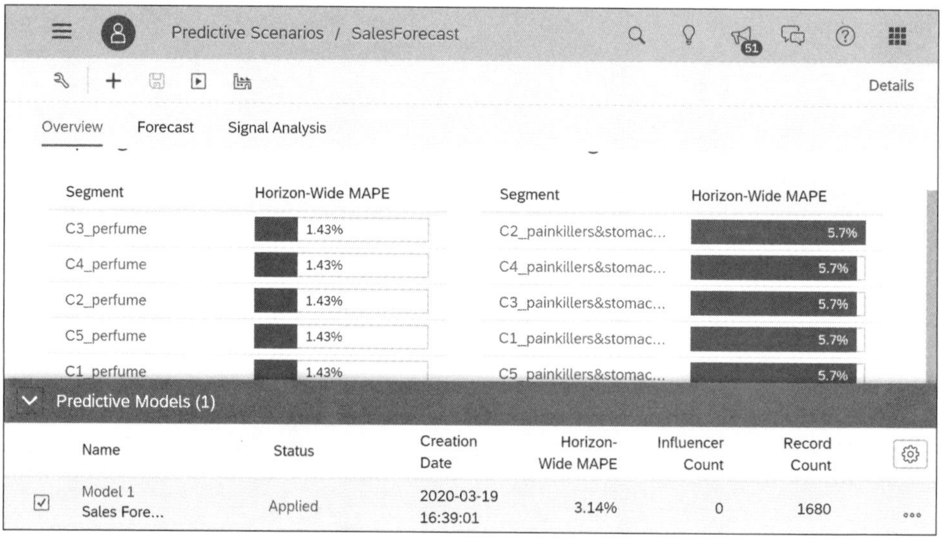

Figure 5.36 Run Prediction Process

Navigate to the folder where the output dataset is saved and click on the dataset to display the results, as shown in Figure 5.37. This is a preview of the dataset. The predictions are shown in the **Forecast** column.

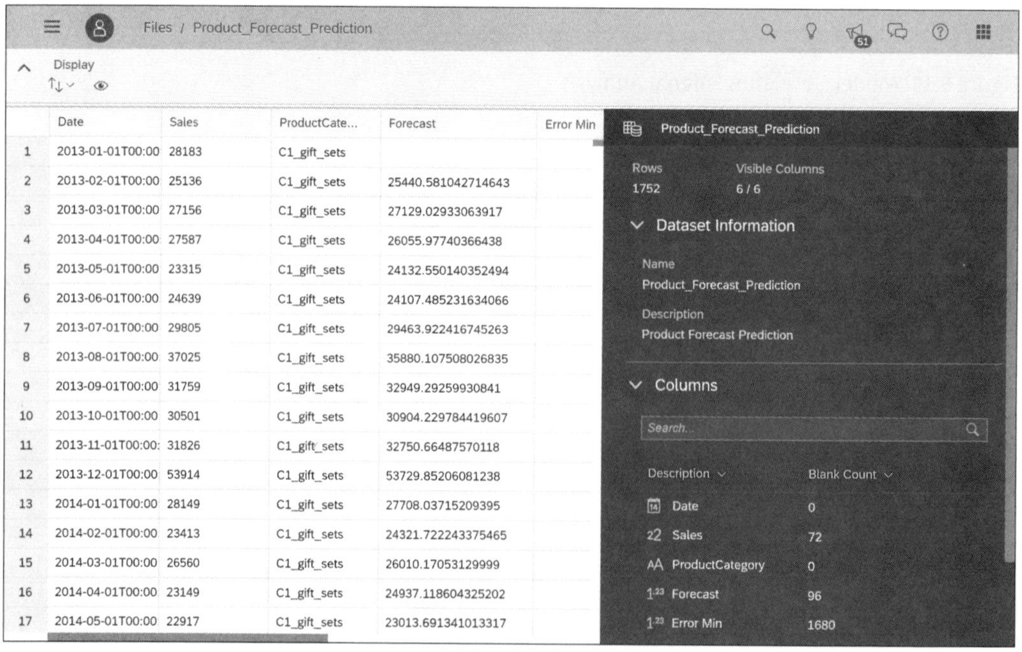

Figure 5.37 Display Prediction Results

The results can be visualized with SAP Analytics Cloud by building models based on the prediction dataset.

5.5 Application Management Processes

When you implement machine learning applications in context of SAP S/4HANA, they must be enterprise-ready. This includes qualities like compliance, lifecycle management, data and process integration, workload management, and performance. In this section, we explain the solution concepts and explore the necessary implementation tasks for each aspect. This will help you to ensure enterprise readiness for your machine learning applications.

Note

As the described approach for predictive analytics is aimed at ad hoc and explorative scenarios, those topics are usually not relevant for this approach. Thus, the main focus here is on the embedded and side-by-side machine learning applications.

As a disclaimer, note that as of the time of writing (summer 2020), SAP does not deliver frameworks for all described components. To enable you to help yourself and implement helpful workarounds, we explain the necessary concepts here.

5.5.1 Lifecycle Management

In context of machine learning, new artifacts and processes are introduced that must be considered from the lifecycle management perspective. For example, machine learning models must be trained, deployed, activated, and monitored. Those aspects are not covered by classic lifecycle management tools and concepts. Therefore, this section explains the business requirements and the necessary technical implementation for lifecycle management of machine learning.

Business Requirements

With machine learning, new aspects must be considered in lifecycle management to provide the operational qualities customers expect from intelligent ERP. The focus of this section is on the phase in which a customer researches a machine learning scenario in SAP S/4HANA to understand its value, its prerequisites, and the required infrastructure to use it. The requirements are described in the context of the steps customers must perform to consume and operate machine learning scenarios: check, setup, train, deploy, and monitor. The focus here is on the machine learning specifics; existing solution approaches for lifecycle management are taken for granted, such as delivery or support processes.

Let's walk through these steps and how machine learning fits into each:

- **Check**
  It's difficult for customers to understand which technical and business prerequisites are required to train and consume machine learning scenarios. For example, sufficient data volume must be in place for training machine learning algorithms. Also, underlying business processes must be activated and configured in order to have a meaningful foundation for the training process. With an increasing number of machine learning scenarios, such evaluations can't be performed by customers manually any longer due to high total cost of ownership (TCO) and huge complexity. Therefore, a prerequisite check capability is required to automatically validate whether the necessary prerequisites for training and consumption are met for each machine learning scenario.
- **Setup**
  Before customers can start using side-by-side machine learning scenarios, the connectivity to SAP Data Intelligence must be set up. Within this onboarding process, a customer account with a service entitlement and a service key is created. The content of the service key provides the information for the initial communication configuration on the SAP S/4HANA ABAP side to access the machine learning services in SAP Data Intelligence. This operation model is manual, time-consuming, and hard to control for customers. Therefore, the setup procedure will be automated based on a wizard that assists customers through the steps of provisioning and connecting SAP Data Intelligence.
- **Train**
  Customers must train machine learning models for consumption. However, the training process is usually manual, time-intensive, and less transparent to customers. Furthermore, the calculated model accuracy KPIs are not sufficient for usage decisions. Therefore, the training process will be provided as a self-service option for the customer for all machine learning scenarios. Customers must be enabled to adjust parameters to make successful training possible. Errors and warnings will be presented in a language that people who aren't machine learning experts will understand. Fully automated training runs based on scheduled jobs will be supported. Event-driven triggering of training jobs will be possible.
- **Deploy**
  Customers require control over the timing of deployment and activation of trained models. However, usually models are deployed in the DevOps modus, which is an expensive and time-consuming task due to manual steps. It's good to automate this process so that customers can deploy the model as a self-service option. Deployment of multiple models simultaneously will be supported so that, for example, A/B testing can be performed before the model is activated. Deactivation of models will also be possible, such as undeploying models with insufficient accuracy. The history of model activation/deactivation will be recorded for monitoring and auditing purposes.

- **Monitor**
  Customers want to evaluate the quality of productively used models at runtime. However, the machine learning infrastructure often provides only technical monitoring, with business process aspects missing. Monitoring gives customers a comprehensive view of the machine learning models used. This will be based on a central cockpit for all machine learning scenarios and will provide, for example, the model status, accuracy KPIs, the state of inference calls, and the amount of processed data. In case of issues, alerts will be raised to inform administrators to act. Solution proposals will be recommended to administrators for resolving issues. Statistics will be offered, like the number of errors, the amount of resource consumption, or the costs incurred.

In this context, different roles must be distinguished:

- **Business users**
  These users have their focus on operating the business processes. They will want, for example, to change the layout of the charts and the format of dates or to hide columns in tables. Those changes are only locally applied and don't affect other people. Such kinds of changes are not machine learning specific and can be taken as granted.
- **Business experts**
  These are specialized key users with profound business and technical knowledge. They are in charge of most of the configuration and extensibility use cases, such as enhancing the data source for model training, replacing algorithms, and extending pipelines. However, usually they must be supported by data scientists, who perform machine learning explorations as a prerequisite. The changes they perform affect people in specific lines of business and in the whole company.
- **Business administrators**
  These technical users are responsible for the administration of machine learning applications in terms of training, deploying, and monitoring machine learning models. In particular, they configure models and schedule training jobs for machine learning scenarios. They consult data scientists and follow the resulting recommendations. The changes they perform affect people in specific lines of business and in the whole company.
- **Developers**
  These technical experts perform complex changes by implementing code. They oversee breakout scenarios in which machine learning logic cannot be modeled graphically but must be coded due to high complexity. Developers are in charge of extending coded machine learning logic. Data scientists must support them with recommendations based on their machine learning explorations. The changes they perform affect people in specific lines of business and in the whole company.

Technical Implementation

In this section, the technical implementation for the business requirements listed in the last section is described. To provide the necessary background, first let's answer the following questions:

- What are the machine learning-specific artifacts that require special lifecycle management considerations?
- What are the lifecycle management processes related to these artifacts?
- How and by whom are these processes triggered?

Artifacts, Processes, and Roles

The machine learning artifacts are illustrated in Figure 5.38. An intelligent scenario is the central design-time entity, covering all artifacts required for the implementation of the machine learning scenario. Semantically, it represents the key questions to be solved with machine learning. For this, the required algorithm and data features are determined by data scientists. For training the algorithm, you need the appropriate application data. Thus, a view is provided that can span multiple application tables and perform preliminary data transformations. The signature for the inference API is also defined during design time as part of the intelligent scenario, so the consuming application and business processes also can integrate the inference API during design time. However, the inference API provides meaningful results once the underlying machine learning model is trained. Before machine learning models are trained, prerequisite checks are performed. These are machine learning scenario-specific checks to validate whether the prerequisites for the training are met—for example, checking that sufficient application data are available or that necessary process configurations have been completed.

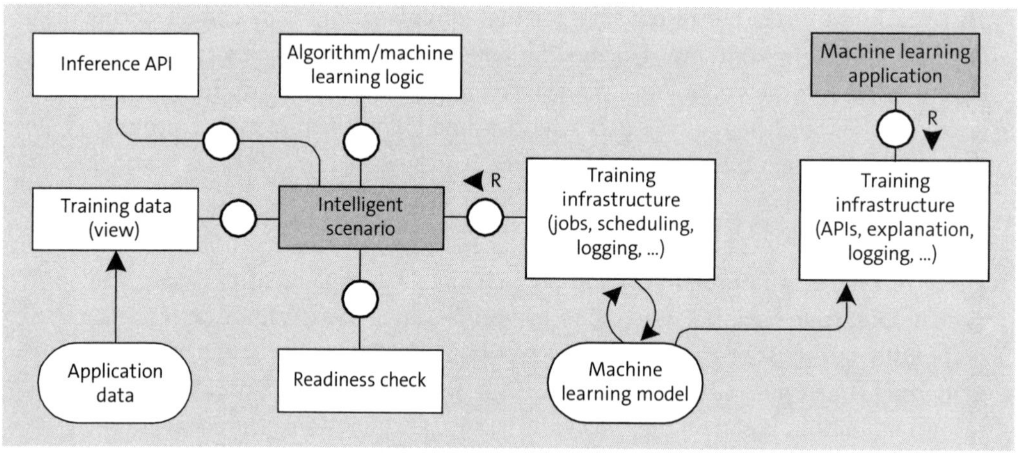

Figure 5.38 Lifecycle Management Artifacts

After the prerequisite checks have been passed successfully, the training is triggered. For this, the training infrastructure is used, which reads the necessary metadata from the corresponding intelligent scenario, runs the training, and stores the trained model. The inference infrastructure provides the consumption API for the specific scenario and model. Thus, the machine learning capability can be integrated into applications and business processes.

Figure 5.39 summarizes the processes, triggers, and roles that are related to the described artifacts and require lifecycle management considerations.

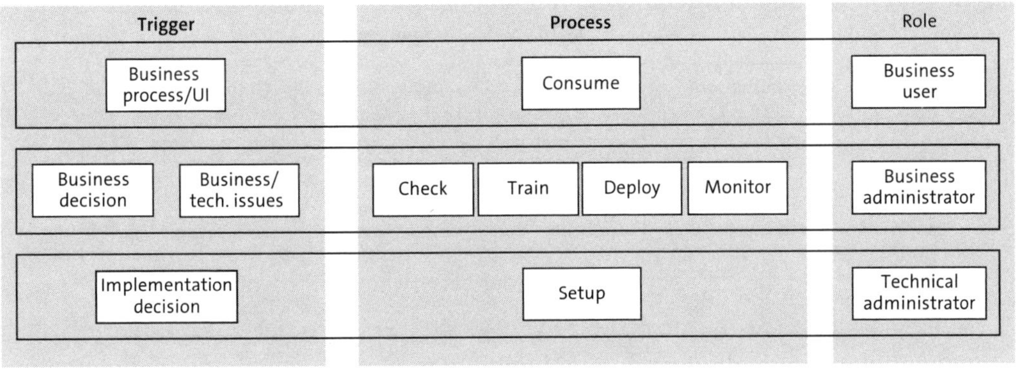

Figure 5.39 Lifecycle Management Triggers, Processes, and Roles

Business users usually consume machine learning capabilities to complete their daily tasks and fulfill their business needs. The prerequisite checks are performed by business administrators. The results are the basis to decide whether the machine learning scenario can be implemented. If the decision is positive, then technical administrators execute the setup while the business administrator performs the training and deployment steps. Machine learning scenarios are continuously monitored. However, typically business administrators only consider the monitoring data when it comes to alerts due to business or technical issues, such as the inference API providing the wrong results or the training job allocating too many system resources.

The lifecycle management of the machine learning application is handled by the ISLM framework in a more generic approach. The framework is offered as a capability in the ABAP application server through the Basis software component (see Figure 5.40). This capability is available for on-premise and cloud versions in Basis to enable the SAP S/4HANA business applications to consume and deliver the intelligent scenarios in a unique way. The initial focus of the framework is to enable and harmonize the operations of intelligent scenarios in SAP S/4HANA areas, but in the future it is expected to be able to extend to other areas that have a dependency on the Basis layer.

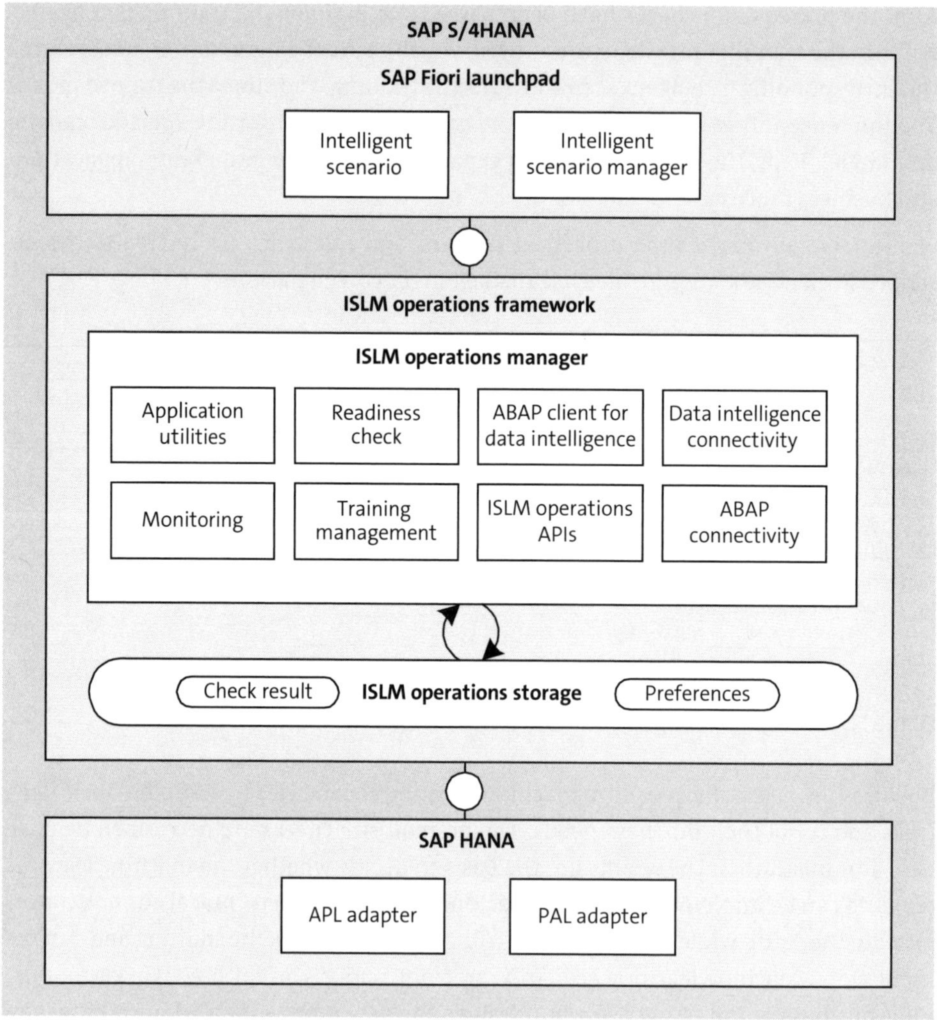

Figure 5.40 ISLM Framework Architecture

The framework harmonizes lifecycle management operations like prerequisite checks, training, deployment, activation, deactivation, monitoring, and inference consumption across various business domain-specific intelligent scenarios. The framework provides simpler functionalities and features for those who are not machine learning experts to perform the lifecycle management operations of an intelligent scenario.

The complexity of handling the intelligent scenarios in different stacks, like SAP S/4HANA and SAP Data Intelligence, is simplified in the ISLM framework. The ISLM framework empowers business administrators to perform lifecycle management operations for intelligent scenarios, like a self-service operation from a central cockpit. The architecture overview in Figure 5.40 illustrates the placement of the ISLM framework in the SAP S/4HANA intelligent scenario application context.

The existing Predictive Analytics integrator (PAi) framework, supporting embedded machine learning scenarios, is also included in the ISLM framework. This inclusion makes the ISLM framework able to offer a harmonized operational experience for embedded machine learning and for the side-by-side machine learning approach based on intelligent scenarios.

For the side-by-side scenarios, the ISLM framework has a built-in, ABAP-based REST client for SAP Data Intelligence. It has the logic to invoke the SAP Data Intelligence-specific REST APIs in a native way. SAP Data Intelligence provides a set of HTTP REST APIs for various aspects, like machine learning scenarios, training, deployment, and metrics. These APIs are consumed and orchestrated by the ISLM framework to provide a simpler view to nonexperts in machine learning for operating the side-by-side intelligent scenarios. The framework assumes the connectivity between SAP Data Intelligence and the SAP S/4HANA system is established up front with the required authorizations and valid authentications.

The ISLM framework is delivered with the following SAP Fiori applications under the Intelligent Scenarios SAP Fiori catalog in the Basis UI software component, as depicted previously in Figure 5.2:

- **Intelligent Scenarios**
  Using this application, developers can create intelligent scenarios in the ISLM framework with the basic details and an ABAP class representing the intelligent scenario. This application will be accessible to the customers as well for the custom-developed intelligent scenarios. For the embedded approach, the ISLM framework generates a class internally during this registration process.
- **Intelligent Scenario Management**
  Using this application, the business administrators and domain experts can perform operations like prerequisite checks, training, deployment, activation, and monitoring of the intelligent scenario for a specific business domain. This application is built with more focus on nonexperts in machine learning to allow them to operate intelligent scenarios.

The ISLM framework for intelligent scenarios covers the customer/business administrator view of operating an intelligent scenario. To offer a simplified operational approach, the framework mandates the development-related artifacts and processes. Figure 5.41 illustrates the phases to be performed by the developer in design time and the phases to be performed by the business administrator on the customer side. The technical details of theses phases differ based on the machine learning approach (embedded or side-by-side).

Creation of an intelligent scenario is a design-time activity to be performed by developers during the development period. The intelligent scenarios from different application areas in SAP S/4HANA apply to the ABAP interfaces designed by the ISLM framework. These rules are detailed based on SAP Data Intelligence for the side-by-side machine learning model. The creation activity helps the ISLM framework to understand the properties and behavior of an intelligent scenario in detail during the runtime. The creation process mandates an ABAP class implementation of the IF_ISLM_INTELLIGENT_SCENARIO interface from the applications delivering an intelligent scenario. This ABAP class represents the intelligent scenario in SAP S/4HANA. IF_ISLM_INTELLIGENT_SCENARIO is the marker interface used by the ISLM framework to identify the intelligent scenario representations in the framework and map it with the content available in SAP Data Intelligence during runtime with a unique SAP Data Intelligence content globally unique identifier (GUID).

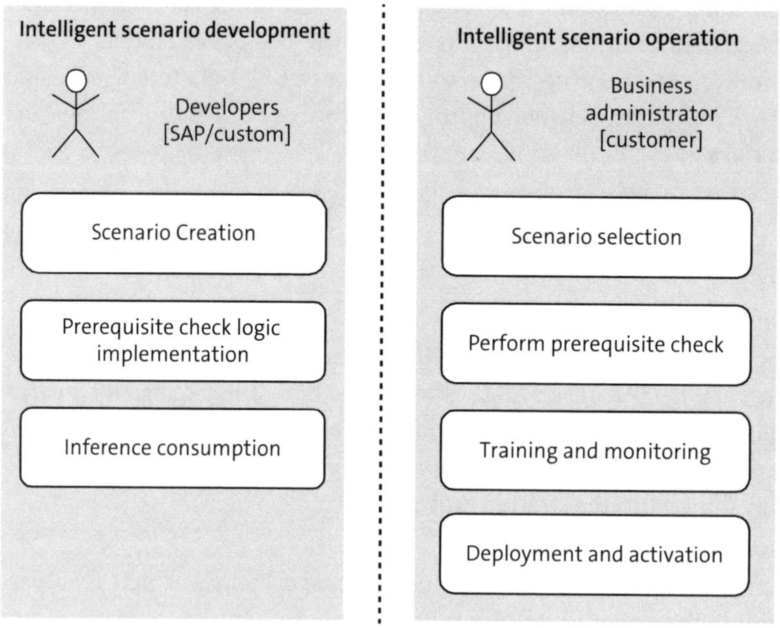

Figure 5.41 ISLM Development and Operations Roles

The IF_ISLM_INTELLIGENT_SCENARIO interface mandates the implementation of the GET_SCENARIO_GUID method. The method returns the scenario GUID to match the content present in the connected SAP Data Intelligence instance. The scenario GUID is a universal unique identifier generated by SAP Data Intelligence.

The ISLM framework does not perform the SAP Data Intelligence content provisioning; rather, it checks for the existence of a machine learning scenario based on a GUID in SAP Data Intelligence. Figure 5.42 illustrates the dependencies among the ISLM framework in Basis, an intelligent scenario from the application areas, and the SAP Data Intelligence content GUID. By default, the registered scenarios are in draft status and are published once the development of the scenario and the integration with the ISLM framework's operational aspects is completed. SAP-delivered intelligent scenarios registered in the ISLM framework are transported and made available for customers to consume in various releases.

The ISLM framework also allows customers to develop their intelligent scenarios and operate those themselves. For the embedded machine learning use case based on APL, the framework itself generates a class during registration. This class is used by the machine learning application for inference consumption.

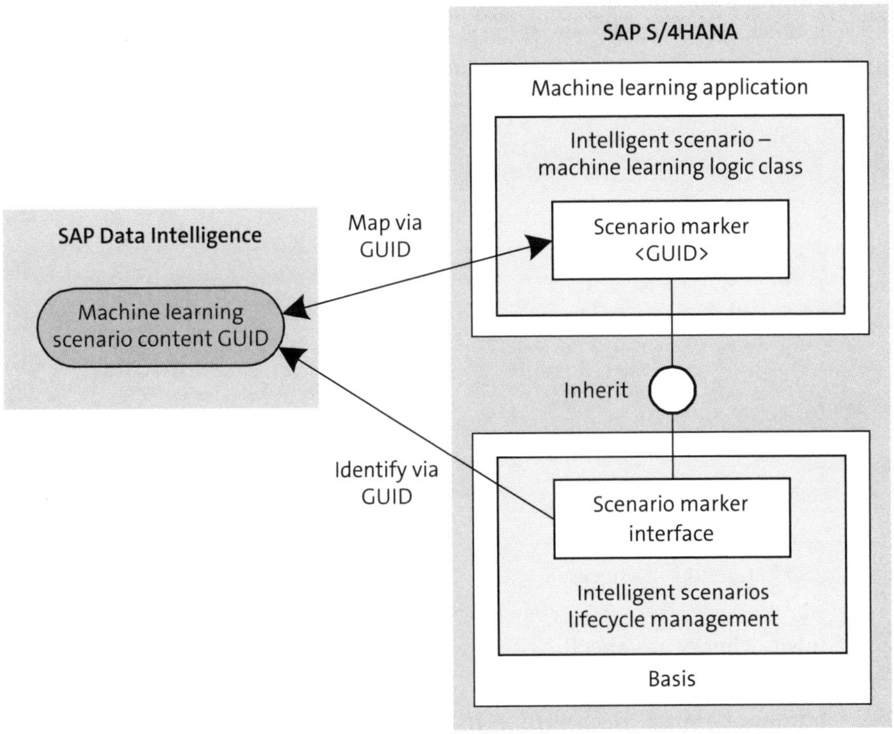

Figure 5.42 SAP Data Intelligence Content Mapping and Identification

Prerequisite Check

As described during the discussion of business requirements, the ISLM framework helps in establishing prerequisite checks for the intelligent scenarios that can be executed before setting up the intelligent scenario in the system, as shown in Figure 5.43. The intelligent scenarios require the training data to be of good quality and of a minimum volume for an optimized result.

Note

This functionality is not yet available as of the time of writing (summer 2020).

In some cases, the intelligent scenarios may mandate the configuration of a business process in the system up front, apart from the data prerequisites. Here the ISLM framework provides a unique way to execute the qualification checks of an intelligent scenario in advance and evaluate the results to decide on the implementation. The business administrator can select an intelligent scenario from the ISLM framework and trigger the prerequisite checks with the required parameters. The result provides decisive information about data quality, data volume, and configuration checks to the business administrators to use the intelligent scenario or perform any further changes required in the system. The ISLM framework also captures the historic evaluations of prerequisite conditional checks of an intelligent scenario to provide traceability and supportability. Usually, the business data present within the SAP S/4HANA application area is considered for checks.

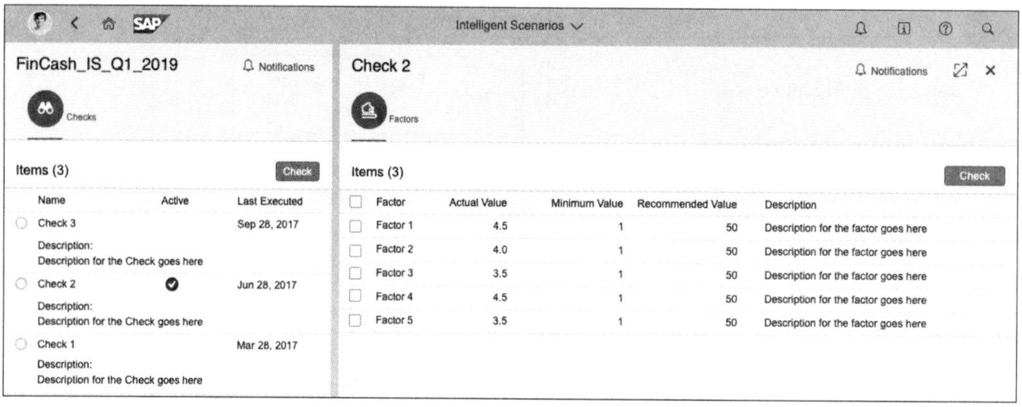

Figure 5.43 ISLM Prerequisite Checks

The prerequisite checks are specific to a business domain, so these checks must be implemented by the intelligent scenario developer and are bounded to the scenario in the ISLM framework with the machine learning logic ABAP class. For example, a finance-related table requires more than 10,000 records for an accurate result in the intelligent scenario inference calculation related to a finance use case. The developer of these checks should follow the design established by the ISLM framework. A scenario may check different conditions at the individual breakdown level in the system and produce a composite result to offer the business administrator about the feasibility of using that scenario.

The respective ABAP structures, data elements, table types, exceptions, and interfaces for the prerequisite checks are defined in the ISLM framework. These objects should be used by the developers for implementation of the prerequisite checks. The business administrator holds the permissions to execute these checks from the business data authorization profile and has access control. To offer prerequisite checks, the ABAP class must also implement the `IF_ISLM_PREREQUISITE_CHECK` interface with the following methods:

- **`GET_CHECKS_METADATA`**
  Returns the list of checks and metadata information like the check name, if it's mandatory or optional, the check classification, the description, and the check ID

- GET_PARAMETERS_METADATA
  Returns the list of parameters metadata required for the prerequisite check's execution
- PERFORM_CHECKS
  Implements the logic to be executed when a prerequisite check is triggered from the UI and returns the check results to be displayed

The prerequisite checks from intelligent scenarios can be improved/fixed over a period due to ongoing learnings and findings from the customer implementation experiences. In such cases, the enhancements to the check logic from the intelligent scenarios can be delivered in the regular fashion without much effort. This approach is similar to delivering SAP S/4HANA readiness checks with SAP Notes.

Training

For an effective inference calculation of a business problem, the intelligent scenario is trained with the respective historical dataset in a regular manner. The business administrator triggers the training of an intelligent scenario in the ISLM framework. The sequence in Figure 5.44 depicts the simplified approach of a training process for the side-by-side machine learning approach.

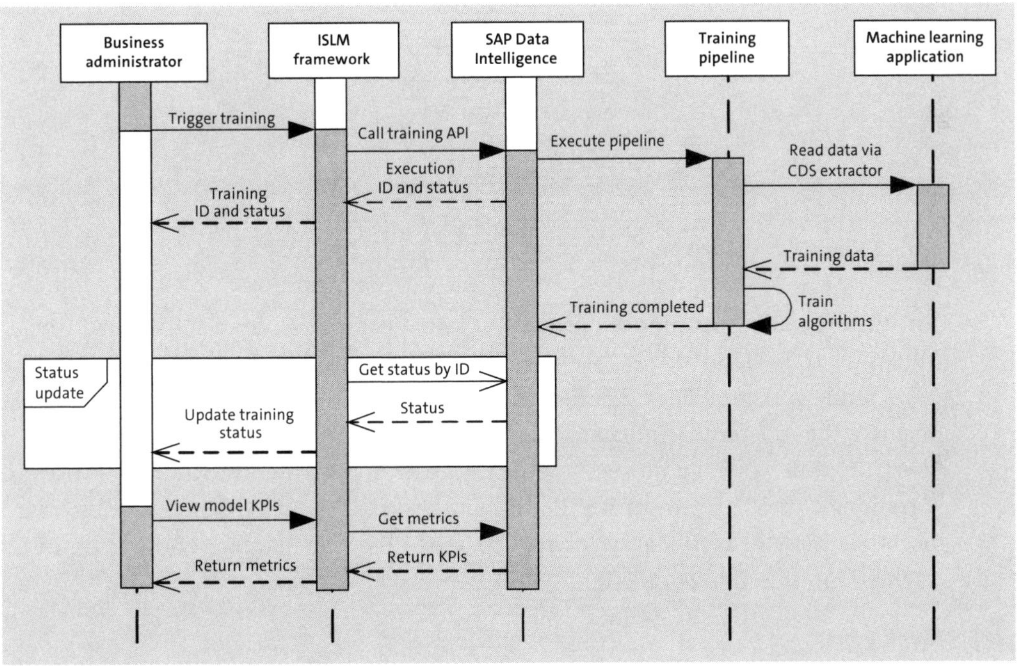

Figure 5.44 ISLM Training Process

After a training operation is completed in SAP Data Intelligence, the respective metrics, key influencers, and quality of the training process are provided to the business administrator for analysis and decision-making purposes, as shown in Figure 5.45.

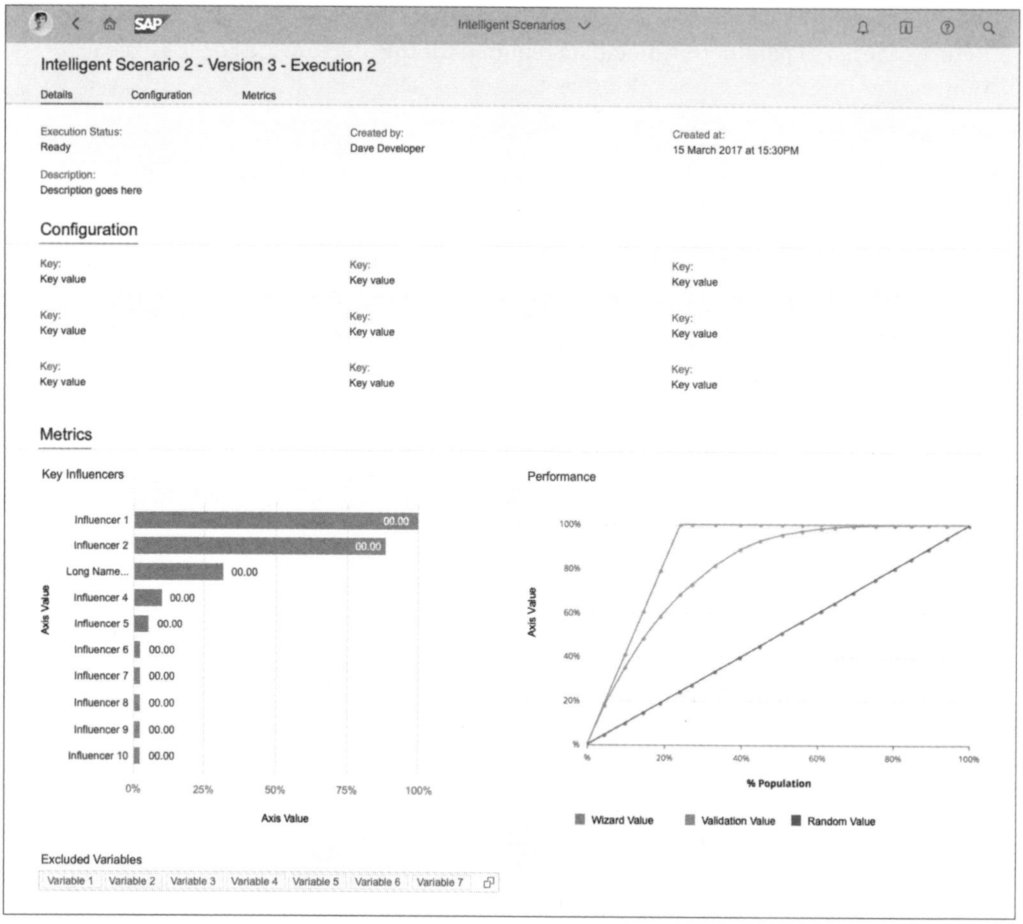

Figure 5.45 ISLM Model KPIs

The business administrator can provide different parameters for the training activity, as designed in the intelligent scenario, and compare the trained model results. The ISLM framework provides a comparative view for the business administrator in which to check the different training executions in order to decide on a trained model to be used for the deployment process. The ISLM framework captures the parameters used in a training activity to templatize the training process. The ISLM framework also supports additional functionality for regular scheduling of training, notifications, and asynchronous training executions.

Deployment

Business applications should be able to use the trained models for inference of a given dataset without disruptions. Business users should not be exposed to the complexity, but they need a way to control the consumption of trained models by business applications. Selecting a trained model is not the concern of the business user; rather, it's the focus of the business administrator.

The ISLM framework helps with a deployment and activation procedure from the business administrator's perspective, as shown in Figure 5.46. Deployment is the process of creating a running server instance out of the trained models from a training activity for the productive inference consumption. The business administrator can decide to create a deployment of a trained model at any point in time in the ISLM framework without any impact on the business applications. Here, the application gains an advantage by changing the deployments in runtime based on the business administrator's analysis.

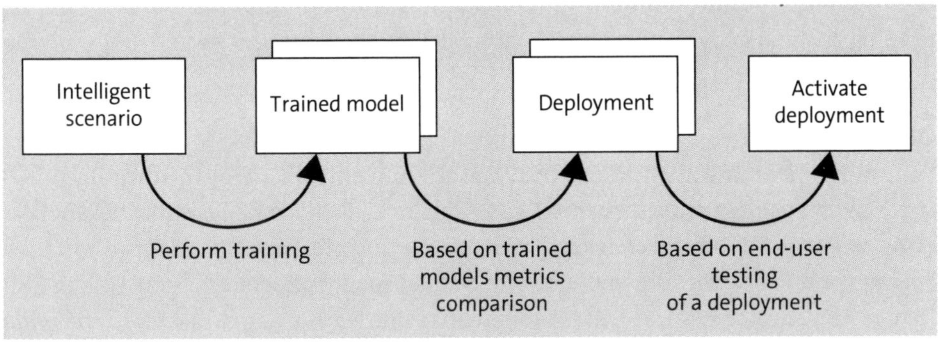

Figure 5.46 Machine Learning Model States

An intelligent scenario can have many trained models and many deployments, but only one active deployment per ABAP client is supported by default. Based on the metrics, the influencers, and the quality of the training models, the business administrator should finalize a trained model for a deployment activity.

The ISLM framework provides an additional capability to test a deployment from the business application perspective and decide to activate a deployment for productive consumption. Activation of a deployment is a feature available in the ISLM framework (see Figure 5.47), which allows the business administrators to perform successful testing of deployment and the respective inference consumption in the business context.

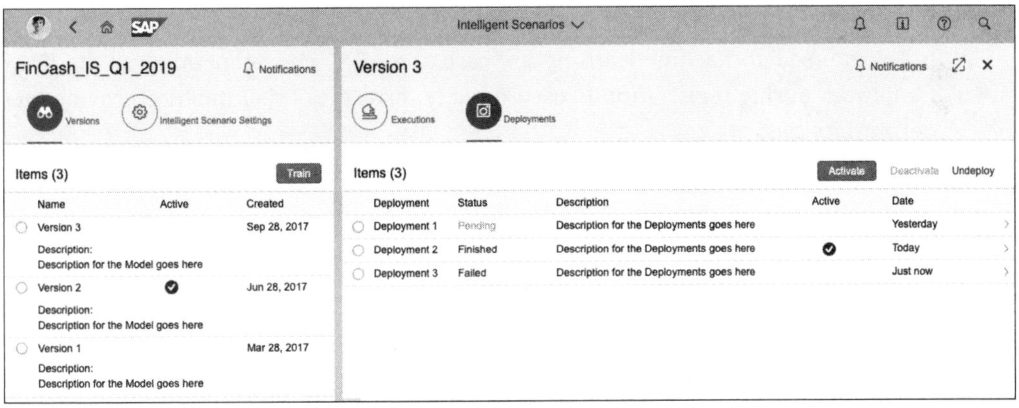

Figure 5.47 Machine Learning Model Activation

Based on the feasibility and accurate results, the business administrator can activate the deployment for all the business users' consumption. This functionality in the framework helps to include intelligence in the business application, with more control from the business administrator's perspective.

The active model of an intelligent scenario is used by the business applications to consume the inference as the ISLM framework hides the complexity of active deployment from the applications. Business administrators can also decide to roll back an active deployment in specific situations and conditions. The ISLM framework also suggests performing an "undeployment" of a trained model in case of nonactive and long-running instances of a trained model.

Inferencing

The machine learning application consumes the inference values by intrinsically utilizing them in business processes and UIs. For side-by-side machine learning, the applications consume the inference results by invoking the inference URL mapped to active deployment. Every new deployment of a trained model generates a new URL in SAP Data Intelligence, so it's problematic for the machine learning applications to consume a stable inference URL. Therefore, the ISLM framework offers a utility ABAP class to consume the inference URL stably. Still, the business applications need to invoke the inference URLs directly.

So, the ISLM framework provides a generic utility and toolset-based ABAP class, `CL_ISLM_INFERENCE_PROVIDER`, to fetch the dynamic inference URL of a specific active deployment with the `GET_INFERENCE_CLIENT` method (see Listing 5.8 and Figure 5.48). The framework returns the `CL_ISLM_INFERENCE HTTP_CLIENT` object, which encapsulates the dynamic URL and other inherited properties from the ISLM framework. Using this return object, applications can proceed to execute the inference and understand the response data. The scenario GUID should be passed to the ISLM framework utility class to identify the respective active deployment in SAP Data Intelligence. The ISLM framework may devise additional parameters like the ABAP client and username in runtime during invocation.

In the embedded machine learning use case, the ABAP class generated by the ISLM framework during registration is used directly instead of this inference provider generic utility class.

```
DATA:
lr_http_client TYPE REF TO if_islm_inference_http_client,
   lr_islm_inference_provider TYPE REF TO
cl_islm_inference_provider.
   lr_islm_inference_provider = NEW cl_islm_inference_provider(
                        iv_scenario_guid =  '<scenario_guid>').
   lr_http_client = lr_islm_inference_provider
->get_inference_client( ).
   IF lr_http_client IS NOT INITIAL.
```

```
lr_http_client->add_request_body( .. ).
lr_http_client->add_request_header( .. ).
lr_http_client->send_and_receive( ).
  ENDIF.
```

Listing 5.8 Coding for HTTP Client Handling of Inference Calls

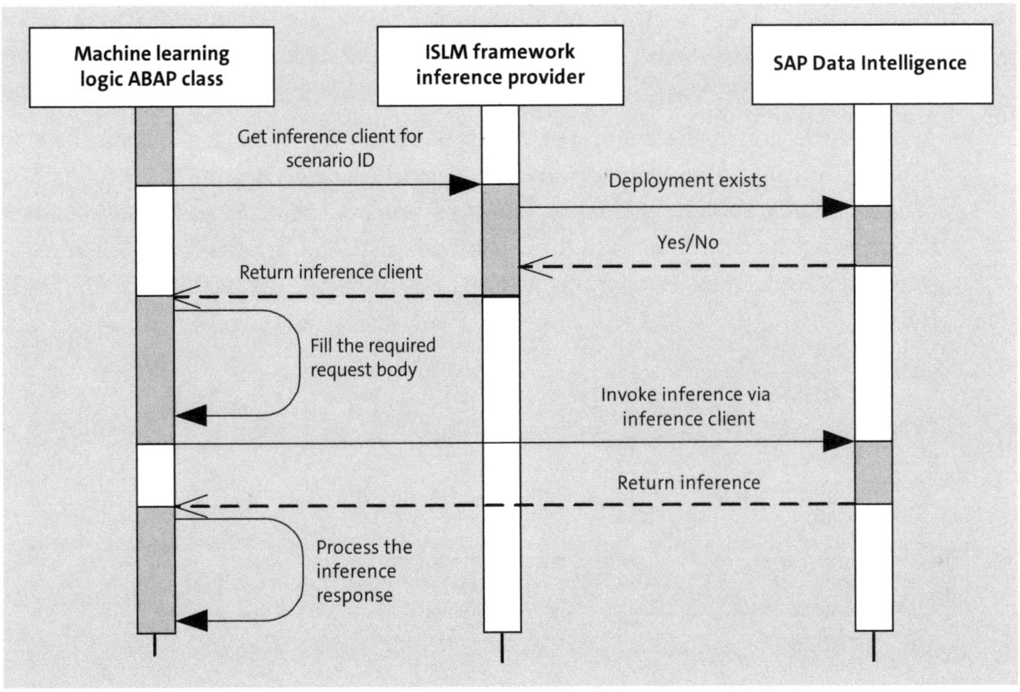

Figure 5.48 ISLM Inference Process

Monitoring

The ISLM framework enables the business administrator to continuously monitor the deployed trained models and the status of asynchronous training executions. The framework has a feature to check the existence of a deployed trained model continuously and notify the business administrator in case of failures. Data ingestion and operating a machine learning algorithm can consume more time and can be executed asynchronously. Hence the ISLM framework continuously monitors the training execution and updates the business administrator at regular intervals. The ISLM framework will also provide advanced monitoring of the degradation of deployed trained models and inference consumption.

5.5.2 Data Integration

For side-by-side machine learning, data integration is required because application data must be extracted from SAP S/4HANA into SAP Data Intelligence for model training. In this section, the business requirements and the necessary technical implementation are explained.

Business Requirements

Data integration involves combining data from several disparate sources, stored using various technologies, to provide a unified view of the data. In the context of machine learning, data integration is required for the training phase. For inference in terms of consuming the machine learning models via APIs, the topic of data integration is not relevant.

As already described, the training process is necessary for embedded and side-by-side machine learning. However, for embedded machine learning, the training data is accessed locally, so data integration is not required. As depicted in Figure 5.49, data integration is usually demanded for side-by-side machine learning, in which application data must be transferred from SAP S/4HANA into SAP Data Intelligence for model training.

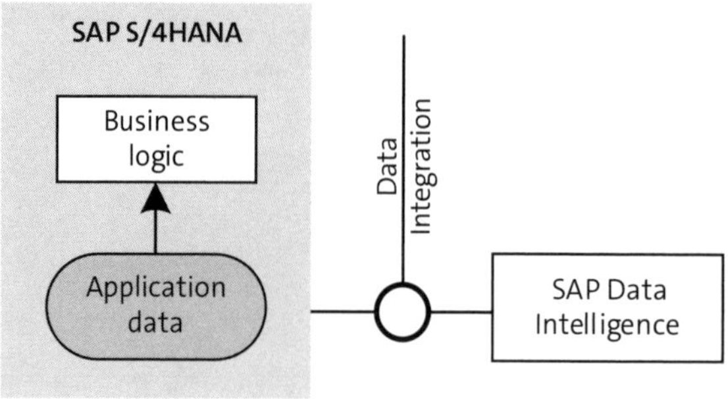

Figure 5.49 Data Integration Requirement

Application data can be structured, in terms of having a well-defined relational schema, or can be unstructured, such as files or emails. Both types will be considered for data integration, although the primary focus is on extracting structured data.

The data integration solution will support an initial load. This is the preliminary process for extracting all the records from SAP S/4HANA into to the target system. Delta loads also will be supported. If the overall size of a data source (the number of rows times the width of a row) is small, then the data consumer can track changes to the data source via a full reload of the source. However, this is only indicated for Customizing-like data sources (e.g., code lists and the like) that are rarely changed. For other sources that change regularly (e.g., tables for transactional data), this approach is not feasible in general. To replicate such data sources, delta handling is required, which accomplishes the initial load by replicating changes that occurred since the last data replication. Long-running data extraction tends to run out of sync. This means that the data source contains different data than the replicated version, but the synchronization process does not show an error. Reasons for this include the following:

- Lost updates in the synchronization process
- Deletion at the receiver side
- Source objects falling out of scope if source object filters are used

Therefore, a resynchronization of the replicated data must be supported, either via complete reload of the data or via comparing and resolving the differences between source and receiver. The second approach requires more complexity, but for large data volumes a complete reload might not be feasible. Usually data replication increases TCO because additional disk space is required, among other factors. For machine learning scenarios in which data replication can be avoided, a live connectivity/data streaming capability will be provided.

SAP S/4HANA provides data for extraction in a provider-driven way; that is, the data sources are modeled to ensure complete coverage of SAP S/4HANA data in a mostly redundancy-free manner. Specific consumer scenarios might require data in a specific structure or with transformations (e.g., aggregations) already performed. Some of these transformations could be achieved via the creation of customer- or consumer-specific data sources. This will be covered by tools for key users to create their own data sources or enhance existing data sources, including the capability to add simple transformations. Some of these transformations might also need access to further data in the source system that are not replicated. In this case, the only option is to use custom data sources.

Other transformations might need access to external data or be so consumer-specific that they won't be implemented in the source system. For these cases, the data integration technology must include the option to add transformations before providing the data to the consumer. SAP S/4HANA stores data in an internal format (e.g., internal code lists or currency-related amounts that always have two decimals). ABAP-based consumers (like SAP Business Warehouse [SAP BW]) might be able to consume the data in the internal format (with the precondition that all relevant Customizing settings are synchronized between the provider and consumer systems). In contrast, most consumers are not aware of the internal format or are unable to perform the needed conversions because they don't support the same conversion routines. Therefore, a consumer must be able to receive the data in an externally understandable format. Furthermore, it must be possible to map the key of the object instance during the extraction process so that sender and receiver instances with different key values/structures can be mapped to each other.

SAP S/4HANA provides one model for data integration scenarios. As CDS views are the internal layer for reading data, the data integration model should be CDS-based as well. Different tools can use this model and provide different integration capabilities and qualities. For machine learning use cases, the following characteristics/capabilities are required in a data integration solution:

- **Consistency**
  Consistency for data integration is required on the CDS level. In general, a CDS view is constructed from several database tables. CDS view consistency requires that the data extraction is aware of the associations of the CDS view to the underlaying database tables and can reliably replicate all changes from one database transaction touching these tables (delta load). In addition, a consistent method for initial load of CDS views will be provided.
- **Read access**
  In specific contexts and for specific data sources, read-access logging might be required to ensure the auditing of each data access in the provider system. Consequently, the data extraction solution needs to enable read-access logging for specific data sources.
- **Data protection and privacy**
  Data in the source SAP S/4HANA system must fulfill the requirements imposed by the legislation and product standards for data protection and privacy (blocking of data, end-of-purpose deletions, retention period handling). Even if the data is extracted to another storage source, it must fulfill these requirements. So the provider system needs to enable the consumer to be informed about any events in the data lifecycle via SAP Information Lifecycle Management (SAP ILM), and the consumer must take these events into account and handle the data accordingly. SAP S/4HANA provides interfaces to get information about SAP ILM events. The data extraction technology must provide this information to the relevant consumers. If a consumer is SAP-delivered, it must ensure the data are handled accordingly, or the customer can add the handling as desired.
- **Implicit selection**
  Because the smallest selectable entity that can be selected for extraction by a consumer is a CDS view, this item requests an option to define a set of CDS views (*1..N*) for extraction. To simplify this selection, an option for an implicit selection of all CDS views belonging to one business object will be provided.
- **Consumer-defined CDS views**
  SAP S/4HANA's data sources (CDS views) are in general not defined for a certain consumer application; they are defined as provider-driven CDS views and might not be specific enough for a concrete customer use case. Therefore, customers must be able to define their own data sources for replication (consumer-defined CDS views). These data sources will bear the same features and functions as defined SAP data sources. A customer might use all available whitelisted artifacts to build his or her own data source. The preferred basis data sources will be stable CDS views. Customer- or consumer-specific data sources might be needed to replicate customer-specific data (custom business objects or fields) or to transform data to fulfill the needs of a specific consumer.

- **Monitoring and analysis tools**
  The data integration solution must provide local tools for monitoring and error analysis to detect and resolve data integration issues. Furthermore, it must support integration into SAP's central tools for monitoring and error analysis, like SAP Solution Manager for cross-system monitoring and error correction.
- **Stability contracts and independent upgrades**
  The communication partners in a data integration scenario must be able to be upgraded independently from each other and without the necessity of downtime on one side during the upgrade of the other one. Therefore, the interfaces for data integration must fulfill specific stability contracts. In addition, the integration technology must ensure the partners can be upgraded independently as well. This includes irregular field length extensions on SAP S/4HANA. It must be ensured that occasional disruptions of the data integration (e.g., network failure, consumer upgrades) don't affect the overall data integration qualities. The technology must provide a means to continue with the data integration at the point when the disruption occurred without the need to perform a complete reload of the data.
- **High data volume/data change support**
  SAP S/4HANA will not be affected by the data integration technology in a way that prevents the normal, operative tasks performed in this system. Particularly, the data integration will not require planned downtime for the initial load, for example, to get consistency. The technology and protocols used must be able to support a high data volume and/or highly frequent data changes. It will be possible for the consumer to get the data in packages of a defined size to prevent overload. The technology and protocols need to support scenarios with a low bandwidth between the communication partners. Depending on the needs of the concrete scenario, it will be possible to configure whether the data changes from the data source are available in the target system in near real time or within a specific time interval.
- **On-premise and cloud support**
  The data integration solution must support data integration between SAP S/4HANA on-premise and cloud versions, including all combinations and integrations with external systems.

Technical Implementation

The overall solution architecture for data integration in the context of side-by-side machine learning is depicted in Figure 5.50. On top of the application tables, CDS views are defined that represent the semantic data model of SAP S/4HANA. The CDS views hide the cryptic database models and wrap them to human-understandable entities. Besides the SQL view definition, the CDS views contain domain-specific metadata called annotations, used, for example, for analytics, data extraction, or search. The CDS views are consumed by the business logic in order to access structured data. To extract structured data from SAP S/4HANA for machine learning training, the operational data provider (ODP) framework is used. Output management and attachment services are used to extract unstructured data from SAP S/4HANA. Key user tools will be provided for customers to define structured and unstructured sources for data extraction, like customer CDS views or the Form Builder.

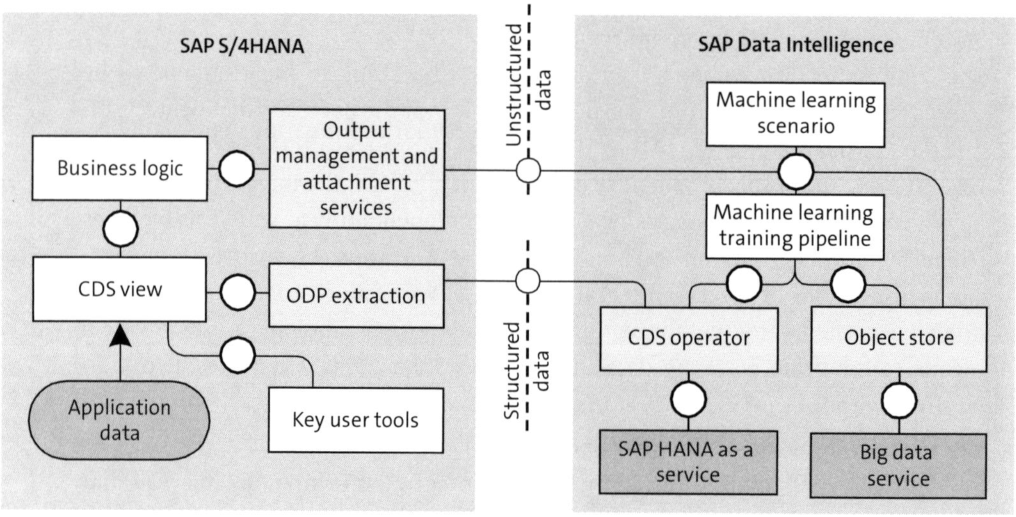

Figure 5.50 Data Integration Architecture

Let's walk through the technical processes involved in data integration with machine learning.

Data Extraction with CDS Views

In SAP Data Intelligence, each machine learning use case is organized by the machine learning scenario artifact (refer back to Section 5.1). This artifact contains all development entities that are required for the implementation of a specific machine learning use case. Inference and training processes are developed as pipelines comprising sequential and parallel tasks. In particular, a training pipeline is provided that receives the training data from SAP S/4HANA and processes the data to train the algorithms for the specific scenario. Structured data is handled by a CDS operator and stored in SAP HANA, while unstructured data is managed by the object store and big data storage. Often, the saved application data is deleted after the training run, although there are use cases with high retraining frequency in which, after an initial upload, the deltas must be received and stored periodically for the next training run. Persisting the data is one option; streaming the data and training the algorithm with live connectivity is also an important capability for various machine learning use cases.

For data extraction, CDS views are used in SAP S/4HANA. The advantage of this approach is that the extracted data is based on the same semantic layer as analytical and transactional applications and provides high quality and consistency. Furthermore, total cost of deployment (TCD) is reduced as existing CDS views can be reused. Let's walk through the initial data extraction steps with CDS views:

1. At design time, CDS views are enabled as outbound data interfaces via the extraction annotation for full delta extraction, as shown in Listing 5.9 ahead.
2. The metadata of these CDS views are exposed and used as structures for data persistency in the target system.
3. As depicted in Figure 5.51, from a technical point of view, an ODP is enabled for CDS views. ODPs support extraction and replication scenarios for various target applications and supports delta mechanisms in these scenarios.
4. In case of a delta procedure, the data from a source, called an ODP provider, is automatically written to a delta queue using an update process or passed to the delta queue using an extractor interface.
5. The target applications, called operational data queue (ODQ) consumers, retrieve the data from the delta queue and continue processing the data.
6. An ODP is used to connect via a remote function call (RFC) or HTTP protocol to the SAP S/4HANA source system. The ODP context allows access to extract an annotated CDS view of SAP S/4HANA. If the CDS view contains the required annotations, it can be used for full extraction or delta extraction.

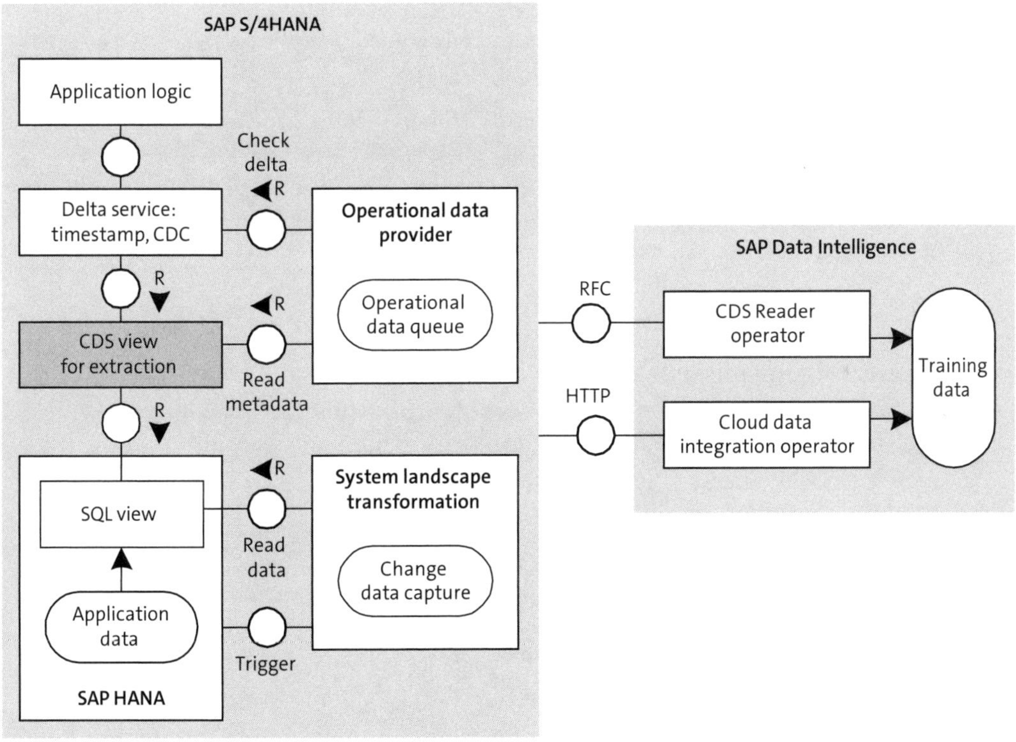

Figure 5.51 Integration of Structured Data for Model Training

Two approaches for delta handling are available: timestamp based and change data capture (CDC) based.

For CDS views containing time-related attributes, the timestamp approach can be applied; based on that information, the engine can calculate deltas. However, there are CDS views without timestamp attributes for which CDC is the appropriate solution.

The CDC approach is based on triggers for the database tables used in a CDS view. From the recorded changes in the table, the resulting changes in the affected data extraction views are calculated and stored in the ODQ for any extraction consumer. This approach doesn't need any field to calculate the delta like a timestamp, but several limitations apply in regard to the view structure. The view must contain the key fields of the underlying tables in order to calculate the changed view rows from the changed table entries. The CDC can be consumed by any component that requires CDC functionality based on database triggers.

The database tables to register the CDS views for delta extraction and to handle the status of the individual recording objects are filled in the source system. All database tables are defined as client-dependent tables with delivery class A (tables containing application data). Using delivery class A has the following implications:

- The data is owned by the customer.
- SAP may not use this table to deliver content.
- During client copy, the data records are only copied if explicitly requested by setting the corresponding parameter option.
- When data is transported between customer systems, the client-specific data is imported into the specified target client.

The annotation schema for enabling CDS views for data extraction is specified as shown in Listing 5.9.

```
annotation Analytics {
dataExtraction : {
enabled : boolean default true;
delta : {
byElement : {
name : RefToElement;
@MetadataExtension.usageAllowed : true
maxDelayInSeconds : Integer default 1800;
detectDeletedRecords: boolean default true;
MetadataExtension.usageAllowed : true
 ignoreDeletionAfterDays : Integer;
};
changeDataCapture : {
automatic : boolean default false;
```

```
nodes : array of {
role : String(20) enum {MAIN; COMPOSITION;
 LEFT_OUTER_JOIN; INNER_JOIN;};
// only one of the next 3 annotation may
// be used
association : associationRef;
entity : entityRef;
table : String(30);
// only used if association is not specified
viewElement : array of elementRef;
// only used if association is not specified
logElement : array of elementRef;
filter : array of {
element : elementRef;
type : String(10) enum { EQ; NOT_EQ;
GT; GE; LT; LE; BETWEEN;} default EQ;
value : String(45);
highValue : String(45);
};
};
};
};
```

Listing 5.9 Extraction Annotations for CDS Views

Let's break down the key elements of this annotation schema:

- **`dataExtraction.enabled`**
  Via this indicator, the application developer can mark those entities that are suitable for data replication (e.g., for mass data, you must provide delta capabilities). This ability is typically lost on the level of consumption views, in which joins, procedures, and aggregations blur the relation to the primary data (in particular, data replication that includes calculated elements leads to the wrong query results in the target system). It's expected that an application will deliberately design a set of data provisioning views that are suitable for consistent and redundant-free replication.
- **`dataExtraction.delta.byElement`**
  The application developer can enable the generic delta extraction by this annotation.
- **`dataExtraction.delta.byElement.name`**
  This is the element that should be used for filtering during generic delta extraction. This element can either be a date (ABAP type `DATS`) or a Coordinated Universal Time (UTC) time stamp.

- **`dataExtraction.delta.byElement.maxDelayInSeconds`**
  There is always a time delay between taking a UTC time stamp and the database commit. This annotation specifies the maximum possible delay in seconds. The default is 1,800 seconds.
- **`dataExtraction.delta.byElement.detectDeletedRecords`**
  By using this annotation, the system will remember all key combinations of the view that were extracted in delta mode. If a key combination does not occur in the view anymore, this annotation will automatically generate a delete image in the extracted data.
- **`dataExtraction.delta.byElement.ignoreDeletionAfterDays`**
  This annotation only makes sense together with `dataExtraction.delta.byElement.detectDeletedRecords`. The extraction will ignore deleted records if they are older than the specified number of days. The main purpose is archiving.
  For example, if records are archived after two years, then this value should be less than 700. In this case, the deletion in the database tables will be ignored if the record is less than 700 days old.
- **`dataExtraction.delta.changeDataCapture.automatic`**
  In the case of simple CDS views like projections, no additional information needs to be provided, but the relevant information about key fields is inferred by the system automatically.
- **`dataExtraction.delta.changeDataCapture.nodes`**
  Defines the list of tables that should be logged.
- **`dataExtraction.delta.changeDataCapture.role`**
  Defines the role of the involved tables. Possible values and semantics include the following:
  - `#MAIN`: The key of the extraction view corresponds exactly to the key of the logging table.
  - `#COMPOSITION`: There is a node with the `#MAIN` role, and the key of the logging table is part of the key of the extraction view.
  - `#LEFT_OUTER_JOIN`: There is a left outer join, but not all requirements of the `#COMPOSITION` role are met.
  - `#INNER_JOIN`: There is an inner join to another table. Entries might disappear in the extraction view due to updates on the fields of the join condition.
- **`dataExtraction.delta.changeDataCapture.association`**
  The name of the association to the CDS projection on the table to be logged.
- **`dataExtraction.delta.changeDataCapture.entity`**
  The name of the CDS projection view on the table to be logged.
- **`dataExtraction.delta.changeDataCapture.table`**
  The name of the table to be logged.

- dataExtraction.delta.changeDataCapture.viewElement
  View elements of the equal join condition between the actual CDS view and the CDS projection view or the logging table.
- dataExtraction.delta.changeDataCapture.logElement
  Elements or fields in the logging table of CDS projection view of the equal join condition. Both arrays, viewElement and logElement, must contain the same number of elements. In addition, the logElement array must contain exactly all key elements of the logging table without the client element.
- dataExtraction.delta.changeDataCapture.filter
  A filter on the logging table is specified with these annotations:
  - .element: Name of the element in the logging table or view
  - .type: Possible values are #EQ, #NOT_EQ, #GT, #GE, #LT, #LE, and #BETWEEN
  - .value: The filter value
  - .highValue: Only used in combination with the #BETWEEN type

Figure 5.52 shows the application of the annotation for CDS views extracting product data. The first section of the coding contains the annotation usage, whereas the second part defines the view structure.

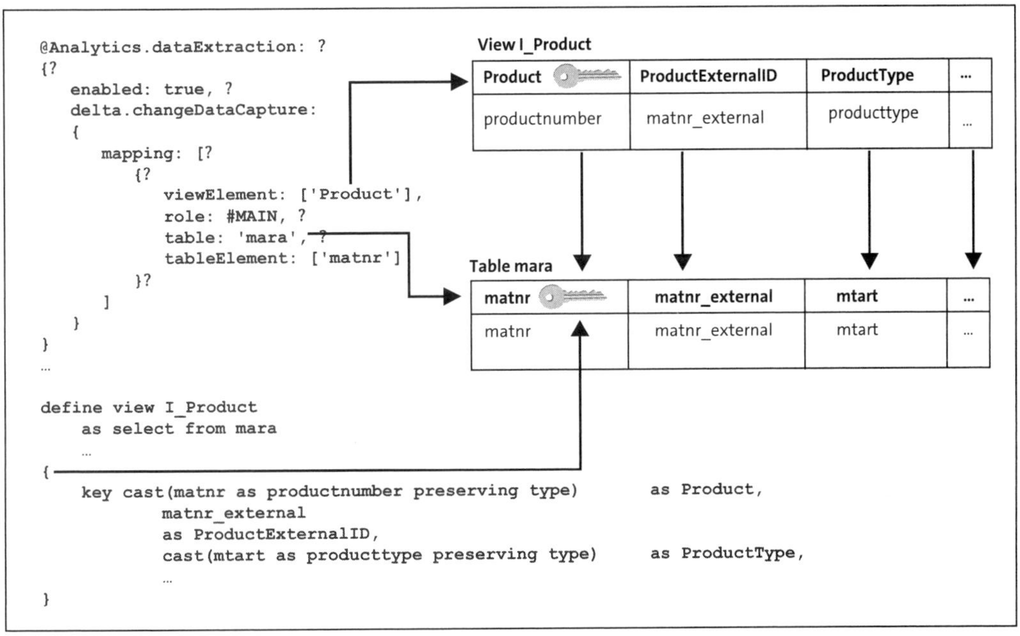

Figure 5.52 Example for CDS Extraction View

Pipelines and Operators

SAP Data Intelligence uses pipelines to process the training of algorithms (refer back to Section 5.3.1). Pipelines represent process logic based on a graphical programming paradigm. Data flows with transformations and persistency are implemented with pipelines. There is an according runtime component to execute pipelines in a containerized environment that runs on Kubernetes.

Pipelines consist of process steps called *operators*. An operator represents a vertex of a graph. An operator is a reactive component and reacts only to events from the environment, which are messages delivered to the operator through its input ports. The operator can interact with the environment through its output ports. The operators are unaware of the graph in which they are defined and the source and target of their incoming and outgoing connections. Operators require a certain runtime environment for their execution. For example, if an operator executes some JavaScript code, then the operator requires an environment with a JavaScript engine. The SAP Data Intelligence modeler provides certain predefined environments for operators, and these environments are made available to users as a library of Dockerfiles. When a graph is executed, the tool translates each operator in the graph into processes. It then searches the Dockerfiles for an environment suitable for the operator execution and instantiates a docker image.

There are predefined operators available to connect SAP S/4HANA and transform structured or unstructured application data into SAP Data Intelligence. For CDS extraction, the ABAP CDS Reader and Cloud Data Integration operators are provided, as illustrated in Figure 5.53. The main difference between both operators is the underlying network protocol. While ABAP CDS Reader is based on RFCs, Cloud Data Integration uses HTTP. As the underlying concepts are identical, we focus here on Cloud Data Integration only. Figure 5.54 shows the detailed configuration properties of the Cloud Data Integration operator, as follows:

❶ **Connection ID**
ID to connect to SAP S/4HANA system.

❷ **Source**
CDS view to based data extraction on.

❸ **Extraction Mode**
Extraction mode to transmit all data (**Full**) or only changes (**Delta**).

❹ **Columns**
Columns of the CDS view extractor to be considered (filtering of columns).

❺ **Filters**
Rows of the CDS view extractor to be considered (filtering of rows).

❻ **Data Preview**
Button to access a review of the extracted data, as shown in Figure 5.55.

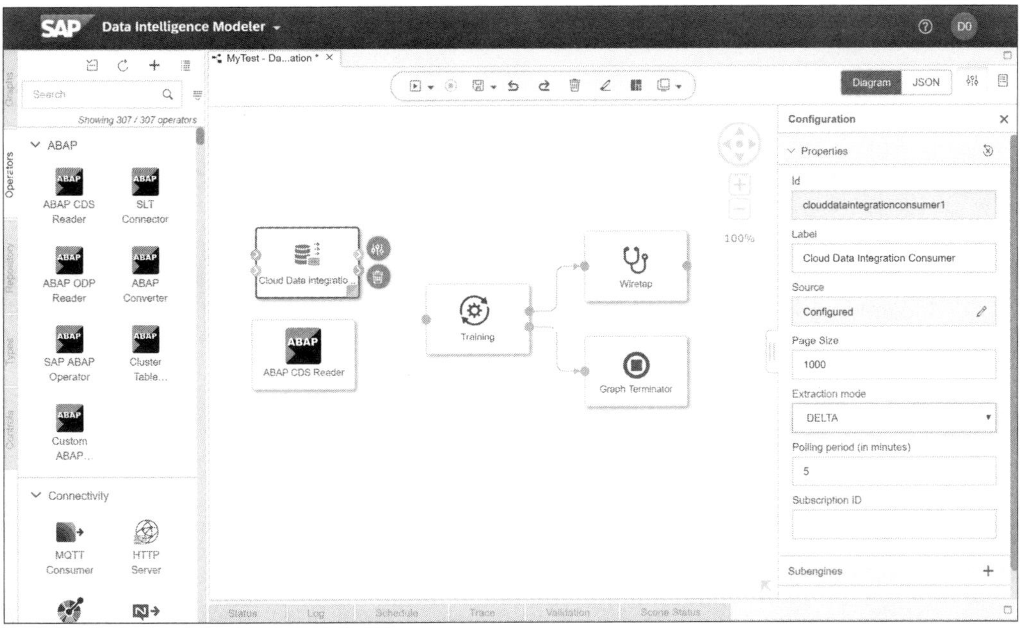

Figure 5.53 CDS Extraction Operators

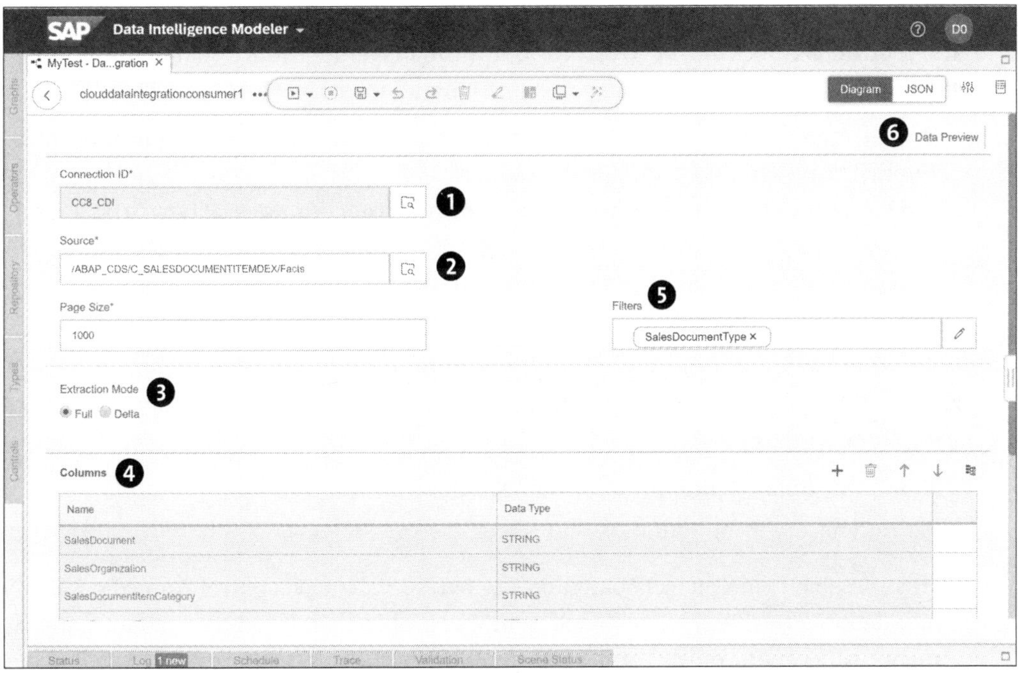

Figure 5.54 Cloud Data Integration Operator: Configuration

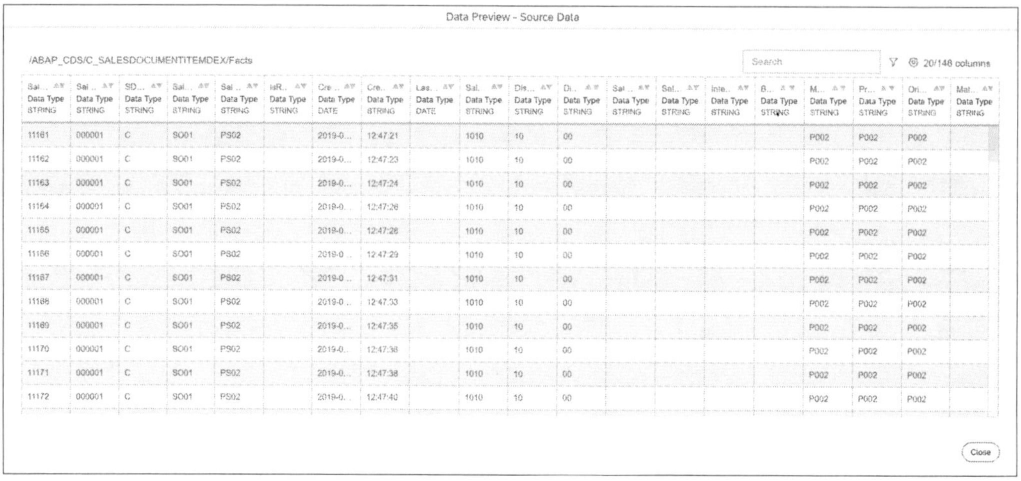

Sal... Data Type STRING	Sai .. Data Type STRING	SD... Data Type STRING	Sal... Data Type STRING	Sai .. Data Type STRING	isR.. Data Type STRING	Cre .. Data Type DATE	Cre.. Data Type STRING	Las.. Data Type DATE	Sal. Data Type STRING	Dis... Data Type STRING	Di.. Data Type STRING	Sal .. Data Type STRING	Sal.. Data Type STRING	Inte.. Data Type STRING	B... Data Type STRING	M... Data Type STRING	Pr... Data Type STRING	Ori... Data Type STRING	Mat... Data Type STRING
11161	000001	C	SO01	PS02		2019-0...	12:47:21		1010	10	00					P002	P002	P002	
11162	000001	C	SO01	PS02		2019-0...	12:47:23		1010	10	00					P002	P002	P002	
11163	000001	C	SO01	PS02		2019-0...	12:47:24		1010	10	00					P002	P002	P002	
11164	000001	C	SO01	PS02		2019-0...	12:47:26		1010	10	00					P002	P002	P002	
11165	000001	C	SO01	PS02		2019-0...	12:47:28		1010	10	00					P002	P002	P002	
11166	000001	C	SO01	PS02		2019-0...	12:47:29		1010	10	00					P002	P002	P002	
11167	000001	C	SO01	PS02		2019-0...	12:47:31		1010	10	00					P002	P002	P002	
11168	000001	C	SO01	PS02		2019-0...	12:47:33		1010	10	00					P002	P002	P002	
11169	000001	C	SO01	PS02		2019-0...	12:47:35		1010	10	00					P002	P002	P002	
11170	000001	C	SO01	PS02		2019-0...	12:47:36		1010	10	00					P002	P002	P002	
11171	000001	C	SO01	PS02		2019-0...	12:47:38		1010	10	00					P002	P002	P002	
11172	000001	C	SO01	PS02		2019-0...	12:47:40		1010	10	00					P002	P002	P002	

Figure 5.55 Cloud Data Integration Operator: Data Preview

Output Management

SAP S/4HANA output control is the central entry point in SAP S/4HANA for functions such as printing, sending emails, or maintaining form templates for the output (see Figure 5.56). It is the key element of SAP S/4HANA output management, which is the framework for all functions and processes related to the output of documents in SAP S/4HANA. Output control structures output management functionality and make it reusable across all business functions that have adopted them.

Output management creates documents based on templates and application data. Templates are kept in a template store and can be authored both by SAP development and key users. Print forms for pixel-perfect print output are a special type of document and are created with the help of Adobe technology. Finally, the document output component handles document output via the UI (frontend output), via backend printing, and via backend email.

The attachment service is a reusable UI component that can be consumed by SAP S/4HANA applications to attach documents. The attachments are stored in the following storage repositories:

- SAP Document Center (in SAP S/4HANA Cloud)
- Content server configuration (in on-premise SAP S/4HANA)

This reusable component is based on the `UploadCollection` control from the SAPUI5 foundation. Attachments are linked to business objects based on document management or via generic object services (GOS). Some of the basic features supported by the attachment service are uploading, downloading, renaming, and deleting an attachment in the parent application. The attachment service supports the concept of drafts for adding a new attachment, changing the attachment name, and deleting an attachment.

The changes are persisted only when an explicit save operation is performed by the user. Attachments can't be created without a consuming application object.

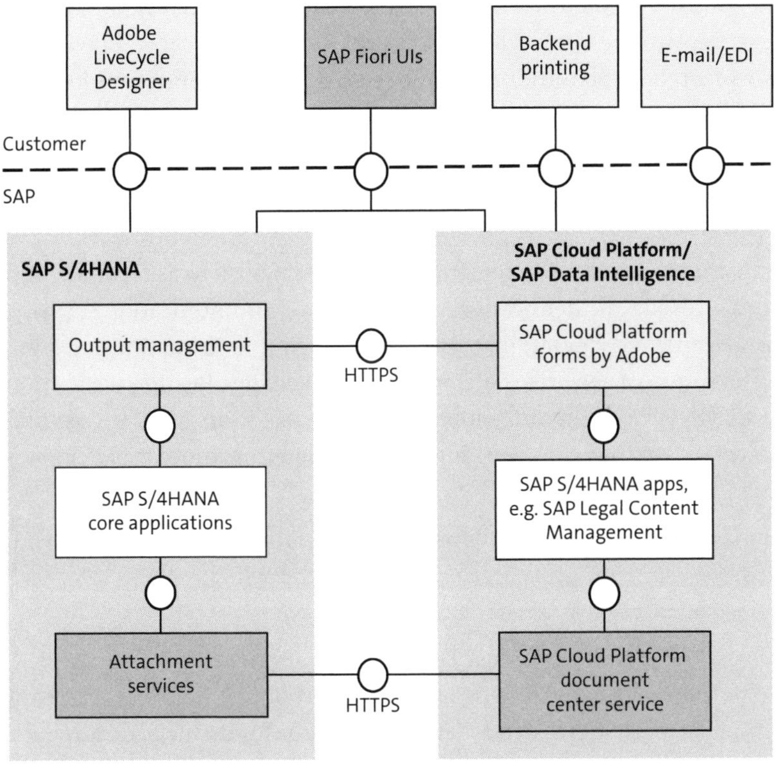

Figure 5.56 Integration of Unstructured Data for Model Training

In context of machine learning, the `ObjectStore` component must enable the training pipeline to access the unstructured data resulting from output management and attachment services. For this, an integration of `ObjectStore` with SAP Cloud Platform Forms by Adobe and SAP Document Center is required.

5.5.3 Data Protection and Privacy

Processing of machine learning must be compliant with data protection and privacy legislations. For example, the training job should only consider personal data for which consent is given. This section explains the business requirements and the necessary technical implementation for General Data Protection Regulation (GDPR) in the context of machine learning.

Business Requirements

GDPR is associated with numerous legal requirements and privacy concerns. In addition to compliance with general data protection acts, it's necessary to consider compliance with industry-specific legislation in different countries. Machine learning must be compliant with those legal requirements and data protection rules.

The term *personal data* is defined within various regulations, standards, and guidelines. Most of them explain the term as broadly as possible to best protect individuals. The common core of all these definitions is that personal data is any information that relates to an identified or identifiable natural person. One of the main GDPR requirements is that processing of personal data is forbidden if no justifying reason is given, such as a contract, other legal reasons allowing for enforcing the processing, or unambiguous consent from the data subject.

Basic requirements that support data protection are often referred to as technical and organizational measures like authentication, authorization, communication security, secure system landscape and operation, read access logging, information retrieval, encryption and decryption of sensitive data, input control change logging, separation by purpose, and masking or anonymization. The implementation of those requirements is well known and not machine learning specific. Therefore, those topics are not considered further in this section.

In the context of machine learning, the following GDPR requirements must be considered:

- **Deletion**
  In certain countries, personal data must be destroyed after the specified, explicit, and legitimate purpose of the processing has ended, but only if no other retention periods are defined in legislation, such as retention periods for financial documents. In some scenarios, personal data also includes referenced data. Therefore, the challenge for destruction is to handle first referenced data and then data such as business partner data. Machine learning training and inference must be able to handle deletion of personal data.
- **Blocking**
  Legal requirements in certain scenarios or countries also often require blocking data, restricting their further processing or use in cases in which the specified, explicit, and legitimate purposes of the processing have ended but the data is retained in the database due to other legally defined retention periods. Machine learning training and inference must not process blocked data.
- **Consent**
  In order to process personal data, a legal basis is required. GDPR identifies six legal options for the processing of personal data, one of which is consent. Other legal grounds are contracts, legal obligation, protection of vital interest, public interest, and legitimate interest. If none of the other legal grounds applies, the consent of the individual for the intended processing of personal data is required. Machine learning training must not process data without consent given. This is relevant only for machine learning scenarios with a consent requirement.

- **Automated decision-making**
  The controller will provide knowledge of the existence of automated decision-making and meaningful information about the logic involved, as well as the significance and the envisaged consequences of such processing for the data subject. A machine learning explanation for automated decision-making and its consequences is required.

Technical Implementation

While the implementation of the GDPR requirements for deletion, blocking, and consent depends on embedded and side-by-side machine learning variants, the automated decision-making can be considered for both variants together and is described in Section 5.5.7 in combination with additional aspects.

We'll discuss the various solutions involved in supporting GDPR requirements in this section, as well as how these requirements are implemented in embedded and machine learning.

Blocking, Deleting, and Consent

Simplified blocking and deletion describe the technical processes for blocking and deleting of personal data in a structured and integrated manner. The functions for deleting data across specific archiving processes and blocking data using authorization-based grouping are enhanced by functions for modeling and applying blocking and deletion rules in SAP S/4HANA system. Simplified blocking and deletion are based on SAP ILM.

The technical components of SAP ILM are illustrated in Figure 5.57 and will be discussed throughout the following sections. To be able to use simplified blocking, residence and retention periods for specific purposes are defined in SAP ILM. Different purposes require different residence and retention periods. Trading peppermint oil, for example, has different purposes than trading cough syrup. The system determines the purposes of the data processing by checking the following technical attributes:

- Line organizational attributes represent the data controller—for example, the company code.
- Process organizational attributes help to differentiate between business processes. For example, by assigning specific order types, processes like the peppermint oil business and the cough syrup business can be differentiated.

Once a document has reached the end of its purpose, you can use the SAP ILM object for archiving and thus blocking the document in the system.

For blocking central master data (business partner, customer, vendor), the end-of-purpose check first determines whether any applications (e.g., sales and distribution, financials) are still using the relevant master data as defined in the original purpose. If this is not the case, the system marks the data as blocked. This prevents the data from any further processing in a legally compliant way. Thus, transactional data is blocked using SAP ILM and master data is blocked using a central blocking indicator. An authorization concept controls both types of blocking. This means that only authorized users—for example, an auditor—can access blocked data. After the retention period ends, both transactional and central master data must be deleted.

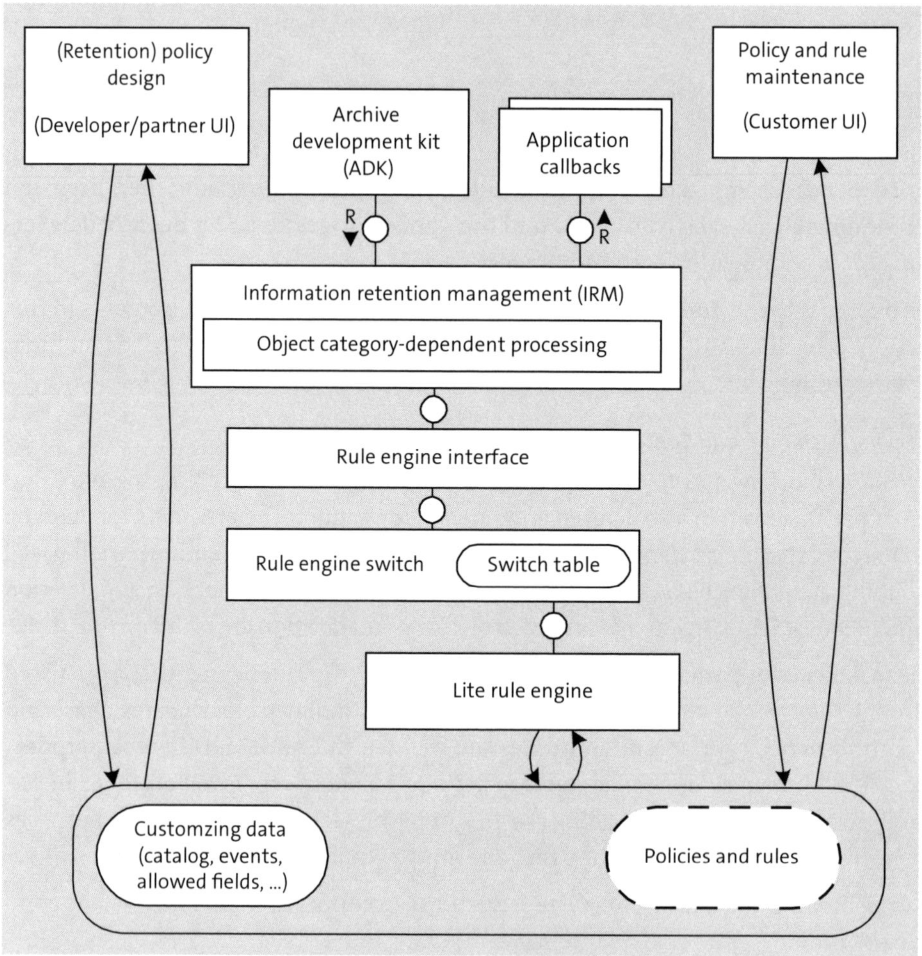

Figure 5.57 SAP Information Lifecycle Management Architecture

If there is no other legal basis for the lawful processing of personal data, consent must be obtained from the data subject to use their personal data. SAP Enterprise Consent and Preference Management enables storing, searching, and displaying consent data, as illustrated in Figure 5.58. Importing consent records as copies from either a file or via the consent repository service is also supported. SAP Enterprise Consent and Preference Management facilitates proving the legality of personal data processing and ensuring the correct processing of personal data according to the purpose for which the consent was given.

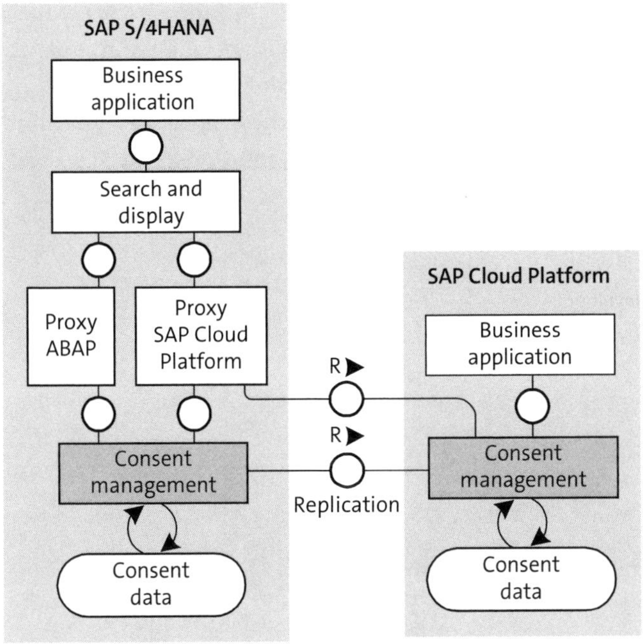

Figure 5.58 SAP Enterprise Consent and Preference Management Architecture

SAP Enterprise Consent and Preference Management aligns consent functions and data structures across applications and products. It provides access to consent data through a local API, which is implemented as a consent façade/proxy. For searching and displaying consent data, generic UIs are provided. This application also builds the foundation for centralization of consent data on SAP Cloud Platform, where front-office applications collect consent documents and back-office applications can access the same for further processing. This is the basis for communication with third parties in order to share not only personal data but also related consent information. In addition, it provides transparency and proof of which personal data and processes/purposes consent was given for.

Embedded Machine Learning

Figure 5.59 shows how deletion, blocking and consent requirements are solved for embedded machine learning. Application data stored in database tables are wrapped via CDS views and processed by machine learning algorithms for model training. The trained models are exposed via APIs for integration into machine learning applications and business processes.

As shown in section ❶ of Figure 5.59, training data is read via CDS views. For training, usually all data is accessed, so no authorization checks are performed. As already described, deletion and blocking are implemented based on SAP ILM. Thus, personal data that must be deleted or blocked is moved from the online database into archive storage by SAP ILM and cannot be accessed via CDS views anymore. Consequently, those records cannot be read during the machine learning training process, which results by design in compliance with the GDPR deletion and blocking requirements. However, there are exceptional objects like business partners or vendors for which blocking is realized by setting a blocking flag. In such cases, the CDS view for machine learning training must consider the blocking flag as a filter condition in the SQL `where` clause, as shown in the following example:

```
define view TrainingInput as select from I_
BusinessPartner ... where BusinessPartnerIsBlocked <> 'x'
```

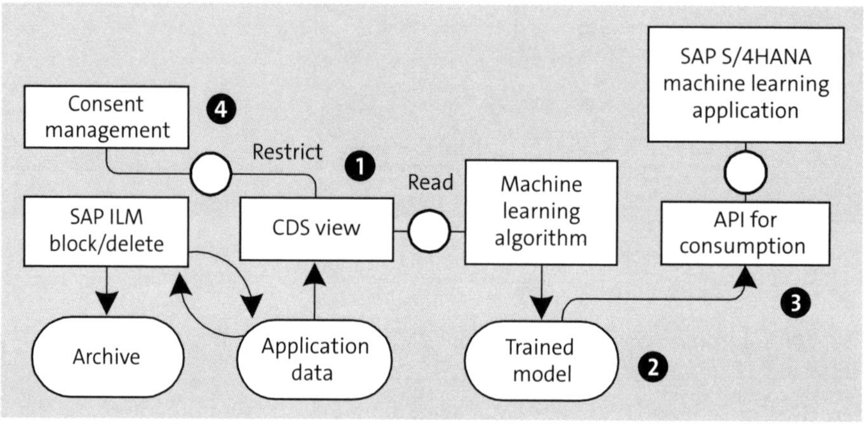

Figure 5.59 GDPR for Embedded Machine Learning

As depicted in section ❷ of Figure 5.59, the trained models are stored in the database. Typically, those models are stored in an encrypted format; they cannot be accessed directly, but only based on controlled APIs with authorization checks. Therefore, it's assumed that the trained models don't require specific treatment from a GDPR point of view. As illustrated in section ❸ of Figure 5.59, the trained models are exposed via APIs for consumption by machine learning applications. There is nothing machine learning-specific concerning the consumption APIs, so existing methods for ensuring GDPR compliance can be applied.

Machine learning applications that have to ensure the GDPR consent requirement must restrict the access to application data as shown in section ❹ of Figure 5.59. Only data with existing consent is allowed to be processed for machine learning model training. To achieve this, CDS views for accessing the application data must be combined with consent data in order to implement the required restriction. To establish a linkage between consent records and application data, which attributes do the following must be defined:

- Contains the consent's data subject and type. For example, if `I_SALESORDER.SoldToParty` is a customer, then `DataSubjectId = SoldToParty && DataSubjectIdType = 3` is the data subject and type.

- Contains the consent's data controller and type. For example, if I_SALESORDER.SalesOrganization is a sales org code, then ControllerName = SalesOrganization && ControllerType = 2 the controller and type.

To match the intended consent purpose, the following qualities need to be ensured:

- The application-specific functional attributes represent the purpose. For example, "I consent that SAP SE uses order data of the past four years to create a profile of my product preferences."

  The term order data matches orders only (not offer data). Thus, purpose X is defined by I_SALESORDER.SalesOrderType = 'OR'.

- The consent-specific functional attributes represent the purpose. For example, "I consent that SAP SE uses payment data of the past quarter to calculate a risk profile of my payment behavior."

  The terms profile/profiling as well as product preferences need to be represented by attributes of the purpose: Purpose.Action = 'Profiling' && Purpose.Aspect = 'Payment'.

Consent management data is stored in database tables. Figure 5.60 shows the static SQL statement to join the application and consent data in order to consider only records with given consent for machine learning model training.

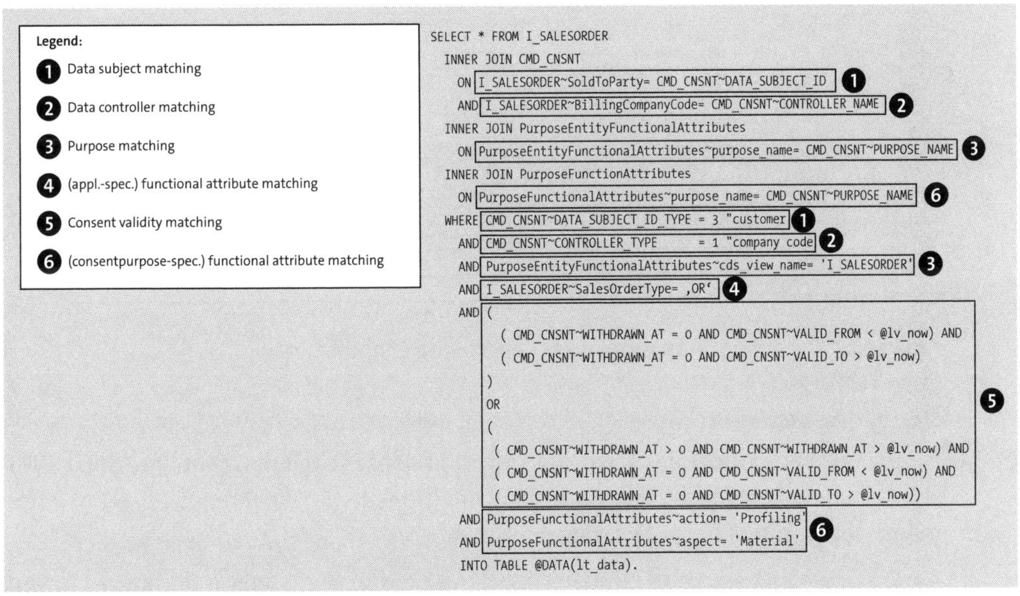

Figure 5.60 Static SQL Statement for Considering Consent Data

However, because the consent management data must be flexible and configurable, the static SQL statement isn't be sufficient. A dynamic approach is needed, which is illustrated in Figure 5.61. The main difference is that this approach is generic and can be applied to arbitrary use cases. However, this dynamic behavior also requires generic database tables for storing the consent data. The structure of those database tables is illustrated in Figure 5.61 as boxes. With those table definitions, the SQL statement to combine application data with consent information can be defined dynamically as shown in the coding in Figure 5.61. Thus, the CDS view for model training considers only the records for which consent is in place. Keep in mind that there is no framework available for combing application and consent data. Therefore, the SQL statements must be provided by the developer of the machine learning scenario.

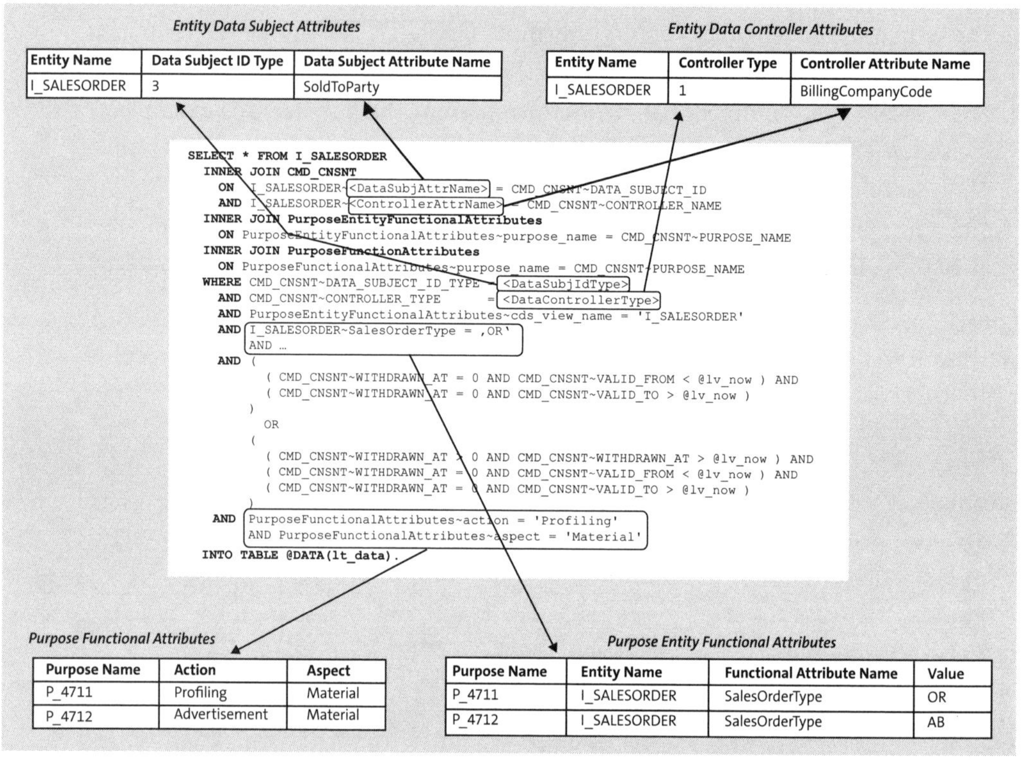

Figure 5.61 Dynamic SQL Statement for Considering Consent Data

Side-by-Side Machine Learning

Figure 5.62 shows how deletion, blocking, and consent requirements are solved for side-by-side machine learning. As already mentioned, the GDPR requirement of automated decision-making is described in Section 5.5.7.

Application data is extracted from SAP S/4HANA based on CDS views and stored in SAP Data Intelligence. Machine learning algorithms are trained based on the extracted data and the trained models are saved accordingly. For consumption of the trained models, SAP Data Intelligence provides REST APIs. Thus, the machine learning capability can be integrated into SAP S/4HANA applications.

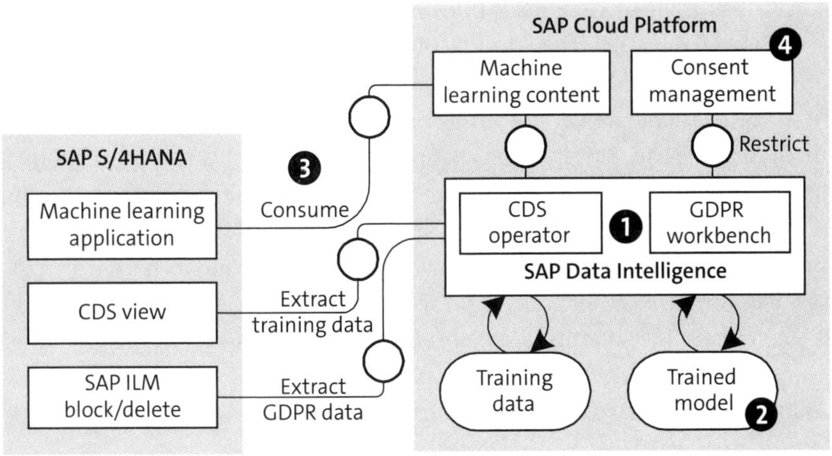

Figure 5.62 GDPR for Side-by-Side Machine Learning

As depicted in section ❶ of Figure 5.62, training data is extracted from SAP S/4HANA into SAP Data Intelligence based on CDS operators covering the initial upload and delta handling (see Section 5.5.2). For the replicated data, GDPR compliance in terms of deletion and blocking must be ensured. However, deletion and blocking of application data depends on the business processes running on SAP S/4HANA. As already described, SAP ILM handles deletion and blocking for SAP S/4HANA. Thus, SAP ILM has all relevant information regarding deleted/blocked records and exposes them with an extraction API to SAP Data Intelligence. An according GDPR workbench is needed, which receives deletion/blocking information based on the extraction API and deletes/blocks the effected records on SAP Data Intelligence. This is how the GDPR requirements for deletion and blocking can be solved for side-by-side machine learning. The details are illustrated in Figure 5.63.

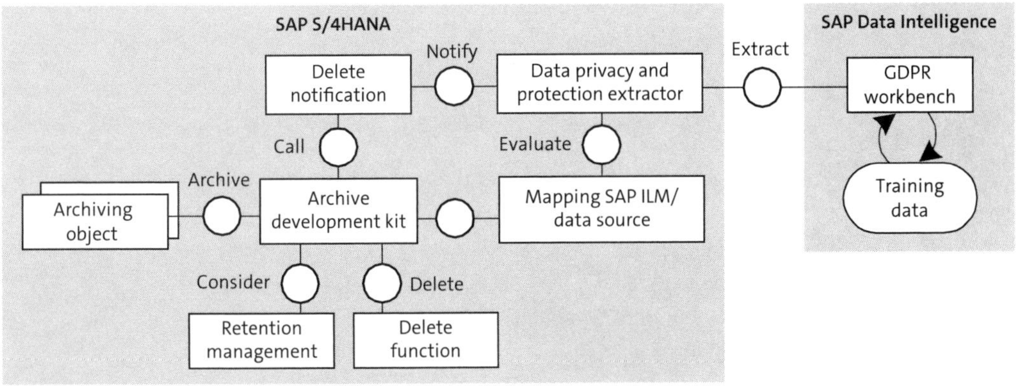

Figure 5.63 SAP ILM Integration for Side-by-Side Machine Learning

A notification API is provided so that applications can call in their delete programs to notify users of deleted personal data. The call of the API is integrated into the Archive Development Kit (ADK) framework of SAP ILM to trigger this automatically in each archiving run. Notifications are sent to targets that previously received the personal data as configured in the usual integration channels (e.g., RFC, application link enabling (ALE), enterprise service) via the same or separate channels. Filter functionality for sending the info to the right recipients is configured. The notification contains the business objects, so mapping from SAP ILM objects is provided. It must check if a recipient also eventually requires other context information. Mapping for the keys is necessary, which is only possible so long as the corresponding data is still in the database. Adequate user authorizations to display/send notifications for the data controller, purpose, and additional attributes are required. Furthermore, delete notifications, including their statuses, are logged.

The data privacy and protection extraction API is implemented as CDS view extractor `I_ILMDataPrivacyNotification`, as shown in Listing 5.10. With the CDS view, the data privacy and protection data can be extracted from SAP S/4HANA to SAP Data Intelligence for further processing.

```
@AbapCatalog.sqlViewName: 'IILMDPNOTIF'
@VDM.lifecycle.contract.type: #PUBLIC_LOCAL_API
@Analytics:{ dataExtraction: { enabled : true  }}
@Analytics.dataExtraction.delta.byElement.name : 'ILMNotifCreationDateTime'

define view I_ILMDataPrivacyNotification as select from
  ilm_notif_cont
  inner join ilm_notif_data on
    ilm_notif_data.guid = ilm_notif_cont.guid
  inner join ilmob_ce_c on
   ilm_notif_data.ilm_object = ilmob_ce_c.ilm_object
  inner join ilmob_cefldmap_c on
   ilmob_ce_c.ilm_object = ilmob_cefldmap_c.ilm_object and
   ilmob_ce_c.cdsextractor  = ilmob_cefldmap_c.cdsextractor and
   ilmob_ce_c.is_active     = 'X'and
    ilm_notif_cont.tabname   = ilmob_cefldmap_c.table_name and
    ilm_notif_cont.fieldname = ilmob_cefldmap_c.field_name
  inner join ilm_bw_modes on
    ilm_notif_data.notif_mode = ilm_bw_modes.ilm_notif_mode
  left outer to one join RSODP_ABAP_CDS_Extraction on       ilmob_ce_
c.cdsextractor = RSODP_ABAP_CDS_Extraction.strucobjn
{
  key ilm_notif_cont.guid as ILMNotificationUUID,
  key ilm_notif_cont.obj_number as ILMNotifDataObjectID,
  key ilm_notif_cont.rec_number as ILMNotifDataObjectRecdID,
```

```
  key ilm_notif_cont.tabname as ILMNotificationTable,
  key ilm_notif_cont.fieldname as ILMNotificationField,
  key ilm_notif_data.ilm_object as ILMObject,
  key ilmob_cefldmap_c.cdsextractor as CDSView,
  key ilmob_cefldmap_c.cdsextractor_fld as CDSViewField,
  @Semantics.systemDateTime.lastChangedAt:true
  ilm_notif_data.notif_tmstmp as ILMNotifCreationDateTime,
  ilm_notif_data.notif_mode as ILMNotificationMode,
  ilm_bw_modes.bw_notif_mode as ILMToSAPBWNotificationMode,
  ilm_notif_data.blocking_date as DataBlockingDateTime,
  ilm_notif_data.expiration_date as DataExpirationDateTime,
  ilm_notif_cont.value as ILMNotificationFieldValue,
  ilmob_ce_c.is_active as CDSViewExtractionIsActive,
  RSODP_ABAP_CDS_Extraction.odpname as
    APBWOperationalDataProvider
}
```

Listing 5.10 CDS View for Extracting Data Privacy and Protection Data from SAP S/4HANA

In SAP Data Intelligence, a kind of GDPR workbench is required that regularly extracts the SAP ILM data protection and privacy data from SAP S/4HANA and deletes/blocks the effected records accordingly.

> **Note**
>
> A GDPR workbench is not yet available in SAP Data Intelligence as of the time of writing (summer 2020), but one could be implemented by the developer of the machine learning scenario following the concepts and building blocks described in this section.

As illustrated previously in section ❷ of Figure 5.62, trained models are stored in an encrypted format without identifiable information and can be accessed not directly, but only based on controlled APIs with authorization checks. Thus, no specific measures must be taken for GDPR.

As shown previously in section ❸ of Figure 5.62, the trained models are exposed via APIs for consumption by machine learning applications. There is nothing machine learning-specific concerning the consumption APIs, so existing methods for ensuring GDPR compliance can be applied.

The GDPR requirement of consent is not relevant for all machine learning applications; for example, if the training data contain no personal information, it doesn't apply. However, if consent handling is requested, section ❹ in Figure 5.62 depicts that the training dataset must be restricted only to the records for which consent is given. In this context, the following questions arise: How can we establish a linkage between consent records and application data? How can we match the intended consent purpose? To solve these challenges, the same concept can be reused that was illustrated for embedded machine learning previously in Figure 5.60 and Figure 5.61.

5.5.4 Configuration

With SAP S/4HANA, predefined machine learning scenarios are delivered. This means that for a business question solved by machine learning, all the needed development is provisioned by SAP: integration to business processes and UIs, defining the data source for training, and delivering predefined models. However, customers or partners might require adapting this machine learning content to their specific needs based on configuration and extensibility. The focus of this section is on machine learning-specific configuration. General concepts are taken as granted, such as configuration for UIs, forms, or analytics. We'll discuss extensibility in Section 5.5.5.

Business Requirements

Configuration is a procedure to set up or adjust systems at a customer site by adapting delivered functions to the customer's business requirements. The process of configuration (or Customizing) is done by utilizing predefined variability of the underlying models. In context of machine learning, the following use cases for configuration must be considered:

- **Multiple model support**
  For the same scenario, active machine learning models for each data segment will be supported. In contrast to having one machine learning model, the prediction accuracy can be improved. For example, having a sales revenue prediction model for all countries usually lacks accuracy power compared to sales revenue models for each country, which incorporate country specifics. The consuming machine learning application must not handle those multiple models individually but based on one stable API. The determination of the correct machine learning model for an according inference request will be managed by the machine learning infrastructure automatically.
- **Model hyperparameter**
  When the machine learning algorithm of a delivered machine learning scenario is tuned for a specific data environment, then hyperparameters of the model are configured. Hyperparameters are estimated without using actual and observed data. Examples for such hyperparameters are the number of clusters in K-means clustering, the number of leaves of a tree, or the number of hidden layers in a deep neural network. Mechanisms and tools are required to maintain, deliver, and apply hyperparameters.
- **Configuration lifecycle management**
  User interfaces are required so that customers can maintain configuration data. It will be possible to maintain the configuration data in a test system and transfer it

safely to the productive system. Configuration data will be stored separately from system and application data. Mechanisms and tools are required to deliver and apply configuration. Only those configuration actions that fit customers' activated SAP S/4HANA business scopes will be suggested. All configuration offered to customers must continue to work after patches and upgrades without manual activities. Changes of core software must not corrupt customer configuration. Extension mechanisms must not break the system's integrity. Also, time-consuming activities before and after the upgrade will be avoided.

- **Model training and validation**
  For training of machine learning models, jobs must be scheduled, either regular jobs or events. To use trained models, customers must explicitly activate and deploy them. In advance of this, customers must validate the machine learning models. They also enable degradation of machine learning models in terms of automated retraining when model accuracy decreases. For those activities, appropriate mechanisms and tools are required.

Although configuration lifecycle management and model training and validation are machine learning-specific, they don't need further investigation here as they are covered by machine learning lifecycle management and tools (see Section 5.5.1). But a solution is required for multiple model support and model hyperparameters, which are considered in this section.

Note

The multiple model support configuration is not yet available as of the time of writing (summer 2020). However, with the described concepts and building blocks, the developer of the machine learning scenario can resolve these requirements.

Technical Implementation

In this section, the implementation for configuration of machine learning use cases are described.

Figure 5.64 illustrates the machine learning-specific artifacts that require special considerations for configuration and extensibility, which we'll discuss further in Section 5.5.5. An intelligent scenario is the central design-time entity covering all artifacts required for the implementation of a machine learning scenario. Semantically, it represents the key question solved with machine learning. For this, the required algorithm and data features are determined by data scientists. For the training of the algorithm, application data is necessary. Thus, a view is provided that can span multiple application tables and perform preliminary data transformations.

From an extensibility point of view, the training data view and the underlying persistency can be extended. Furthermore, the algorithm can be exchanged and the hyperparameters reconfigured. The machine learning logic in terms of data transformations and feature engineering can be extended to meet a specific customer situation. The signature for the inference API is also defined during design time as part of the intelligent scenario. Consequently, the consuming applications and business processes can integrate the inference API during design time. However, meaningful results are provided by the inference API once the underlying machine learning model is trained. The signature of the consumption API can be extended with optional fields.

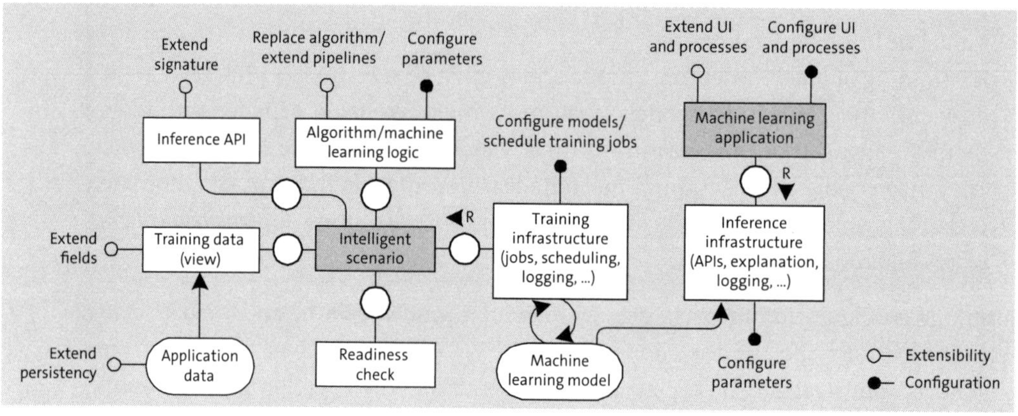

Figure 5.64 Machine Learning Artifacts: Configuration and Extensibility

Before machine learning models are trained, readiness checks are performed. These are machine learning scenario-specific checks to validate whether the prerequisites for the training are met, such as sufficient application data being available or necessary process configurations being completed. After the readiness checks have been passed successfully, the training is triggered. For this, the training infrastructure is used, which reads the necessary metadata from the corresponding intelligent scenario, runs the training, and stores the trained model. Scheduling training jobs and activating machine learning models are in the scope of the configuration. The inference infrastructure provides the consumption API for the specific scenario and model. Thus, the machine learning capability can be integrated into applications and business processes.

The configuration and extensibility of the UIs and business processes of the machine learning application are resolved with already established concepts and frameworks. Because they are not machine learning-specific, those topics are not considered here.

As already mentioned, the focus of the configuration implementation is on multiple model support and model hyperparameters; the supplementary requirements are provisioned by machine learning lifecycle management and do not need further investigation in this section.

5

Multiple Model Support Configuration

Figure 5.65 shows the solution architecture for multiple model support. Depending on the machine learning scenario, having only one gross granular machine learning model can result in poor model accuracy and long-running training jobs. To improve the prediction power of machine learning models (see Section 5.5.6), it's necessary to operate multiple models for the same machine learning scenario in parallel based on different data segments. To resolve this requirement, the training infrastructure and runtime must be enhanced for configuration.

The intelligent scenario design-time artifact contains metadata regarding the view for accessing the training data and the consumption API. The list of intelligent scenarios is the entry point for the business administrator to schedule training jobs. In this context, filters will be defined to segment the training data. The training process uses those filter criteria during runtime and trains models for each of the segments. An important constraint is that the filter criteria must be based on attributes of the API inference request structure only.

With those fine-grained, trained models, the next challenge is how to determine the relevant model for consumption. Here the *model dispatcher* component comes into play. Whenever the machine learning application calls the inference API for consumption, the underlying request is routed to the model dispatcher. This forwards the request to the appropriate segment model by matching the request to the defined filter criteria. This is possible because the request and the filter are based on the same set of attributes. For example, if a sales organization is part of the request signature of the API inference, then the sales organization can be used to segment the training data and provide machine learning models for different sales organizations.

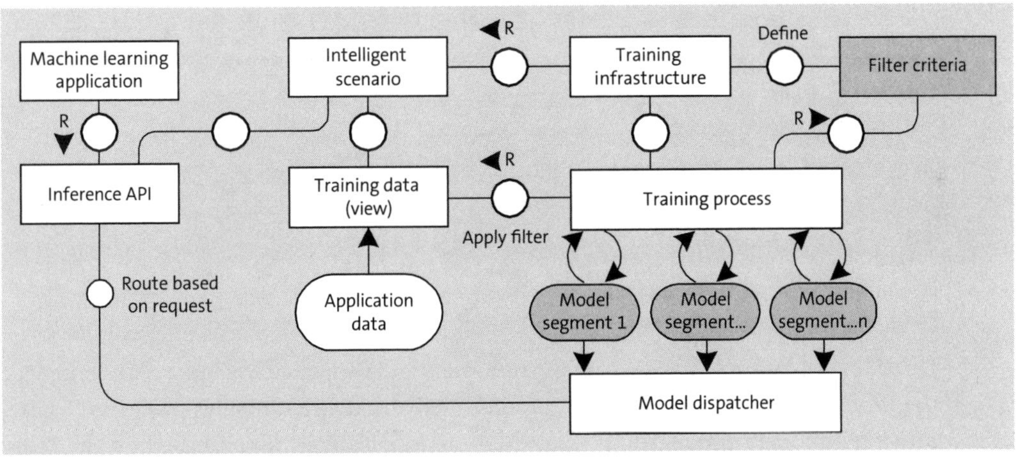

Figure 5.65 Multiple Model Support

The interplay of these components is illustrated in Figure 5.66. As already described, the training jobs are scheduled by the business administrator. For each segment, the business administrator needs to define the filter criteria and run the correct training job, which results in finely granular models.

The consumption of system resources in terms of memory and CPU time can be coordinated better with those finely granular models compared to having one model for all segments. For consumption, the correct model must be determined. This is managed by the model dispatcher, which routes the inference requests from the machine learning application to the relevant model. The necessary correlation happens based on attributes of the request structure, which are used also for model segmentation.

Note

This approach is also helpful for performance and workload management of the training process.

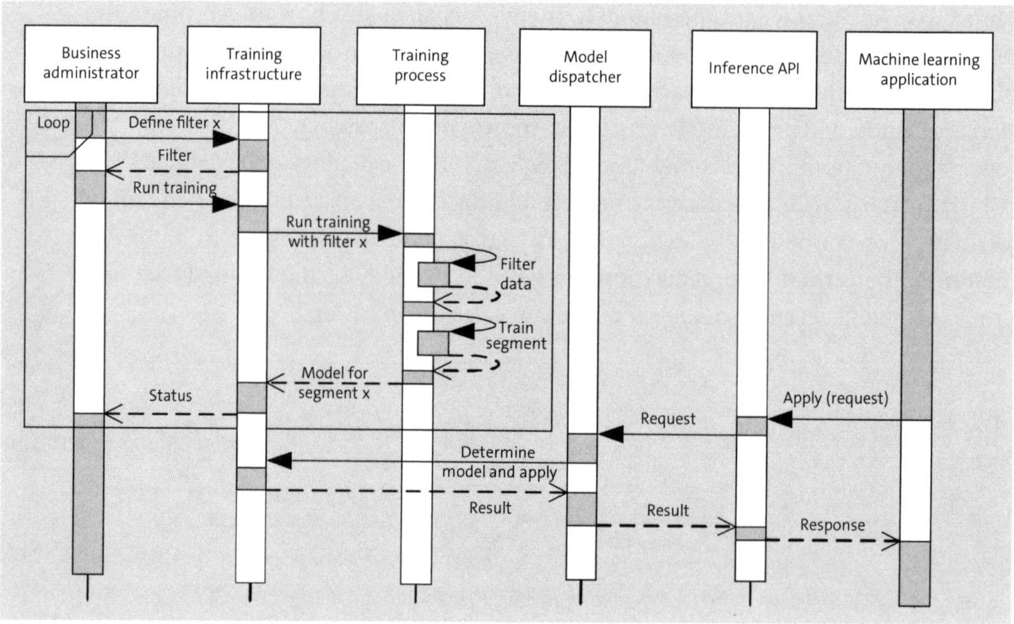

Figure 5.66 Multiple Model Processing

The UI of the multiple model support requirement is depicted in Figure 5.67. It depicts the training run for a use case that predicts delivery delays for sales products. To improve the model's prediction power, multiple models are trained based on the sales organization and sales group segmentation criteria. Figure 5.67 illustrates the training run for the Asia sales organization and the **01** to **05** sales groups.

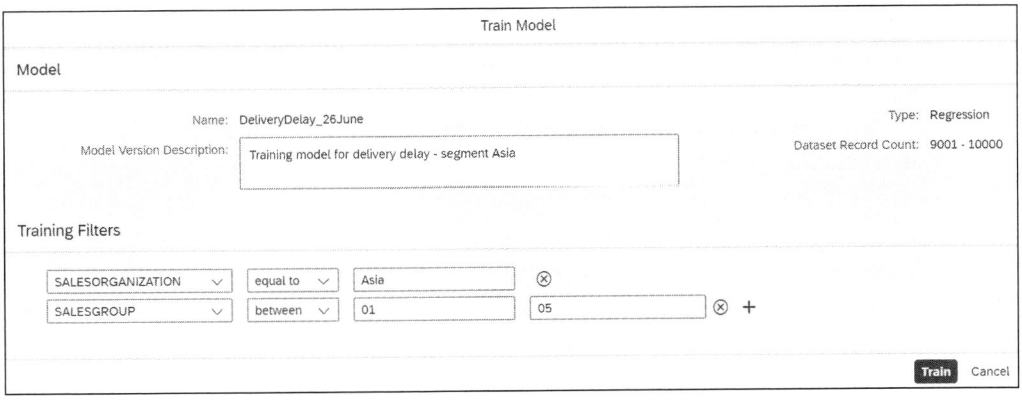

Figure 5.67 Multiple Model User Interface

Model Hyperparameter Configuration

Now, let's move on to the solution for the model hyperparameter configuration requirement, as shown in Figure 5.68. Usually there are various parameters for configuration to consider, such as country codes, sales order types, limits for CPU time, or connectivity variables. Because most of these are universal technical and process-related parameters, they are not considered here. Our focus is on machine learning-specific configuration parameters. In this context, hyperparameters of machine learning algorithms during the training run are considered in particular. However, the concept is defined more generally so that arbitrary machine learning-specific configuration parameters are taken into account. This particularly includes configuration parameters that are required for the inference process.

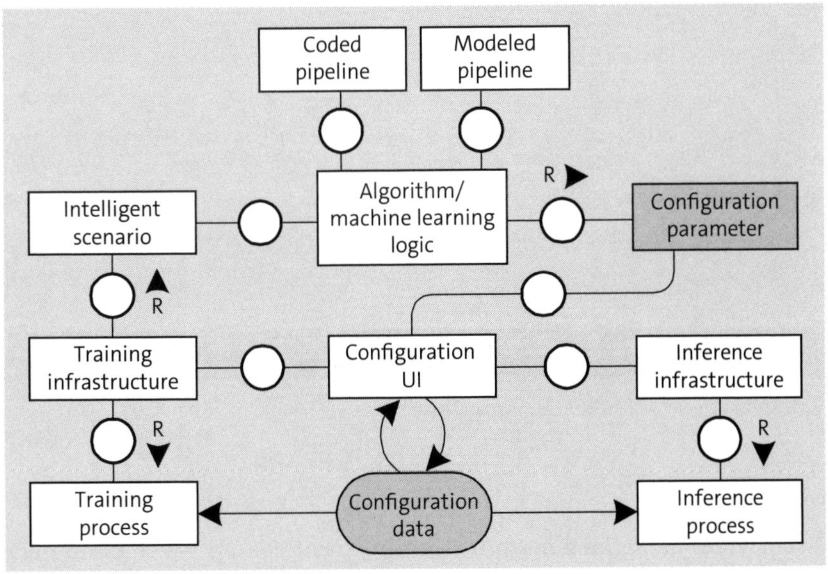

Figure 5.68 Configuration Parameter

The machine learning logic consists of data transformation, feature engineering, launching algorithms, and postprocessing steps. To provide flexibility, parameters are integrated into the machine learning logic instead of having static values. Thus, different values can be assigned to the parameters during configuration time. This is especially the case for hyperparameters of the machine learning algorithms, which are tuned to a concrete customer situation to improve the model accuracy. The machine learning logic can be coded or modeled graphically, depending on the requirements of the underlying use case. Coding is more expressive, whereas graphical models are more comprehensive. Wherever configuration flexibility is necessary in the machine learning logic, parameters are introduced accordingly. The concrete values for those parameters can be maintained based on a generic configuration UI consisting of key-value pairs.

An exemplary UI for hyperparameters of a machine learning scenario is depicted in Figure 5.69. The configuration parameters are associated through the machine learning logic to an intelligent scenario that is the logical bracket. The entered configuration values are incorporated by the training and inference processes. Thus, the runtimes can be impacted according to specific needs.

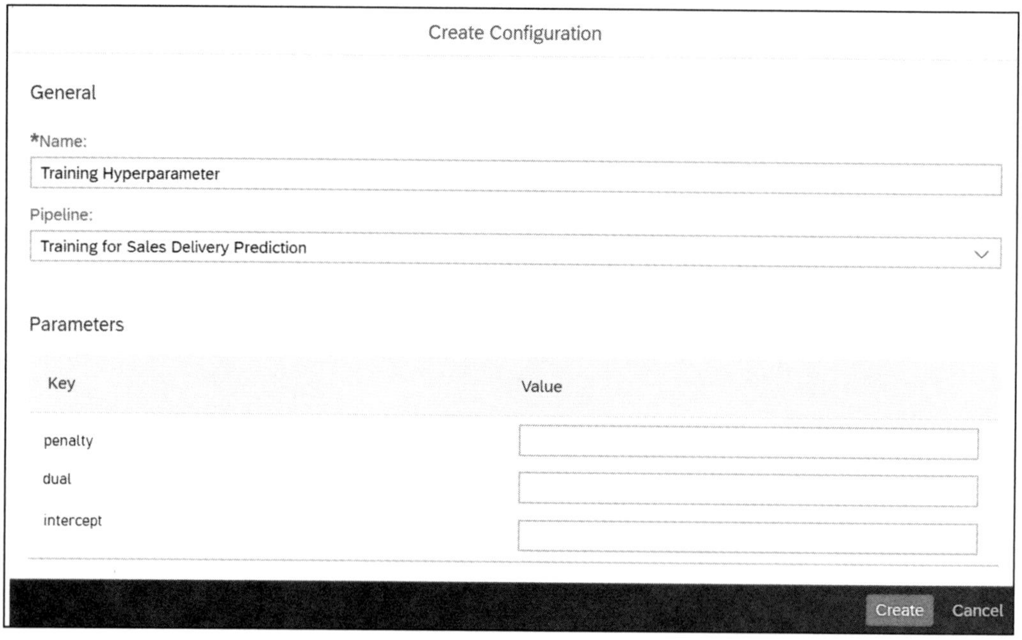

Figure 5.69 Configuration Parameter User Interface

5.5.5 Extensibility

SAP S/4HANA includes predefined machine learning scenarios. However, customers or partners might require adopting this machine learning content to their specific needs based on extensibility. This might be also necessary for use cases you develop.

Therefore, this section explains the business requirements and the necessary technical implementation for extensibility in context of machine learning.

Business Requirements

The following definition of extensibility can be considered as a basis for all extensibility use cases: Extensibility is the adaptation of standard software by partners, customers, or SAP, including the related integration into system landscapes. Its goal is to add functionality for individual or industry-specific needs that cannot or should not be covered by the standard software. Its main obligation is to enable business experts to create simple extensions by themselves. Customers usually extend machine learning use cases to improve the accuracy of the machine learning models. Better models can be achieved in a couple of ways, as follows:

- **Training data source extension**
  The accuracy of machine learning models can be improved by using more attributes. For this, customers will be enabled to extend the SAP-delivered CDS view for model training with additional fields. Different options must be supported:
  - Add fields from extended applications (custom field extension)
  - Add fields from lower-level CDS views (data source extension)
  - Add fields from new CDS views (custom CDS view joining other SAP CDS views)
  - Add fields from external datasets (custom CDS view joining custom business objects)
- **Algorithm exchange**
  The prediction power of machine learning models can be improved by changing the algorithm. Thus, it's requested that customers be able to exchange the predefined algorithm with a new algorithm of the same type—for example, substituting linear regression with exponential regression. For this, a customer's data scientists have to perform explorations and experiments to identify the algorithm that's best tailored for the specific customer situation and data environment.
- **Machine learning logic extension**
  Machine learning logic covers aspects like data preparation, feature engineering, and transformations. By extending the predefined pipelines of machine learning logic, the accuracy of models can be improved. Thus, customers need to be able to extend the predefined machine learning logic. As a prerequisite, a customer's data scientists have to perform explorations and experiments to identify what's best tailored to the specific customer situation and data environment.
- **Consumption API extension**
  The consumption API wraps the machine learning model and exposes it for integration into applications and business processes. This API contains the necessary request-and-response signature to interact with the trained model. To avoid breaking the consuming applications and business processes, the API must be kept stable

and extended in a compatible manner. Therefore, customers need to be able to extend the consumption API with optional fields.

- **New machine learning application**
  Customers and partners should be able to build a new machine learning application from scratch. This includes modeling of CDS views for training, creating intelligent scenarios, defining consumption APIs, and integrating these into the business logic. All the follow-up development activities are not machine learning-specific, so SAP S/4HANA concepts and tools can be reused.
- **Extensibility lifecycle management**
  All extensibility capabilities offered to customers must continue to work after patches and upgrades without manual activities. Therefore, extensibility should follow the whitelisting approach and offer only measures or objects for extensibility that are designed for those purposes. Changes of core software must not corrupt customer extensions. There will be a clear separation between standard functionality and extensions (e.g., by a namespace concept). Extension mechanisms must not break the system's integrity. Also, time-consuming activities before and after the upgrade will be avoided. All extensions are created and transported with standard mechanisms in such a way that a transport contains an extension completely without the need for further activities and tools.

Technical Implementation

The technical implementation for the extensibility requirements is described in this section.

Training Data Source Extension

The training data source extension requirement is about extending the data source with additional fields to improve the prediction power of the machine learning model. Depending on which additional data fields are required, an appropriate extensibility approach must be applied:

- Custom field extensions add fields from extended applications.
- Data source extensions add fields from lower-level CDS views.
- Custom CDS views joining other SAP CDS views add fields from new CDS views.
- Custom CDS views joining custom business objects add fields from external datasets.

Custom field extensions offer the capability of adding customer-specific fields to the business context of an application in a one-to-one relation or adding SAP fields that are available in tables and structures of the application. This approach is based on extension includes. An extension include is an almost empty include with a dummy field due to data dictionary (DDIC) restrictions. It's shipped by SAP as an anchor point for enhancements. The include is part of all necessary structures, such as databases, API structures, or service implementations. All enhancement fields are part of DDIC appends, which is a modification-free DDIC extensibility approach.

The extension includes are the basis for the custom field extensions, which must be prepared explicitly by the application. In this context, database tables and DDIC structures must be incorporated as extension includes to provide a stable anchor point for DDIC appends. Custom fields will be then added via DDIC appends. ABAP code in the application must generically transfer extension fields between structures/internal tables via `move-corresponding`.

Extensible CDS views for training data must provide a so-called extension include association to an extension include view. This is a CDS view on the database table containing only the key fields and later the custom fields from the extension include. The extension include view acts as a stable anchor point for CDS view extensions and makes the extension fields accessible on extensible CDS views via the extension include association. Custom fields will be added to the extension include view as soon as they are added to the persistency. The extensible CDS views will be extended when selected in the **Where Used** dialog of the custom field UI because not all consumers are able to traverse the extension include association to access the custom fields (see Figure 5.70).

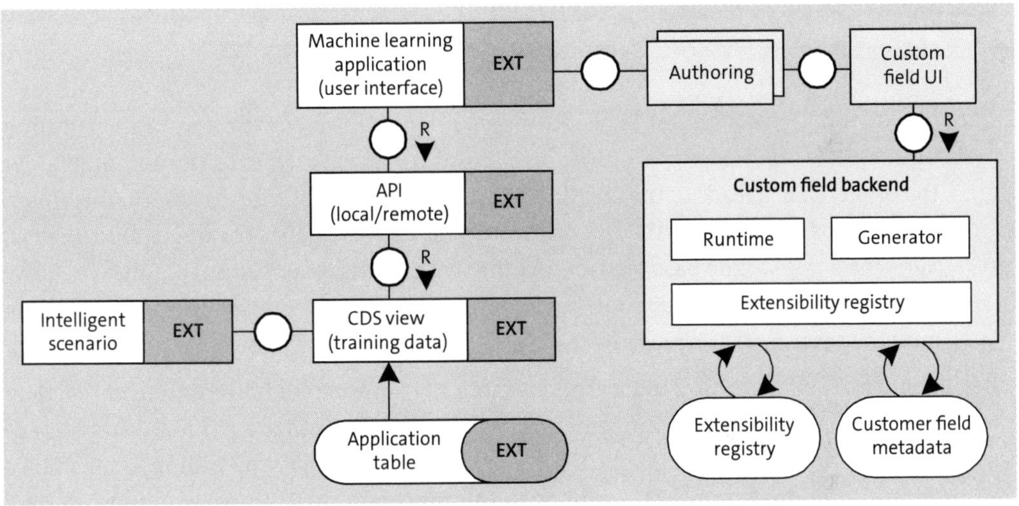

Figure 5.70 CDS View: Customer Field Extension

Listing 5.11 illustrates an example of the extension of CDS views.

```
// SAP-defined View:
Namespace sap.orgmgmt;
view EmployeesView as SELECT from Employee
{
ID, name,
salary,// returns nested structure
address,// returns the association as itself
```

```
{// resolves :1 association and creates a structured type
manager,
name as orgunitName,
} as orgunit
}
// Extension of SAP-defined View by customer "acme":
namespace acme.orgmgmt;
//adds a new label to the view
@EndUserText.label: 'Extended Employee View'
extend view sap.orgmgmt::EmployeesView
{
// A new customer-defined field is added:
acmeFlags,
// the SAP-defined element "name" is extended with a new label text:
@EndUserText.label: 'Name of Employee'
extend name,
// SAP-defined structure "orgunit" is extended by the SAP-
defined field "costcenter":
extend orgunit { costcenter }
}
```

Listing 5.11 Example for CDS View Extension

The `extend view` statement leads on the one hand to the enrichment of the runtime object of the `sap.orgmgmt::EmployeesView` base artifact with the extension elements. On the other hand, it leads to the creation of the `acme.orgmgmt::EmployeesView` name, which acts both as the name of the extension artifact and as an alias for the extended view, allowing access to the base artifact together with all the extensions. This alias is equivalent to the effect of the following statement: view `acme.orgmgmt::EmployeesView` as alias to `sap.orgmgmt::EmployeesView`.

The structural definition of local and remote APIs must contain the extension include as well. Depending on the implementation variant, hand-coded parts in the implementation of an API must be explicitly prepared for extensibility by calling extensibility function modules, such as to enhance the metadata or to transfer extension data between internal and external structures. At a customer site, a key user launches the custom field UI, which allows you to specify the technical details of a custom field (label, type, length, etc.) and provides a where-used list. The tool can be consumed standalone or, preferably, integrated into the runtime authoring of the UI.

Further Resources

General SAP S/4HANA extensibility falls outside the scope of this book. For more information on the custom field UI tool, refer to *http://help.sap.com/s4hana* and search for "Custom Fields and Logic".

The where-used list is based on the extensibility registry, which identifies all APIs and CDS views and exposes the extended persistency. When selecting an entry in the where-used list, the underlying API and CDS view is extended. Thus, the corresponding consumers have the extension field available in their field catalog and they can utilize it like any standard field delivered by SAP.

Next, the data source extension allows customers to add new SAP standard fields from the underlying CDS views. The CDS view for model training is usually built on top of basic CDS views. Those can have fields that could be incorporated into the training process in order to increase the prediction power of the machine learning model.

Figure 5.71 illustrates the basic idea of data source extension. The CDS View 1 that will be enhanced has to be exposed in the extensibility registry. The additional fields can be selected from the CDS View 2.1 (Field1) subview or from the CDS View 2.2 (Field2)/CDS View 2.3 (Field3) views that can be reached via 0..1 associations. From a key user perspective, the data source for the extension has to be selected and enhanced in the custom field application. Adding additional standard fields might enable end users to access restricted data in the extended data source without having the required authorization. Furthermore, depending on the field path and the nature of the selected fields, data retrieval might become slower after a data source extension has been added.

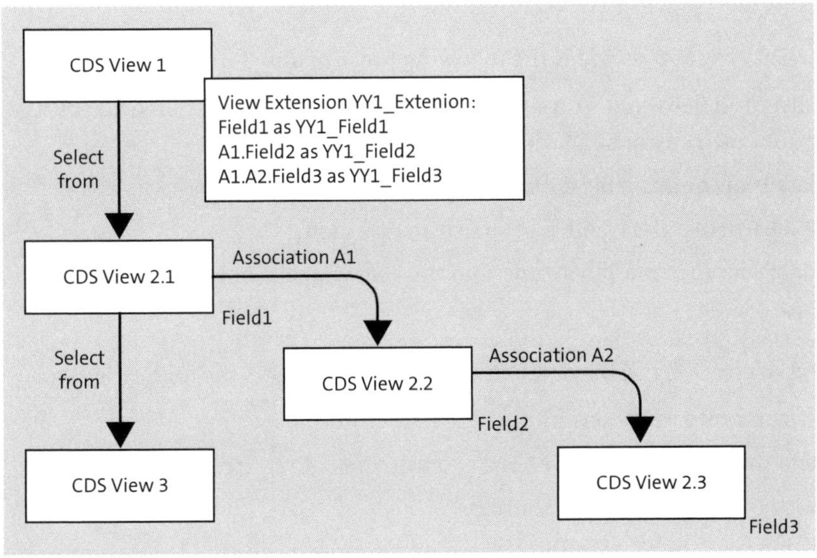

Figure 5.71 CDS View: Data Source Extension

New CDS view modeling is required if the already described approaches of customer field and data source extension are not sufficient or cannot be applied at all. Examples of such circumstances are that fields from CDS views of other view hierarchies are required or the training CDS view is not registered for data source extension. A new training CDS view can be defined to combine fields from other CDS view hierarchies, to access other data tables, or to reduce the fields of the existing training CDS views by SQL projection. With the Custom CDS View app, shown in Figure 5.72, key users can model customer-specific data access on top of public CDS views. In addition to this, already created custom CDS views can be used as a basis for new custom CDS views.

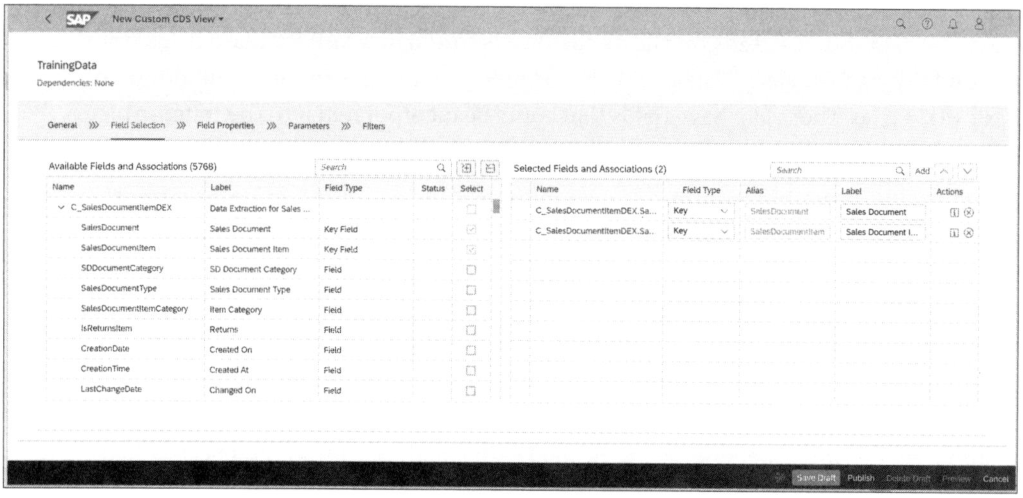

Figure 5.72 Custom CDS View

The Custom CDS View app provides the following functionality:

- Display a list of all delivered data sources and already existing custom data sources, including your own custom CDS views.
- Display details of the available data sources and custom CDS views.
- Search for a particular data source or a custom CDS view.
- Filter by data source name, label, type, and the user who last published the view.
- Create a CDS view.

In the context of creation of CDS views, the application supports the following steps:

1. Define the name and the description of the new custom view.
2. Select a data source that should be used as a primary data source.
3. Add one or more associated data sources.
4. Assign the mapping fields of the associated data sources to the fields of the primary data source.
5. Select the fields and associations that should be used in the new custom CDS view.
6. Expose the custom CDS view via an OData service for external usage.
7. Perform changes on the semantics of the selected data sources for usage within the new custom CDS view.
8. Display the parameters of the selected data sources and check the filter criteria.

To improve the accuracy of machine learning models, it might be also necessary to extend the training data source with customer-specific tables. This table extensibility denotes the capability of adding customer-specific fields to a business context of an application in a 1:1 or 1:*n* relation. New customer-specific tables are created in the database and accompanied with CDS views. Those CDS views can be combined with the delivered CDS view for model training to a new data source. In context of customer-specific tables, two types of enhancements can be distinguished:

- New standalone custom tables that are not child tables of SAP tables are fed through a UI or data load from other customer systems. These standalone tables might be used as code lists, for process control, or as facts or dimensions for transactional and analytical purposes. A further enhancement might be to bundle several standalone custom tables into a hierarchy, creating a new application with simple business logic. This functionality is available as a custom business object.
- Custom tables are used to add fields to SAP business contexts in a 1:*n* relation (e.g., hobbies of a customer) or to resolve the technical constraints of field extensibility in case of extensions in a 1:1 relation. In this use case, the custom data behaves like the standard SAP data (e.g., the authorization inherited from the SAP parent), and custom data is deleted when the parent is deleted. This use case requires that the SAP applications are prepared for this kind of extensibility, which is usually not the case.

With the Custom Business Object SAP Fiori app (see Figure 5.73), customers can create and manage business objects. Such a custom business object is a hierarchical set of database tables with an API for creation, retrieval, update, and deletion of data. Business logic can be provided in a web editor using the cloud-enabled version of ABAP.

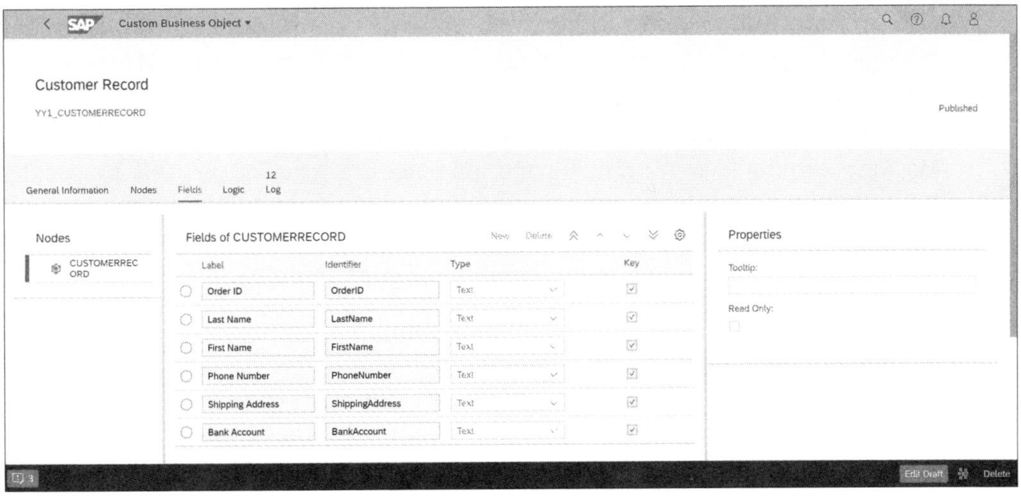

Figure 5.73 Custom Business Object App

The following features are supported with the Custom Business Object app:

- Create business objects and the corresponding database tables.
- Add fields to business objects.
- Delete fields and business objects that have not yet been transported to production system.
- Create CDS views, OData services, and a UI to enter data or load data from other customer systems.
- Create multiple subnodes for one business object.
- Define fields of type association to other custom business objects and to standard SAP business objects.
- Implement custom logic on the node level.
- Write to custom business objects from custom logic.
- Publish business objects.
- Edit business objects that have already been published.
- Reset business objects to their last published version.
- Copy or delete an existing custom business object.
- Perform a trace.

Depending on how the data source is extended, this is reflected automatically by the intelligent scenario because it's just keeping a reference to the CDS view. Particularly, the data source extension must not cause an incompatible change of the consumption API.

Algorithm Exchange and Machine Learning Logic Extension

Now, let's move on to the algorithm exchange and machine learning logic extension. Machine learning logic refers to the implementation of the steps required for processing of training and inference. Such steps include data validations, statistic calculations, transformations, feature reductions, and computing of model accuracy measures. Machine learning logic for training and inference are provided as scenario-specific. Extending of machine learning logic means adopting the SAP standard-delivered content with customer enhancements that survive upgrades. As part of the training logic, the algorithms could be principally exchanged to improve the prediction power of the model. However, only replacement into an algorithm of the same type is usually meaningful, such as linear regression to robust linear regression.

Depending on the underlying technology, different approaches are required to enhance the machine learning logic, which we first discussed in Section 5.2:

- **Automated Predictive Library (APL)**
  Machine learning logic for training and inference are incorporated in the library implementation itself. Therefore, the necessary machine learning logic is provided automatically by APL. Enhancements are metadata-driven and are also provisioned

without explicit development activities. This works for SAP S/4HANA Cloud and on-premise.

- **Predictive Analysis Library (PAL)**
  This library is used for break-out scenarios, which are usually more complex compared to APL-based ones. The machine learning logics for training and inference are implemented as an AMDP class with interfaces as shown in Figure 5.74. Extensibility of the ABAP class is based on BAdIs. These are the basis for object plug-ins that serve to enhance the functions in ABAP programs without having to make modifications. BAdIs enable the creation of enhancement options in the form of interfaces that can be appropriately implemented later in the same system or in an external system by customers. Future upgrades of the original business function can be applied without losing the customer-specific enhancements and without the need to merge the changes. The two code lines (the original one and the customer-specific coding) are strictly separated but still integrated.

  As with all enhancements, BAdIs make a distinction between the definition and individual implementations. The definition of a BAdI is usually created by SAP and, together with its calling points in ABAP programs, it forms an explicit enhancement option in such programs. The definition of a BAdI contains a BAdI interface, a set of selection filters, and some settings that affect runtime behavior later. A BAdI interface forms the entire interface or part of the interface of an object plug-in. *BAdI implementation* is the term used in the enhancement framework for an enhancement implementation element. It consists of a BAdI implementation class that implements the BAdI interface and a condition imposed on the filters specified in the BAdI definition. These can be used to select the BAdI implementation. The calling points of a BAdI are defined through the `GET BADI` and `CALL BADI` ABAP statements, which form the enhancement spot element calls of the explicit enhancement option.

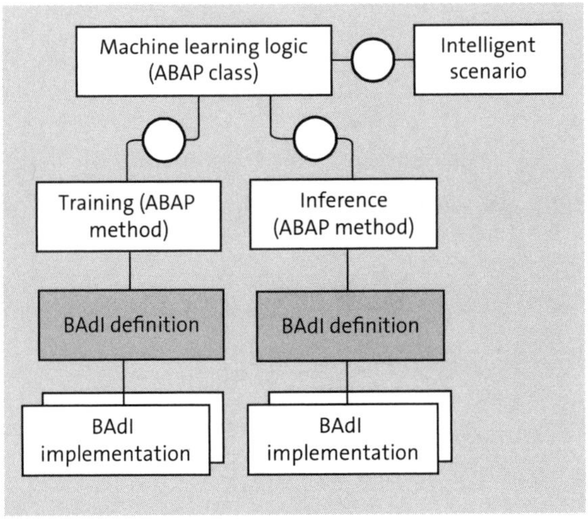

Figure 5.74 PAL Extensibility

For SAP S/4HANA on-premise, the standard development environment can be used to implement BAdIs; for SAP S/4HANA Cloud, a web editor for ABAP is provided to easily add business logic to existing enhancement options. The web editor completely hides the technical details of the underlying BAdI technology for the key user. It provides an easy mode of exploration for APIs with key user documentation and features such as syntax check, syntax highlighting, and code completion. Testing capabilities are included. The tool is designed according to cloud qualities; it allows neither usage of obsolete ABAP statements nor direct database access or task handling. However, the web editor currently supports ABAP classes only and must be enhanced for AMDP.

- **SAP Data Intelligence**
  For advanced scenarios like image recognition or language processing in which neuronal networks, GPUs, and external data are required, SAP Data Intelligence is used. The machine learning logic for training and inference are modeled graphically based on pipelines. Also, CDS extraction views can be incorporated into those pipelines for accessing training data. The SAP Data Intelligence design-time tools are already cloud-enabled and can be used by SAP S/4HANA Cloud and on-premise customers.

  As depicted in Figure 5.75, enhancement points must be provided in those pipelines by SAP development in order to enable customer-specific extensibility. Of course, the enhancement must be placed where it makes sense semantically, such as after the delivered data validation steps so that additional validations can be added. Those extensions are implemented as subpipelines by customers and will remain stable after upgrade of the SAP-delivered parts of the pipelines. The pipelines are exposed via REST APIs to SAP S/4HANA for consumption. Therefore, it's important that the pipeline extensibility results in stable and compatible REST APIs.

Note

There is no framework for extensibility of pipelines available in SAP Data Intelligence as of the time of writing (summer 2020). However, with the concepts presented here, developers can resolve this.

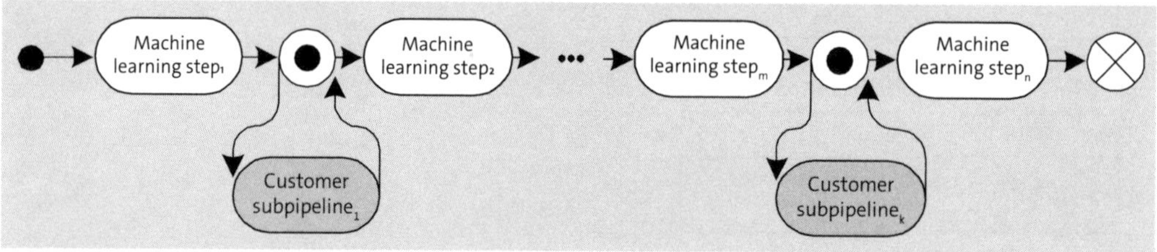

Figure 5.75 Pipeline Extensibility

Consumption API Extension

Next, we'll discuss consumption API extensibility. In the context of machine learning scenarios, different APIs are provisioned and consumed as illustrated in Figure 5.76.

Depending on the way the machine learning scenario is implemented, remote or local APIs must be considered. Local APIs are called within the SAP S/4HANA system, while remote APIs leave the system boundaries toward SAP Data Intelligence.

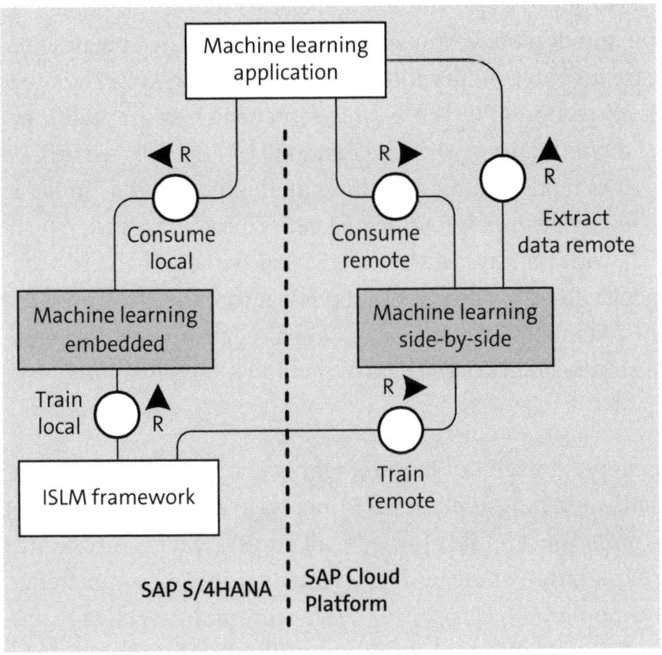

Figure 5.76 Consumption API Extensibility

The consumption API is called by the machine learning application for integration into business processes and UIs. Depending on whether the use case has been implemented in embedded or side-by-side mode, the consumption is local or remote. To train machine learning models, a training API is provided that is local or remote depending on the machine learning implementation paradigm. However, the orchestration of the training APIs is managed by the ISLM framework, which is in charge of lifecycle management of machine learning models (refer to Section 5.5.1). For embedded machine learning, the training data can be accessed locally through the SAP S/4HANA platform, but this isn't possible for side-by-side machine learning. Therefore, the API for extracting application data from SAP S/4HANA to SAP Data Intelligence must be provided.

Depending on the underlying technology (APL, PAL, CDS view, SAP Data Intelligence), there are different tools for implementing those APIs. From an extensibility point of view, it's important that all the mentioned APIs are enhanced for compatibility. This means that the consumer of the API must not break after extensibility. This is critical for the consumption APIs that are integrated into the machine learning applications and business processes. Thus, local and remote consumption APIs must be extended with optional fields only. The existing fields of the signature will not be changed, not deleted, renamed, or their data types replaced. To ensure this kind of stability and compatibility, governance measures and automatic checks are required.

As the training APIs are orchestrated by the ISLM framework, there is more freedom with regard to extensibility. However, to provide this generic handling, the signatures of training APIs are uniform and don't have any business process-related parameters. Therefore, there is no use case for extensibility concerning the training APIs. The situation is different for the data extraction APIs. Those can be extended because additional fields will be considered for model training. The data extraction APIs are based on CDS views for data extraction. To extend such an API, the underlying CDS view must be extended first. As the CDS view extraction is founded on generic data integration technology (refer to Section 5.5.2), with the extension of the CDS view the API and the consuming SAP Data Intelligence functionality are adopted automatically. Thus, the consumer is served with the extended fields while the data integration processes continue to work. With that, the extensibility of the data extraction API is ensured.

New Machine Learning App

The new machine learning app extensibility requirement is about enabling the customer to implement a machine learning application from end to end. As illustrated in Figure 5.77, for this the customer must be able to create all necessary design-time artifacts. It's assumed that the exploration phase by data scientists is already completed so that the focus is on the development objects. SAP S/4HANA (on-premise) already provides the development environment and tools to create all the depicted design-time artifacts. Thus, here we'll target SAP S/4HANA Cloud.

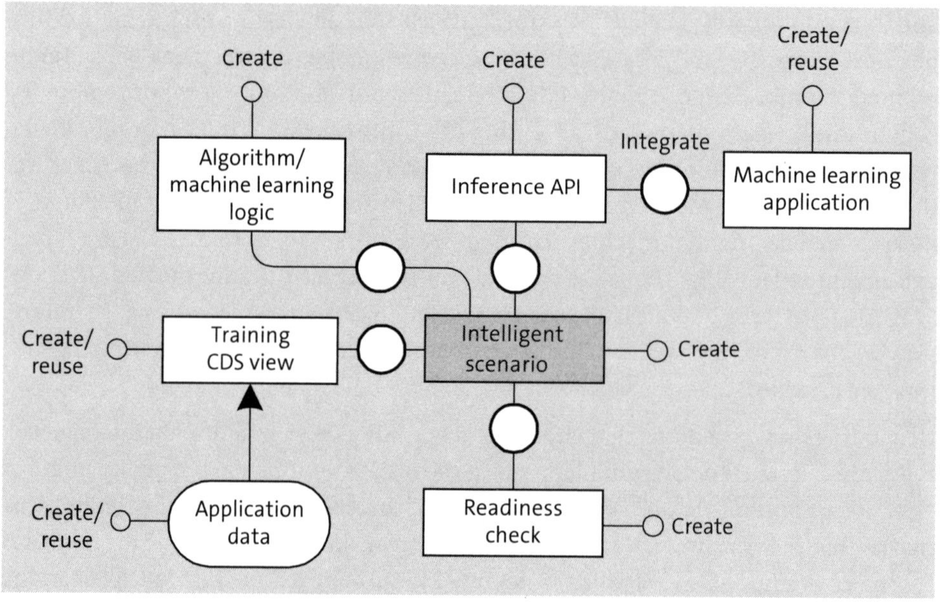

Figure 5.77 New Machine Learning Application

The starting point is the application data. Usually the standard SAP tables are sufficient and can be reused. If customer-specific tables have to be created, then the already described Custom Business Object app can be used. On top of the application tables, new CDS views can be created with the Custom CDS Views app. In this context, existing CDS views can be reused. Intelligent scenarios are the central artifact to keep all development objects together in terms of lifecycle management. This artifact can be created with the Intelligent Scenarios SAP Fiori app (refer back to Figure 5.2). Existing integration scenarios cannot be reused as an integration scenario always represents one machine learning use case. Depending on the underlying library, the machine learning logic (e.g., data transformations, feature reduction) and inference API have to provided differently:

- **Automated Predictive Library (APL)**
  The inference API is specified in context of the intelligent scenario and generated during the activation step. CDS table functions and an according AMDP class are provisioned as result of the generation. Thus, no customer coding is required. This is also true for the machine learning logic in terms of transformation or feature reduction as APL incorporates those data science activities into the library implementation itself.
- **Predictive Analytics Library (PAL)**
  This library is used for break-out scenarios that are usually more complex than APL-based ones. The machine learning logic and inference API are implemented as an AMDP class with a given interface. This implementation is already possible for on-premise implementations. For SAP S/4HANA Cloud, an editor must be provided to implement SQLScript code. The allowed SQLScript statements must be restricted to prevent side effects for the application server. For machine learning, read-only statements are sufficient.
- **SAP Data Intelligence**
  For advanced scenarios like image recognition or language processing in which neuronal networks, GPUs, and external data are required, SAP Data Intelligence is used. The machine learning logic and inference APIs are modeled graphically based on pipelines. Also, CDS extraction views can be incorporated into those pipelines for accessing training data. The SAP Data Intelligence design-time tools are already cloud-enabled and can be used by SAP S/4HANA Cloud and on-premise customers.

Readiness checks are performed in advance of training jobs to ensure that the necessary prerequisites are met, such as whether enough application data is available for model training. The readiness checks are developed as ABAP classes. For SAP S/4HANA Cloud, an ABAP editor is available for implementation of enhancement points. This editor can be reused and might need to be enhanced for creation of ABAP classes. In SAP S/4HANA on-premise, ABAP classes for readiness checks can be implemented already as of the time of writing (summer 2020). The inference API can be integrated immediately into the business processes if the existing applications provide the necessary enhancement points. For this, a particular BAdI implementation must be provided, which is supported for SAP S/4HANA Cloud and on-premise.

Completely new machine learning applications can be created in the cloud using SAP Web IDE. For on-premise SAP S/4HANA, existing tools can be used for this task.

Extensibility Lifecycle Management

Extensibility lifecycle management is in the scope of solutions and tools for general lifecycle management. There are no machine learning-specific aspects to be resolved. Therefore, only some general remarks are provided in this context. All key user extensibility items follow the principles of decoupled extensions:

- All extensibility capabilities offered to customers must continue to work after an SAP software update without manual work; this implies that SAP software updates do not depend on adoptions by the customer.
- Extensibility objects must never block an update of the core SAP software.

To support these core principles, the custom artifacts must comply with the following principles:

- Custom artifacts must be technically modification-free; they are created as own objects that refer to the base object.
- Custom artifacts must be technically clash-free. Also, extensions must not clash with parts of the core objects that are added later. Further, extensions of different extending parties must not clash. This requirement is fulfilled by applying a unique namespace that is considered in every layer of the architecture.
- Custom artifacts use released, stable SAP extension points and APIs only. This is enforced by object/code checks. On the SAP side, checks prevent incompatible changes to objects that are marked as released/extensible and have been delivered once. A deprecation mechanism is required as in the course of time incompatible changes might be necessary for the API, such as due to adoption of the underlying business processes or database tables. Because such incompatible changes break existing consuming applications, they must be informed upfront that the current API is going to be deprecated and replaced with a new one. Thus, the consumers have enough time to react to the incompatible changes.

For business-critical applications, extensions are typically created and tested before the extension is active for all users in the production environment. Here we distinguish between SAP S/4HANA Cloud and on-premise editions. In SAP S/4HANA Cloud, the transport of extensibility objects from the test to the production system is performed by the key user without interaction with the service provider and outside of the maintenance window of the service provider. In the on-premise environment, customers have much greater freedom for development and for setting up the system landscape and quality assurance processes. Therefore, customers can manage SAP updates and customer transports for key user extensibility with traditional transport tools. Transports of extensions created with key user tools can be merged with transports with custom development done with classical ABAP development tools.

5.5.6 Model Degradation

Over the course of time, the prediction power of machine learning models decreases due to changes in the data environment. Determining this time point and triggering retraining is the objective of model degradation. This section explains the business requirements and the necessary technical implementation for model degradation.

Business Requirements

When designing an intelligent ERP system based on a machine learning algorithm, the data model is typically dynamic. After being trained with an initial set of data, the model needs to be retrained on an ongoing basis as new data become available during usage, as shown in Figure 5.78. This new data also include feedback from end users.

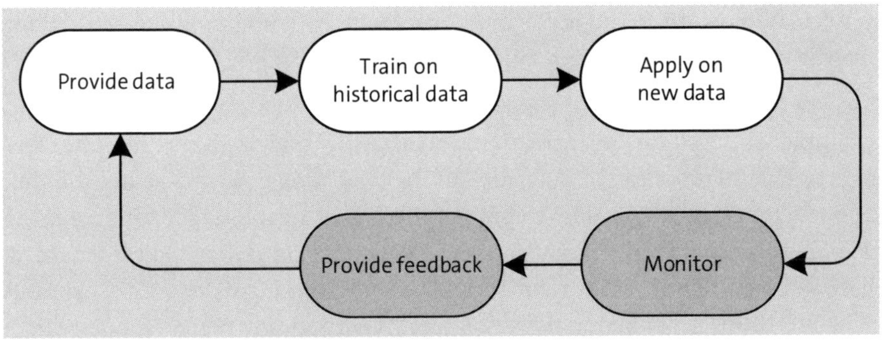

Figure 5.78 Machine Learning Model Lifecycle

The process of retraining is one of the differences from classical systems. It's a recurring process that brings specific requirements to the user experience and, depending on the structure of the feedback loop, requires the involvement of new roles (such as the data scientist) and specific UIs (for monitoring and data analysis). Particularly over time, the predictive performance of machine learning models decreases as the environment evolves and so retraining or even remodeling is required. This is known as model degradation. The key question is: How can you detect and resolve model degradation?

Technical Implementation

To face the challenge of model degradation, the accuracy of the machine learning model must be monitored regularly. Feedback loops must be incorporated, and the machine learning models have to be validated with datasets reflecting changes in the environment. If model performance in terms of prediction accuracy decreases, then retraining or even remodeling must be triggered. To determine model degradation, selected strategies are described in the next sections.

Accuracy KPIs

Calculating the accuracy of a machine learning model helps to determine model degradation. A machine learning model is a function that maps a given set of values for the *x*-columns to the correct corresponding values for the *y*-column. Finding such a function for a given dataset is called training the model. Good models not only avoid errors for *x*-values they already know, but, in particular, create predictions for situations that are only somewhat similar to the situations already stored in the existing data table.

For all machine learning methods, it's essential to know how well the machine learning model performs, whether it's a decision tree or a deep learning model. How can the accuracy of a machine learning model be calculated? The basic idea is to train a machine learning model on a given dataset and then use that underlying function on data points for which the value of *y* is already known. This results in two values of *y*: the actual one and the prediction from the machine learning model. It's now relatively easy to calculate how often the predictions are wrong by comparing the predictions with the true values. This is the foundation of various static methods to calculate accuracy KPIs.

As an example for classification, regression, or clustering models, the prediction power and prediction confidence can be determined. The *prediction power* measures the ability of the machine learning model to predict the values of the target variable using input variables present in the input dataset. The prediction power indicator takes a value between 0% and 100%. This value should be as close as possible to 100%, without being equal to 100%. A prediction power of 1 is a hypothetically perfect model, where the explanatory variables are capable of accounting for 100% of information in the target variable. In practice, however, this is usually an indication that an explanatory variable that is 100% correlated with the target variable was not excluded from the dataset analyzed. A prediction power of 0 is a purely random model with no prediction power. To improve the prediction power of a model, try adding new variables to the training dataset and combining explanatory variables. For example, a model with a prediction power of 79% is capable of accounting for 79% of the variation in the target variable using the explanatory variables in the dataset analyzed. No exact threshold exists to separate a good model from a bad predictive model in terms of prediction power as this depends on your business case.

The *prediction confidence* indicates the robustness of the model to achieve the same performance when it is applied to a new dataset that has the same characteristics as the training dataset. The prediction confidence indicator takes a value between 0% and 100%. A model with a prediction confidence that is equal to or greater than 95% is considered robust. It has a high capacity for generalization. A prediction confidence less than 95% should be considered with caution. There is a risk of generating unreliable results if the model is applied to a new dataset. To improve the prediction confidence, usually more observation rows are added to the training dataset.

The accuracy KPIs must be monitored continuously. If, for example, the prediction power and/or the prediction confidence of the model on the control dataset are significantly low, it means that the relation between the input variables and the target variable has changed. When this happens, it is recommended to retrain the machine learning model using new data.

Drift and Skew Detection

While the strategy of accuracy KPIs described previously aims to monitor the relation between the input variables and the predicted target variable, drift and skew detection focus on controlling the distribution of the input and prediction variables. On live data, the predictions of the machine learning model are tracked. If the distribution of these predictions starts to change, then it could be a sign of model degradation or at least be worth investigating. For instance, say the model is classifying images as urban versus nature, and for the first month it was predicting urban for 50% of the images and nature for the rest. Then in the next month, if the distribution suddenly became 90% urban, one might suspect degradation. Similarly, one can track the input distribution. For example, if you have a model for classifying texts and suddenly start getting text documents with new words or some words' frequency starts to spike, then it means the data distribution shifted and likely the model degraded.

Feedback Loops

The feedback of an end user is an important source for reinforcing the underlying data model. Negative feedback indicates a decrease of model performance and can result in retraining or even remodeling. Different types of feedback can be distinguished as shown in Figure 5.79.

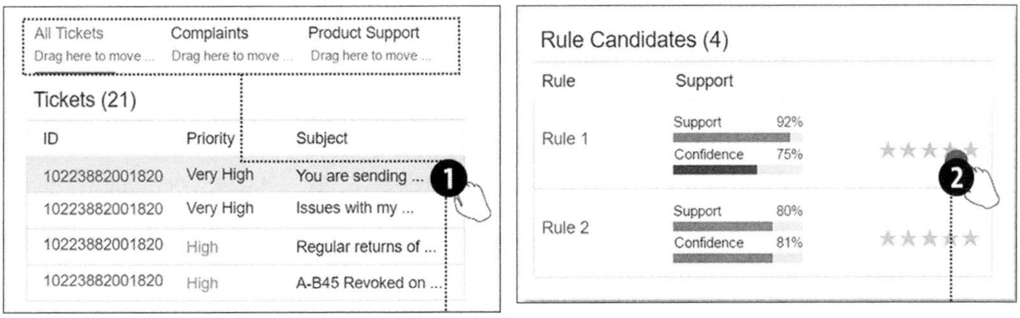

Figure 5.79 Types of Feedback

Implicit feedback ❶ is derived by observing the user's activities, including navigation paths and search queries, but also business process-related interactions like overruling the predictions, as illustrated in Figure 5.79. With implicit feedback, the user can change the classification output by dragging **Subjects** into different categories. Collecting implicit feedback results in dense data, which is usually aggregated for groups of users. Implicit feedback is more representative than explicit feedback because it's based on what users actually do. The feedback can be gathered in the background and does not disrupt the user experience. However, this also means that the process is not transparent to the user. One of the disadvantages of implicit feedback is its low correlation with the long-term user behavior.

Explicit feedback ❷ is derived from direct interaction with the user (e.g., through ranking, voting, flagging, polling, commenting, and so on). In this case, the data gathered can be more biased, especially if it isn't restricted to expert users. Explicit feedback is also subjective by nature because it relies on the user's opinion, which may differ from actual behavior. Explicit feedback is transparent to the user but may be very disruptive and impact the user experience, which is one of the major design challenges. On the other hand, it's a valuable input for optimizing long-term strategies. In some cases, explicit feedback can also reduce the effort involved in retraining models.

Delayed feedback (not pictured in Figure 5.79) is a way of gathering explicit feedback from users without breaking their flow. If the system needs feedback from the user (e.g., to understand why an approved amount exceeds the system recommendation), it doesn't interrupt the user with a feedback form. Instead, the user is just notified that feedback is required and can choose whether to react immediately or continue working and give feedback later. At any time, the user can call up a list of all outstanding feedback questions that still need answering and work on them whenever it's convenient. The main design challenge of the delayed feedback system is to reconstruct the context of the user's decision when the user provides the feedback to the system at a later point in time. To address typical feedback issues, feedback data may require additional processing before it can be used for model degradation. For example, implicit feedback might be subject to event-related bias, or explicit feedback might only reflect the opinion of a specific group of users. In these cases, the feedback may first need to be assessed by a user in a different user role (such as a data scientist) with a separate UI.

Solution Architecture

The model degradation strategies of accuracy KPIs, drift and skew detection, and feedback loops can be technically implemented with the solution described in Figure 5.80.

> **Note**
>
> As of the time of writing (summer 2020), there is no generic framework available. However, the proposed approach can be implemented by developers of the machine learning application.

A machine learning application is represented by the machine learning scenario artifact. With the concept of machine learning scenarios, consistent lifecycle management is ensured and monitoring on the business process level is enabled. On top of the application tables, CDS views are defined, which represent the semantic data model of SAP S/4HANA. The CDS views hide the encrypted database models and wrap them in human-understandable entities. Besides the SQL view definition, the CDS views contain domain-specific metadata called annotations. For machine learning applications, a specific CDS view for model training is provided. Machine learning algorithms read the application data during the training process. The trained machine learning model is stored in the database system.

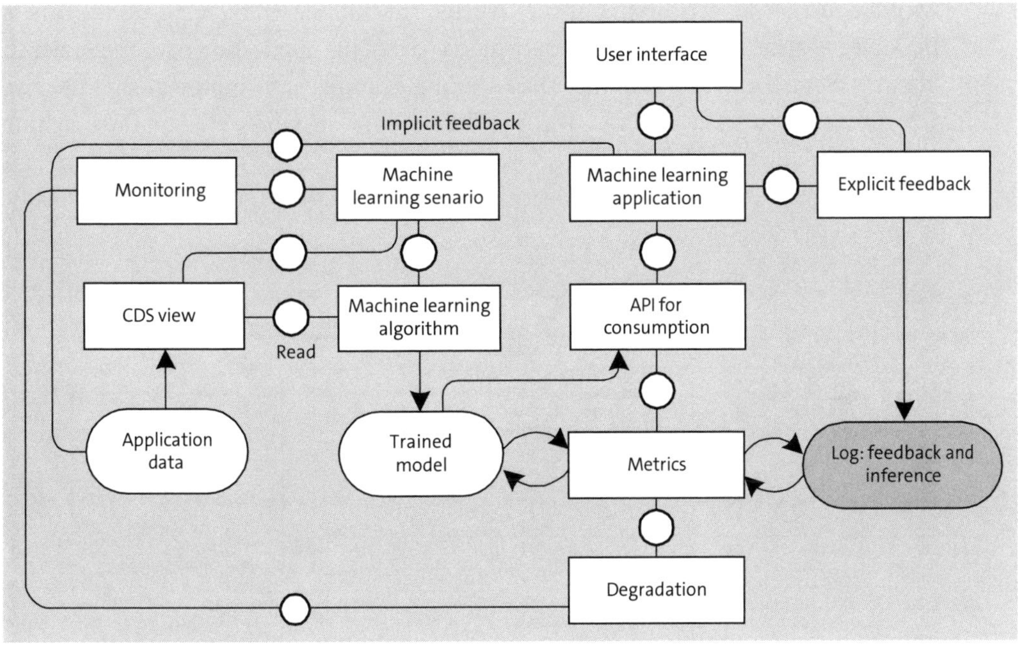

Figure 5.80 Model Degradation Solution

For consumption, the trained machine learning model is wrapped with an API. Thus, machine learning capabilities can be integrated into business processes and their respective UIs. The *metrics* component calculates accuracy KPIs during the training process by dividing the application data into training and validation datasets. The validation datasets can be used to measure the performance of the trained model. The *degradation* component interprets the accuracy KPIs and provides recommendations to the *monitoring* module, such as to retrain with more data or adjust the model definition. Drift and skew detection can be also provided by metrics component as the stream of inference calls in terms of input and prediction can be continuously analyzed. Depending on the drift and skew detection method, the inference data could be temporarily stored or processed in real-time without the need to persist them.

Implicit feedback is usually stored as part of the application data and is reflected accordingly by the training process per the metrics component, as shown in Figure 5.81. Handling of explicit feedback is typically more complex and is not automatically part of the application table design. Therefore, it's suggested to have a generic persistency for the explicit feedback in order to reduce the TCD for application development. With the explicit feedback and the related inference data in terms of input and predictions, the degradation component easily can produce evidence for the machine learning model's performance. As the point of giving explicit feedback is later than the execution of the related inference call, correlation between those is usually challenging.

Figure 5.81 shows how model degradation is determined based on implicit feedback without further investigations of the affected machine learning application. This is possible because implicit feedback is stored as part of the application data and exposed via CDS views for model training. The ticketing example shown previously in Figure 5.79 illustrates this: the user can overrule the ticket priority proposed by the machine learning model. Such a change of priority is saved and can be considered as implicit feedback. Overruling the machine learning recommendation too often is an indicator of decreased model accuracy.

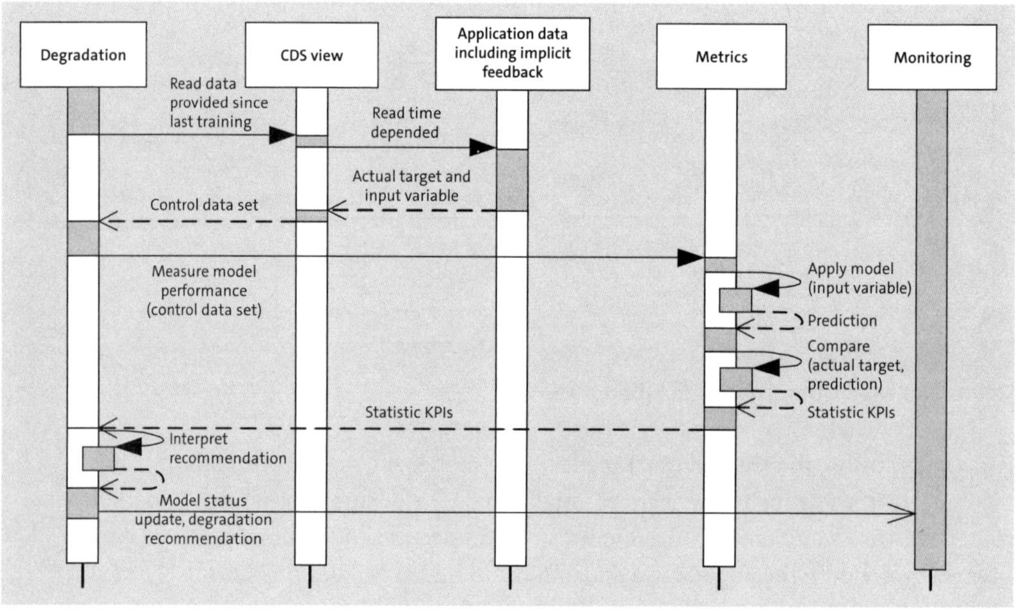

Figure 5.81 Model Degradation Process

The ticket priority represents the target variable and is therefore part of the CDS view for model training (see Listing 5.12). Those CDS views are usually time-dependent in terms of having fields for selecting data records for a specific period of time. To determine whether the machine learning model performs well, the degradation component uses the CDS view for model training to read new data available since the last training run. Because implicit feedback is part of the application data, this control dataset contains the actual values for the target variables and the value of the input parameters. The current trained model version is used to calculate the predictions based on the input values. This is the task of the metrics component, which also compares the actual and predicted values for calculating static KPIs. The degradation component interprets those KPIs and recommends related actions (e.g., retraining or remodeling), which are displayed in the monitoring view.

Listing 5.12 offers a time-dependent training CDS view.

```
@AbapCatalog.sqlViewName: 'TicketTraining'
@ClientHandling.algorithm: #SESSION_VARIABLE
@VDM.viewType: #COMPOSITE
@AbapCatalog.preserveKey: true
@AccessControl.authorizationCheck: #NOT_REQUIRED
@EndUserText.label: 'C_TicketPriorityTraining'
define view C_TicketPriorityTraining as select from ticket
association [1] to actionlog as _actions
  on ticket.incidentid = _actions.incidentid {
  key incidentid,
  description,
  status,
  component,
  priority,
  reporter,
  processor,
  pt as processingteam,
  sid as systemid,
  lastupdate,
  created,
  link as attachment,
   _actions
}
```

Listing 5.12 Example of Time-Dependent CDS View

5.5.7 Explanation of Results

To build trust between human and machine, it's important to explain the results provided by machine learning models. Transparency and traceability of machine learning models are also needed for statutory reasons. Therefore, this section explains the business requirements and the necessary technical implementation for explanation of results computed by machine learning.

Business Requirements

By providing sufficient information about the underlying model and explaining the reasoning behind the results of an algorithm, trust between human and machine is built. Assessing trust is fundamental for human beings to take action based on a prediction. This aspect is even more critical for business applications, in which users are held accountable for all decisions they made due to legal requirements. *Explainable machine learning* means the reasoning behind intelligent system proposals can be explained in context and at the right time. Thus trust can be built, and legal requirements of automated decision-making can be anticipated. Machine learning explanations will be considered if one or more of the following conditions apply:

- **Criticality**
  There is a high risk associated with making the wrong decision. Once applied, actions are hard to reverse. By contrast, if the risk is low and actions can easily be rolled back, users might not need an explanation for the system proposal.
- **Complexity**
  It's hard for users to assess the impact and quality of their decisions right away. By contrast, if users can easily tell when a proposal fits (without training), they probably don't need additional input.
- **Transparency**
  The business process is subject to extensive auditing requirements. Auditors must be able to trace transactions back and see the underlying reasoning for each execution step. By contrast, if there are no auditing requirements, explanations might not be necessary (assuming they are also not required by end users).
- **Volatility**
  To adapt to changing conditions or requirements, the machine learning application relies on continuous feedback. By contrast, if the feedback has little or no effect on the output of the algorithm or the user experience, it's more distracting than helpful to provide an additional explanation.

Technical Implementation

In the context of SAP S/4HANA applications, two main questions must be solved regarding explainable machine learning:

- How will UIs be designed to include explainable machine learning?
- What must be implanted in the backend to provide explainable machine learning?

Note

The methods of calculating explanations for specific machine learning algorithms are out of scope for this book. They are well known in the data science community and are assumed to be reused here.

User Interface

The amount of information users need to understand a system proposal can vary. This depends on the applied machine learning technique, usage context, user task, and user role. The following three explanation levels can be differentiated:

1. **Indicator (What?)**
   This is the minimum explanation level. An indicator is required whenever machine learning output is provided. The indicator is also the access point for the next explanation level.
2. **Abstract (Why?)**
   This is a condensed view of the relevant properties, amounts, and contextual information. An abstract helps users to better understand the machine learning proposals. It can contain links to the last and most detailed explanation level.
3. **Detail (How?)**
   This is an extended report specifically for advanced users. It covers all aspects processed by the intelligent system, the machine learning performance, and any further context and conditions that help users to monitor machine learning operations.

Explanations can be provided with different UI techniques, from just textual information to natural language engagement through SAP Conversational AI. The example in Figure 5.82 illustrates how an application can explain the machine learning-based ranking of the suppliers for a specific material.

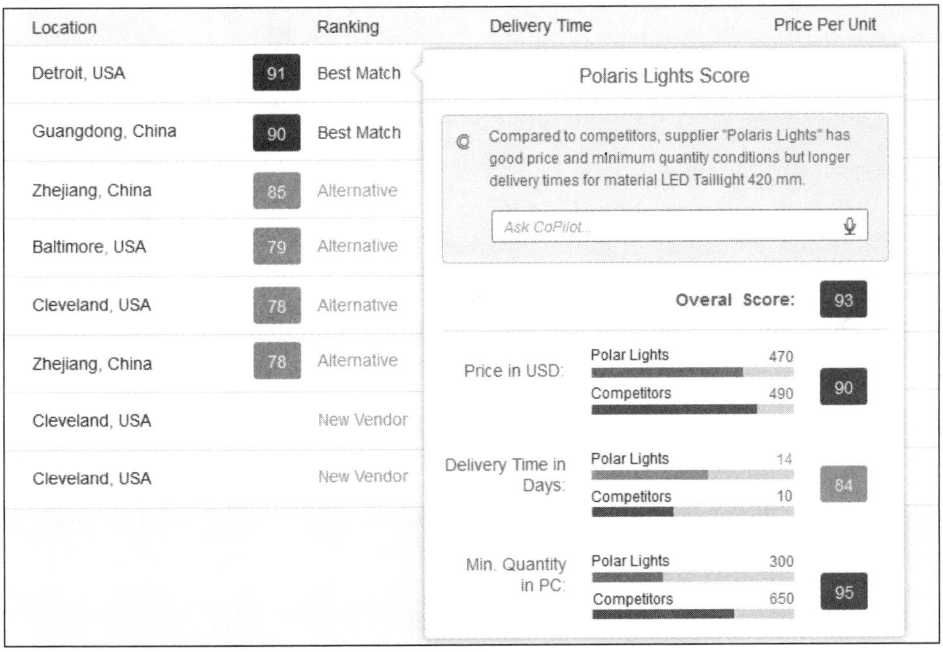

Figure 5.82 Example for Machine Learning Explanation

At the line item level, the user can see how the system arrived at the score for a particular object. Explanations can include the following elements:

- A short explanation in natural language as a trigger for navigation to explanation details
- An option to interact with the digital assistant (SAP CoPilot) in the context of this explanation
- *Optional:* A list of the most important parameters that affected the results being explained
- *Optional:* A comparison with average results for other objects

The progressive disclosure technique is used to avoid overwhelming the user with too much information at once. The most frequently used information is shown in a concise form on the main screen, as shown in Figure 5.83, with an option to drill down to more detailed information on secondary screens. The benefit of this approach is that the users enjoy a more simplified and compact UI and only need to concern themselves with the details if they actually need them.

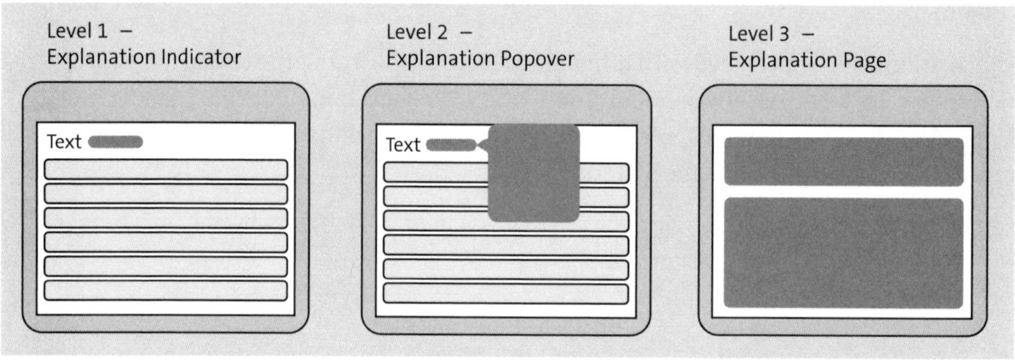

Figure 5.83 Machine Learning Explanation UI Illustration

Let's break down the UI levels shown in Figure 5.83:

- Level 1 shows a short explanation without the option to get more detailed information. This is used in simple cases in which a short explanation clarifies the issue. There is no deeper knowledge required from the user. He or she is guided by the quality indicator as to how he or she needs to handle the provided information, what is important, and on which level.
- Level 2 presents more detailed information including tables, charts, and other UI elements that offer more insight into the issue. On this level, the current situation is brought into perspective and into context. The user can view the criteria on which the prediction is based on and the impact they have had on this prediction. To read that information properly, the user requires good domain and business knowledge, but no additional skill beyond that.

- Level 3 is not really part of the explanation but gives the user the option to engage SAP CoPilot to get more information (via the **Ask CoPilot** button or input field). This level requires strong knowledge and a deep understanding of the technology to manage the given information. The user must be trained to be able to make profound decisions and take action.

Figure 5.84 illustrates, based on an example of a linear regression model, how an explanation for machine learning results can be calculated.

$$\text{Claim complexity} = 0.47 + \underbrace{10^{-6}\ \text{Capital}}_{+0.1} + \underbrace{0.03\ \text{Loan Seniority}}_{+0.21} - \underbrace{0.01\ \text{Interest Rate}}_{-0.03} - \ldots$$

Capital = 100,000€
Loan Seniority = 7 years
Interest Rate = 3%

Figure 5.84 Example of Machine Learning Explanation for Regression Algorithm

The prediction is the linear combination of the feature's values, weighted by the model coefficients. Actually, the weights represent the contribution to the prediction and provide the explanation. However, as shown in Figure 5.85, the highest accuracy is often achieved by complex machine learning models, which even experts struggle to explain.

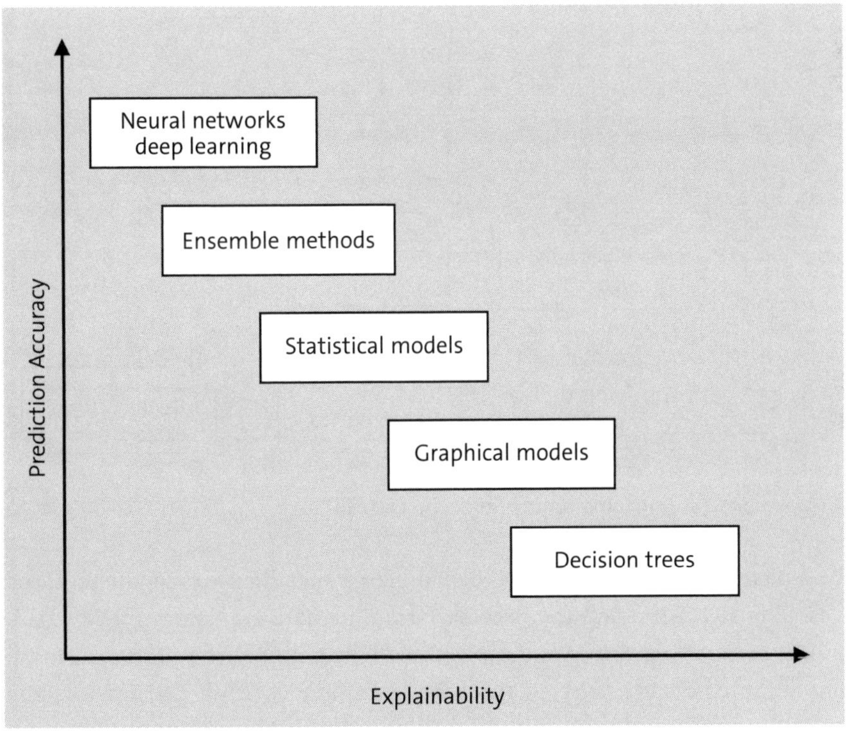

Figure 5.85 Degree of Explanation of Machine Learning Algorithms

Backend Processes

Explainability is achieved by restricting the complexity of the machine learning model (intrinsic) or by applying methods that analyze the model after training (post hoc). *Intrinsic interpretability* refers to machine learning models that are considered explainable due to their simple structure, such as short decision trees or sparse linear models. As already mentioned, for machine learning algorithms, corresponding explanation methods exist that are well known in the data scientist community and are not in focus here. In the context of business applications, you must choose methods that explain the individual prediction and the entire model behavior.

Figure 5.86 outlines the overall machine learning solution. During the machine learning training process, basic explanation information is calculated and stored in addition to the trained model. Those are statistical values for the overall accuracy of the machine learning model, which are referred to as the *global explanation*. For incoming inference calls, individual explanation numbers are computed, which are referred as the *local explanation*. The *explanation* component provides the global/local explanations and enriches the inference results with those values exposed to the consuming business logic. For the illustration of explanation data, UI controls are required as specified previously.

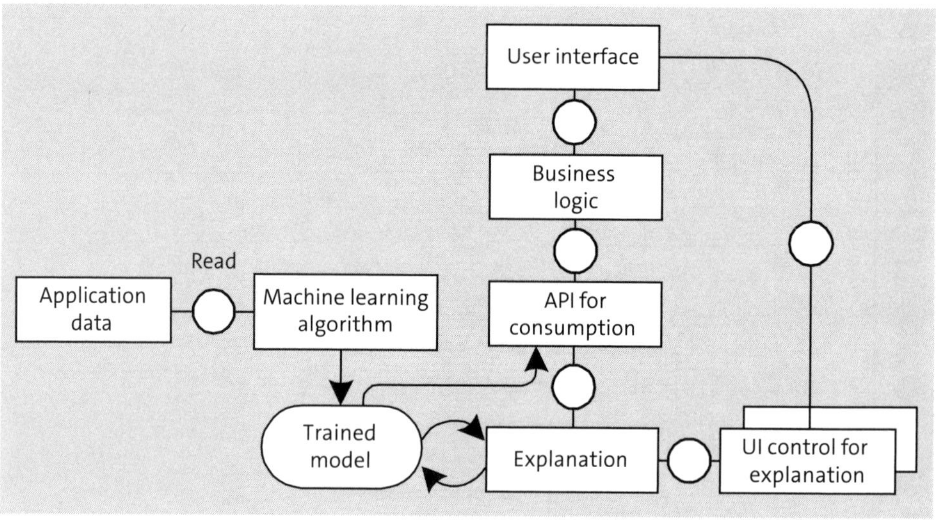

Figure 5.86 Machine Learning Explanation Solution: Overall

Figure 5.87 depicts the solution one level deeper. For a specific use case, the required machine learning algorithm and the necessary training data are determined by data scientists after completing the exploration phase. The training job processes those application data to train the selected machine learning algorithm and to store the trained model for consumption. During the training job, global explanations also are computed by providing accuracy measures based on a specific explanation method. Calculating the accuracy of a machine learning model helps explain the model.

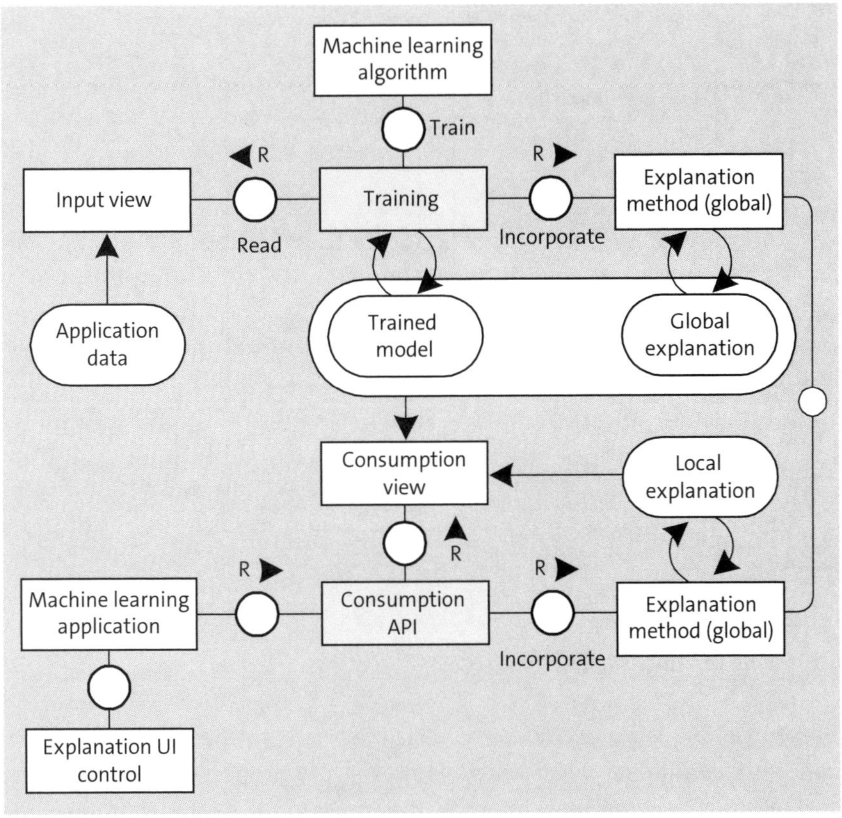

Figure 5.87 Machine Learning Explanation Solution: Detailed

How can the accuracy of a machine learning model be calculated? To review, the basic idea is to train a machine learning model on a given dataset and then use that underlying function on data points for which the value for *y* is already known. This results in two values of *y*: the actual one and the prediction from the machine learning model. It's now relatively easy to calculate how often the predictions are wrong by comparing the predictions with the true values. This is the foundation of various static methods to calculate accuracy KPIs for global explanations. The trained model is exposed to the machine learning applications by consumption APIs. The consumption request is used to calculate a local explanation based on a selected explanation method. For legal auditing reasons, the local explanation must be stored on the database as is the case for the global explanation. The consumption view is the foundation for the consumption API. It determines the inference results and enriches this with the local and global explanations. Those are provided to the machine learning application as part of the response structure and can be displayed in the UI as suggested in the previous section. This interplay is illustrated in Figure 5.88.

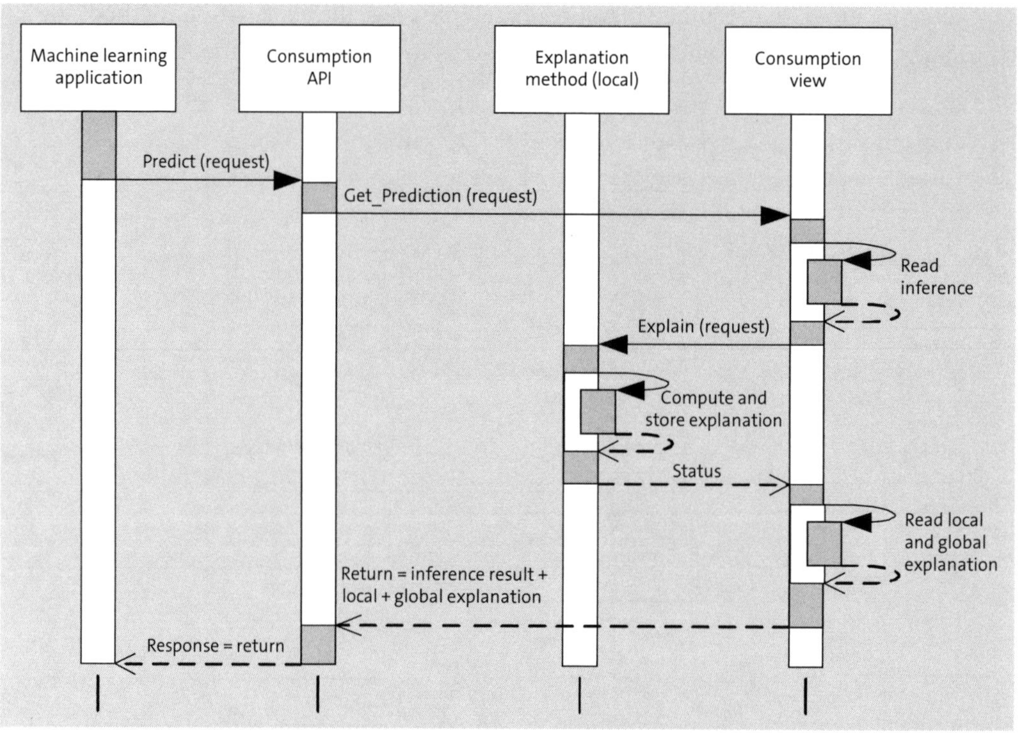

Figure 5.88 Machine Learning Explanation Process

Explanation of machine learning models is usually specific to the use case. Thus, a generic framework can be hardly provided. However, the proposed approach helps developers of machine learning applications to overcome this. The solution implementation is explained with an example regarding predicting the delivery time for products. For this, regression is used as the underlying machine learning model. As depicted in Figure 5.89, for the global explanation, the prediction power and prediction confidence were computed; for the local explanation, the variable contributions of the inference calls were calculated.

For the local explanation, the method for variable contributions was applied, which shows the relative importance of each variable used in the model or category influence. The *category influence* is an analysis of the influence of different categories of a variable on the target, computed from basic information. The higher the absolute value of the influence, the stronger the influence of the category: categories with values equal to 0 or close to 0 are categories with no influence on the target. The influence of a category can be positive or negative:

- Categories with positive values are categories in which observations are more likely to be in the positive category of the target. The percentage of positive targets within such a category is above the percentage of positive targets in the whole dataset.

- Categories with negative values are categories in which the observations are more likely to be in the negative category of the target. The percentage of positive targets within such a category is below the percentage of negative targets in the whole dataset.

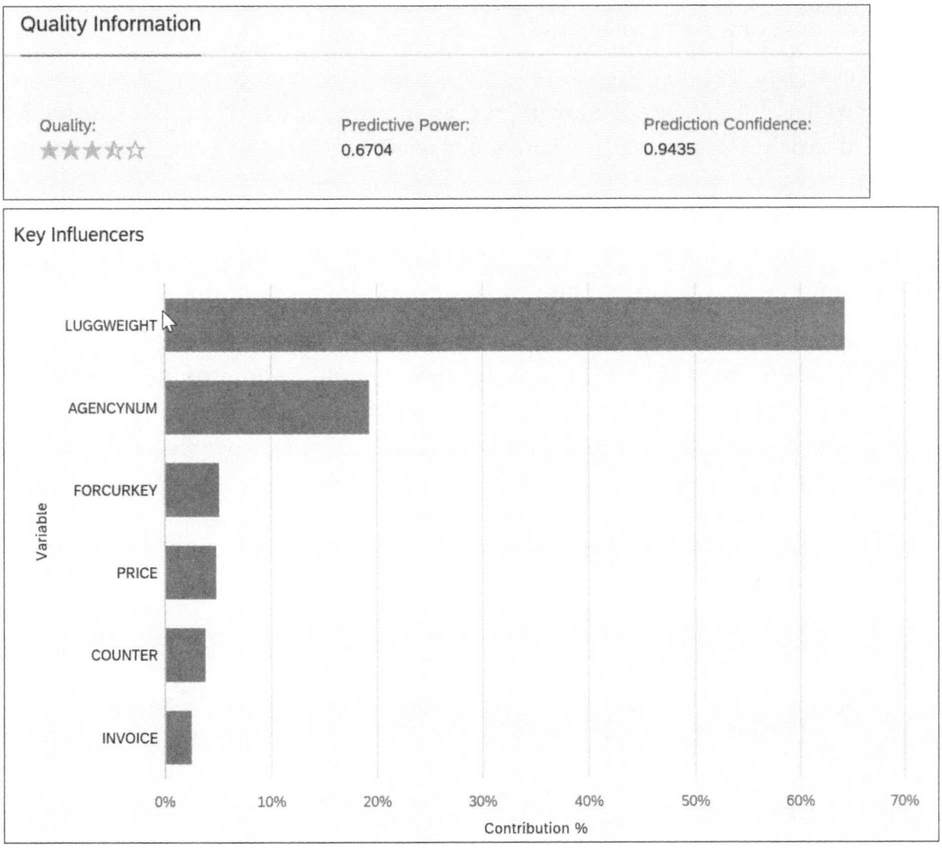

Figure 5.89 Example Implementation for Local and Global Explanations

5.5.8 Workload Management and Performance

Depending on the underlying algorithms, training of machine learning models can have high requirements for hardware. Inference calls must have very short response times. Thus, measures must be put in place to resolve those challenges. This section explains the business requirements and the necessary technical implementation for workload management and performance in the context of machine learning.

Business Requirements

The process of training a machine learning model involves providing a machine learning algorithm with training data to learn from. The training data must contain the correct answer, which is known as a *target* or *target attribute*. The machine learning algorithm finds patterns in the training data that map the input data attributes to the target, and it outputs a machine learning model that captures these patterns. Machine learning training is an asynchronous process usually performed yearly, monthly, or weekly. Therefore, performance is not critical, but the allocation of memory, CPU time, and disk space is. *Performance* refers to the total effectiveness of a system, including throughput, individual response time, and availability.

Inference is the process of using the machine learning model to get predictions on new data for which the target is not known. For example, say that a machine learning model is trained to predict if an email is spam or not spam. In this case, SAP Data Intelligence would be provided with training data that contains emails for which the target is known (i.e., a label that tells whether an email is spam or not spam). SAP Data Intelligence would train a machine learning model by using this data, resulting in a model that attempts to predict whether new email will be spam or not spam.

Thus, training and inference of machine learning can have negative effects on the transactional processes of SAP S/4HANA because typically inferences focus on performance, while memory, CPU time, and disk space are neglected. Those implications must be avoided. However, as illustrated in Figure 5.90, they depend on the machine learning infrastructure used, the required machine learning algorithms, and the underlying programming model.

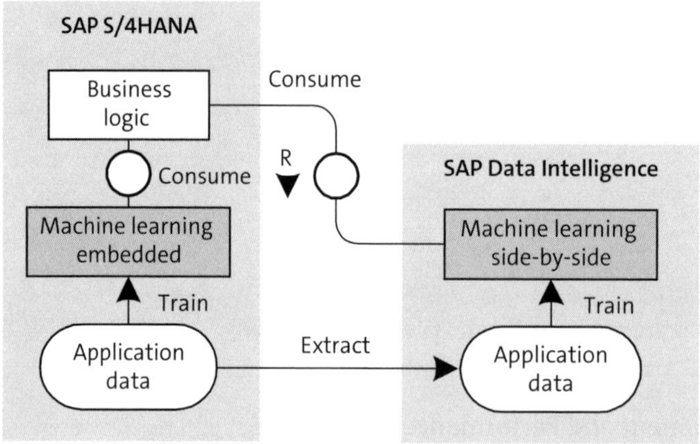

Figure 5.90 Workload Required for Training and Performance for Inference

As already mentioned, use cases that require algorithms with low memory and CPU time are implemented in embedded in SAP S/4HANA, while resource-intensive algorithms like deep learning are scaled out to SAP Data Intelligence. However, embedded machine learning still can slow down SAP S/4HANA processes, so the following requirements must be considered:

- End-to-end response time for consumption of a machine learning model must be below one second because machine learning capabilities are integrated into transactional (mass) processes and UIs.

- Machine learning training must not impact the performance of transactional processes by inadequate allocation of memory, CPU time, and disk space.

For side-by-side machine learning, the model training is processed on SAP Data Intelligence, which is a scalable and distributed platform—so no negative impact is expected. Nevertheless, for the consumption of the model, the related KPI must be considered because (mass) transactional processes are involved on the SAP S/4HANA side.

Technical Implementation

As described in the last section, for machine learning training, the consumption of memory, CPU time, and disk space must be controlled and optimized. To solve this requirement, workload and quota management is necessary.

Workload and quota management are actions that cover optimization of infrastructure resources (memory, CPU time, disk space) to maintain or improve performance or throughput. These actions include rescheduling, moving, technical virtualization, or limiting a particular service or workload. For example, an administrator is enabled to set limits for jobs concerning CPU time and memory consumption. For successful workload and quota management, ongoing diligent monitoring is necessary. In the context of machine learning, quota management typically isn't relevant for the following reasons:

- During embedded machine learning, training the application data is read in real time from the SAP S/4HANA database.
- For side-by-side machine learning, SAP S/4HANA's training application data are replicated to SAP Data Intelligence but usually destroyed after the training run.

However, read access of training data should be optimized by packing and pipelining mechanisms. This means that the training data are not read all at once, which this could result in out-of-memory issues, but are processed in packages by the training process. As outlined in the last section, for machine learning inference, optimum performance in terms of response time is the main requirement. Besides the well-known techniques of improving API performance, caching and bulking inference results helps to fulfill this requirement.

Let's see how workload and performance management implementation techniques differ depending on your machine learning approach.

Embedded Machine Learning

For embedded machine learning, inference calls are processed locally in real time. Thus, from a performance point of view, this is already the optimal solution and no further investigations are necessary. As training is an occasional background task, no extraordinary performance KPIs must be considered. In particular, measurements of training for embedded machine learning algorithms showed that response time is below five minutes and exceeds by far the performance KPIs.

Let's walk through some exemplary measurement results in seconds. Table 5.2 shows performance measurements for logistic regression. For this, tables with different numbers of columns and rows were considered as input for training. The training runtimes in seconds are provided in the table, such as a dataset with 700 columns and 300,000 rows that resulted in a 25.740 second training time.

	50 Columns	100 Columns	200 Columns	500 Columns	700 Columns
10,000 rows	0.159	0.259	0.308	0.968	1.278
50,000 rows	0.367	0.840	1.673	4.085	5.475
150,000 rows	0.925	1.669	3.953	9.819	13.494
300,000 rows	1.155	3.380	7.317	17.868	25.740

Table 5.2 Logistic Regression

Alternatively, Table 5.3 shows performance measurement for K-means algorithms. Different datasets are considered here also.

	50 Columns	100 Columns	200 Columns	500 Columns	700 Columns
10,000 rows	0.2476	0.327	0.4814	0.9158	1.2896
50,000 rows	0.407	0.6458	1.9714	3.1444	3.6752
150,000 rows	0.7828	1.9548	3.4436	9.8342	9.201
300,000 rows	1.453	4.083	7.0614	19.748	28.9356

Table 5.3 K-Means

The situation is different for training runs concerned with consumption of system resources, which could negatively impact SAP S/4HANA's business processes. For this, workload management measures must be applied as illustrated in Figure 5.91.

Scheduling allows for running training automatically at specified times. To schedule training, recurrence patterns have to be provided, such as weekly on Mondays or monthly on the first of the month. Thus, additional parameters can be specified in order to control exactly when and how often training will be run. When training is scheduled, the system will create a scheduled instance that solely contains object and schedule information. However, it doesn't contain any data. When the system runs the training, it creates machine learning models as an output. Event-based scheduling provides additional capability in terms of triggering training runs; for example, if model accuracy decreases below 60%, then retraining occurs. The first line of defense restricting resource consumption is to process scheduled training runs in background jobs and limit the number of parallel background jobs; for example, the number of jobs must be less than five. Scheduling must be integrated into the Intelligent Scenario Management embedded training app.

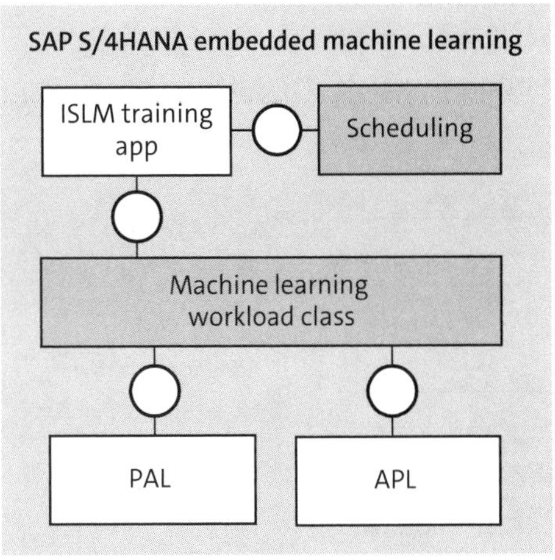

Figure 5.91 Workload Management Measures for Embedded Machine Learning

The second line of defense restricting resource consumption is to limit the memory and CPU time usage by the machine learning algorithms. This is realized by workload class and mapping technology provided by the SAP HANA database. The goal of workload class and mapping is to avoid overbooking system resources such as CPU time and memory by invoking predefined mapping rules, following these general steps, as shown in Figure 5.92:

1. The administrator defines a workload class, which specifies the amount of system resources that can be exploited by group of applications, and a workload mapping, which specifies how to match an application workload with a workload class.
2. When a request from an application arrives at the session layer of SAP HANA, the corresponding workload class is determined based on the information about the session context, such as the application name, application user name, and database user name.
3. Once the corresponding workload class is determined, the application request can have its resources regulated according to the workload class definition.

As of the time of writing (summer 2020), all workload management is performed at the statement level, but this has yet to be extended to the application level.

The workload class supports three kinds of resource properties to regulate:

- **Statement thread limit**
  The statement thread limit represents the maximum number of parallel threads (more specifically, Basis job executors) to execute a statement.

- **Statement memory limit**
  The statement memory limit represents the maximum amount of memory allocated per statement.
- **Statement priority**
  The statement priority represents the execution priority for a statement to be executed in the Basis job execution framework.

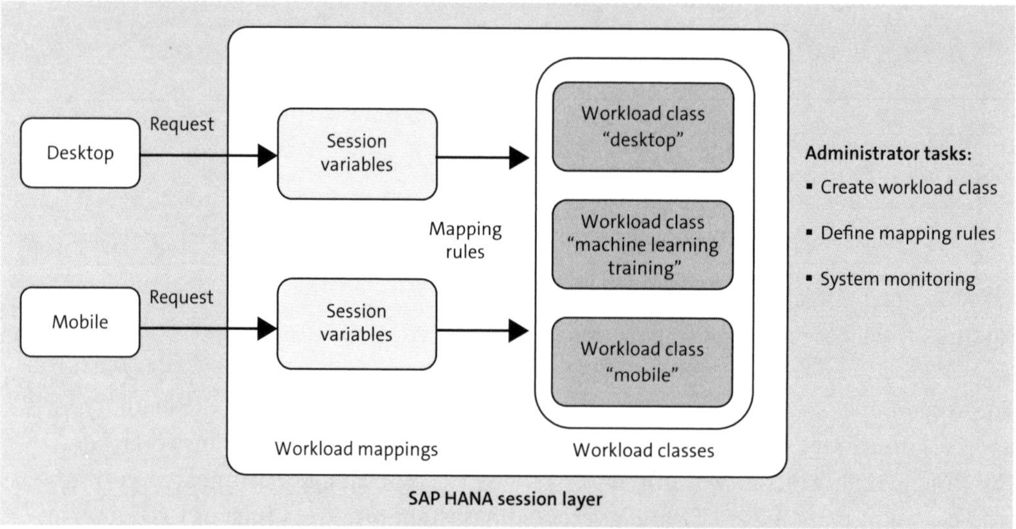

Figure 5.92 Workload Classes for Embedded Machine Learning

All property regulations are considered by the Basis job framework. Listing 5.13 illustrates the classic way of invoking the random decision tree embedded machine learning algorithm, whereas Listing 5.14 also shows the application of workload classes.

```
-- Setup a temporary table with parameters
create local temporary column table #pal_parameter_tbl (
       "PARAM_NAME" varchar (100),
       "INT_VALUE" integer,
       "DOUBLE_VALUE" double,
       "STRING_VALUE" varchar (100)
);
insert into #pal_parameter_tbl values ('TREES_NUM', 10, null, null);
insert into #pal_parameter_tbl values ('SPLIT_THRESHOLD', null, 1e-3, null);
insert into #pal_parameter_tbl values ('SEED', 2, null, null);
insert into #pal_parameter_tbl values ('CALCULATE_OOB', 0, null, null);
insert into #pal_parameter_tbl values ('NODE_SIZE', 50, null, null);
insert into #pal_parameter_tbl values ('THREAD_RATIO', null, 0.9, null);
-- Call the random decision trees algorithm
```

```
call _sys_afl.pal_random_decision_trees
("SAPR3"."airline2008_label_no_null",
#pal_parameter_tbl, ?, ?, ?, ?)
-- Result: The execution takes around 20 seconds.
```

Listing 5.13 Call Random Decision Tree Algorithm without Workload Class

Listing 5.14 shows how thread, memory, and time limitations are applied using the `thread`, `memory`, and `timeout` statements.

```
-- The statement memory limit requires to enable following configuration
parameters as SYSTEM:
alter system alter configuration('global.ini', 'SYSTEM') SET ('resource_
tracking', 'enable_tracking') ='true' with reconfigure;
alter system alter configuration('global.ini', 'SYSTEM') SET ('resource_
tracking', 'memory_tracking') ='true' with reconfigure;
-- Create a simple workload class on the statement timeout
create workload class "TIMEOUT_10" set 'STATEMENT TIMEOUT' = '10';
create workload class "THREAD_LIMIT_2" set 'STATEMENT THREAD LIMIT'='2';
create workload class "MEMORY_LIMIT_1GB" set 'STATEMENT MEMORY LIMIT' = '1';
select * from workload_classes;
and the mapping on the user name:
create workload mapping "TIMEOUT_MAPPING" workload class "TIMEOUT_10" set
'USER NAME' = 'SAPR3';
-- Execute the random decision tree algorithm to check that the timeout is
triggered:
call _sys_afl.pal_random_decision_trees_new ("SAPR3"."airline2008_label_no_
null", #pal_parameter_tbl, ?, ?, ?, ?);
-- Result:
Could not execute 'call_sys_afl.pal_random_decision_trees_new ("SAPR3".
"airline2008_label_no_null", #pal_parameter_tbl, ?, ...' in 10.232 seconds .
[613]: execution aborted by timeout

-- Execute the random decision tree algorithm with thread limit:
call _sys_afl.pal_random_decision_trees_new
("SAPR3"."airline2008_label_no_null", #PAL_PARAMETER_TBL, ?, ?, ?, ?) with hint
(workload_class(THREAD_LIMIT_2));
-- Execute the random decision tree algorithm with memory limit:
call _sys_afl.pal_random_decision_trees_new
("SAPR3"."airline2008_label_no_null", #PAL_PARAMETER_TBL, ?, ?, ?, ?) with hint
(workload_class(MEMORY_LIMIT_1GB));
```

Listing 5.14 Create Simple Workload Classes and Mapping

Side-by-Side Machine Learning

The workload management situation looks different for side-by-side machine learning based on SAP Data Intelligence.

Scalability is the degree to which a business scenario, component, or system can be expanded (or reduced) in size, volume, or number of users served and still continue to function properly and predictably. In other words, scalability refers to the predictable resource consumption of a software application under different system loads (increasing multiuser or parallel load) while keeping response time within a reasonable range.

SAP Data Intelligence provides a scalable inference and training infrastructure based on Kubernetes technology. Kubernetes automates the deployment, scaling, maintenance, scheduling, and operation of multiple application containers across clusters of nodes. Containers run on top of a common shared operating system on host machines but are isolated from each other unless a user chooses to connect them. Kubernetes can be used with container runtimes and with the container runtime interface. It contains tools for orchestration, secrets management, service discovery, scaling, and load balancing. Kubernetes technology includes automatic bin packing to place containers with the optimal resources for the job, and it applies configurations via configuration management features. It protects container workloads by rolling out or rolling back changes and offers availability and quality checks for containers.

With Kubernetes, containers run in pods, which are the basic scheduling unit for Kubernetes and which add a layer of abstraction to containers. Pods are comprised of one or multiple containers located on a host machine, and they can share resources. Kubernetes finds a machine that has enough free compute capacity for a given pod and launches the associated containers. To avoid conflicts, each pod is assigned a unique IP address, enabling applications to use ports. A node agent, called a kubelet, manages the pods, their containers, and their images. A node, also known as a minion, is a worker machine in Kubernetes. It can be either a physical machine or a virtual machine. Nodes have the necessary services to run pods and receive management instructions from master components. Services found on nodes include Docker, kube-proxy, and kubelet.

Tenant namespaces and content are deployed and deleted as Helm releases. A release is an instance of a chart running in a Kubernetes cluster. One chart can often be installed many times in the same cluster. The Helm tool installs charts into Kubernetes, creating a new release for each installation. The tenant ID and related configuration are also easily injected by Helm. With appropriate templates, different logic is handled, for example feature flags or a cloud provider. Upgrade of tenants with new releases also is supported. Templating is also provided for creating Kubernetes job specs. Furthermore, mounting of tenant data and access to GPUs are enabled.

Performance of remote machine learning consumption must be high as (mass) transactional processes are involved on the SAP S/4HANA side. Therefore, measures must be taken depending on the concrete requirements of specific machine learning use case, as illustrated in Figure 5.93. Inference results can be easily cached on the server side to improve response time. However, this might be applicable for a small number of scenarios. Thus, the bulking of inference calls is helpful in terms of combining multiple requests into one inference call.

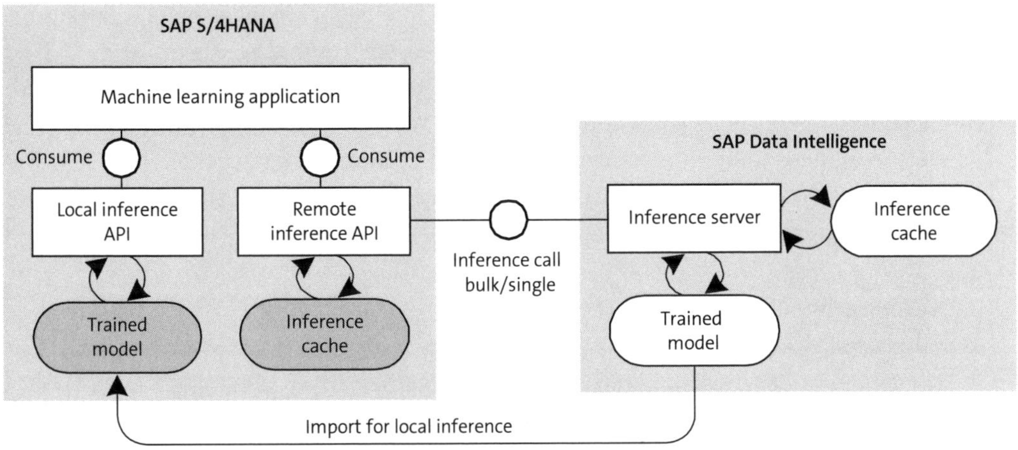

Figure 5.93 Performance Improvement for Inference Calls

Furthermore, on the SAP S/4HANA side, inference results could be cached by having batch jobs, making the results available in advance for local consumption. Specific types of side-by-side trained models could be also imported into SAP S/4HANA and consumed locally. This would improve performance of inference calls dramatically because local APIs are usually 10 to 100 times faster than remote APIs.

5.5.9 Legal Auditing

Machine learning training and inference processes must be traceable. For this, proper logging is necessary. These logs are also the basis of auditing of machine learning, which is a legal obligation. This section explains the business requirements and the necessary technical implementation for logging and legal auditing in the context of machine learning.

Business Requirements

Logging is the process of keeping track of what's happening in the system with the purpose of transparency and reproducibility. System, security, and application logs must be considered for enterprise software. System logs are about technical events like errors, warnings, or user locks to enable problem detection and resolution. Security events like failed logins, unauthorized access, or anomalies are logged to prevent attacks or reconstruct intrusions. Application logs are provided to track and rebuild the steps of business processes, such as sales order 4711 created by Mr. Smith on July 3 in the approved status.

Auditing is a systematic and independent examination of enterprise systems to validate compliance with applicable rules, regulations, and laws. For this purpose, logging information is used, among other info.

Machine learning is a new paradigm of infusing intelligence into business applications. In this context, additional artifacts are developed, such as machine learning models or accuracy measures. For auditing and logging, various aspects must be illuminated: What must be logged? For whom is it logged? Where will it be logged? How will it be audited? These topics are resolved in the next section. However, the focus is on the machine learning specifics; existing solution approaches for auditing and logging are taken for granted.

Technical Implementation

System and security logs are usually covered by the underlying infrastructure generically and therefore are not considered further here. To explain the machine learning-specific logging and auditing aspects, first the entities for logging must be identified. Figure 5.94 classifies the entities for logging and auditing into data and action categories.

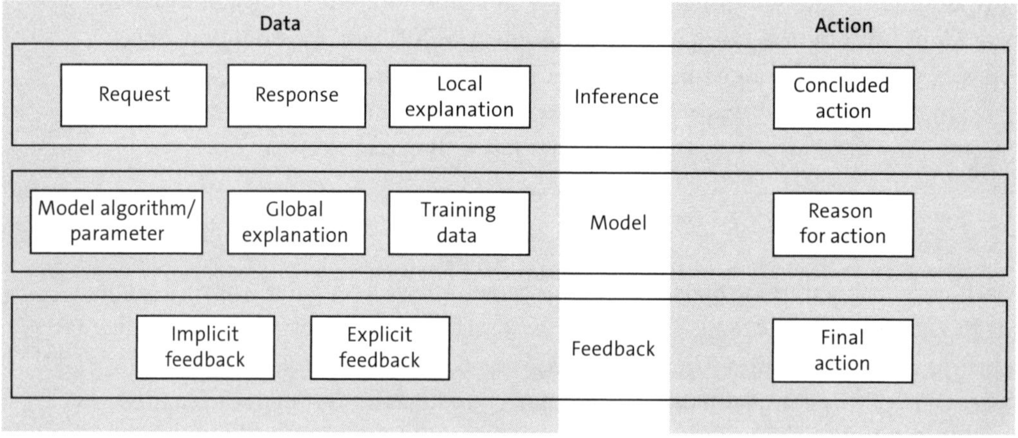

Figure 5.94 Entities for Logging and Auditing

The inference call consists of the request and response. The request contains the machine learning inquiry, such as the product and month for a sales prediction, whereas the response represents the calculated machine learning result, such as 5 million euros in sales revenue predicted for bicycle product sales in June. For transparency reasons, the inference result must be explained to the end user in terms of how the value was determined and the confidence level (refer to Section 5.5.7). Depending on the local explanation, the end user accepts or ignores the prediction/recommendation and performs actions accordingly, such as triggering a marketing campaign to increase sales revenue for bicycles in June.

Machine learning models are provided by training the algorithm with data and determining specific parameters. During the machine learning training process, basic explanations are calculated and stored in addition to the trained model. These are statistical values for the overall accuracy of the machine learning model, which are referred to as the global explanation. On the action side, the reason that the specific action was performed should be captured.

End user feedback is an important source for reinforcing the underlying data model. Negative feedback indicates a decrease of model performance and can result in retraining or even remodeling. After specifying the content for logging machine learning scenarios, the involved actors consuming the logs are discussed. Table 5.4 summarizes the relevant actors and the required log information. The main scenarios for consuming the machine learning logs are to verify legal compliance by auditors, to provide transparency for end users, and to resolve failure situations.

	Auditor	Support	End User	Developer	Key User
Request	X	X	X	X	
Response	X	X	X	X	
Local explanation	X	X	X	X	
Model algorithm	X	X		X	X
Model Parameter	X	X		X	X
Global explanation	X	X	X	X	X
Training data	O	O		O	
Concluded action	X	X	X	X	
Reason for action	X	X	X	X	
Final action	X	X	X	X	
Implicit feedback	X	X		X	
Explicit feedback	X	X		X	
X = Access restricted with authorization; O = Access depends on availability					

Table 5.4 Information Required by Actors

An auditor is a person who executes an audit. To act as an external auditor of the company, a person usually must have a certificate of practice from a regulatory authority. The auditor will have access to all the machine learning-specific logging entities described previously. However, the access is restricted with authorization checks; for example, only data for the auditing timeframe will be visible in read mode. A snapshot of the data used for the training could be stored with additional metadata (e.g., timestamp, machine learning scenario) for logging. As the data amount of those snapshots could rapidly increase the balance between disk space and TCO, logging requirements must be kept in mind. The recommended solution is to make the logging of the training data configurable. Thus, customers can decide according to their needs whether no snapshot, only the last snapshot, or snapshots for a given period of time will be captured (by default, no snapshot of the training data is logged).

An end user is a person who uses the machine learning application for the purpose of executing a business process, not for the purposes of carrying out an administrative or developer task; end users might include purchasers, sales representatives, or product planners. End users need access only to the logging data required for comprehensibility, traceability, and reproducibility of the business processes.

A developer is a person who programs the machine learning application to solve business problems. Like support persons, developers are involved in failure situations. These actors need access to all logs because they usually are relevant for the resolution. However, access to the log data will be protected with authorizations. As key users schedule and perform training jobs, access to the logging data for this purpose will be provided. Again, access will be restricted based on authorizations.

Figure 5.95 depicts the provisioning and consumption of machine learning logs. Before the machine learning application can perform inference calls, the underlying machine learning model must be provided by a training job.

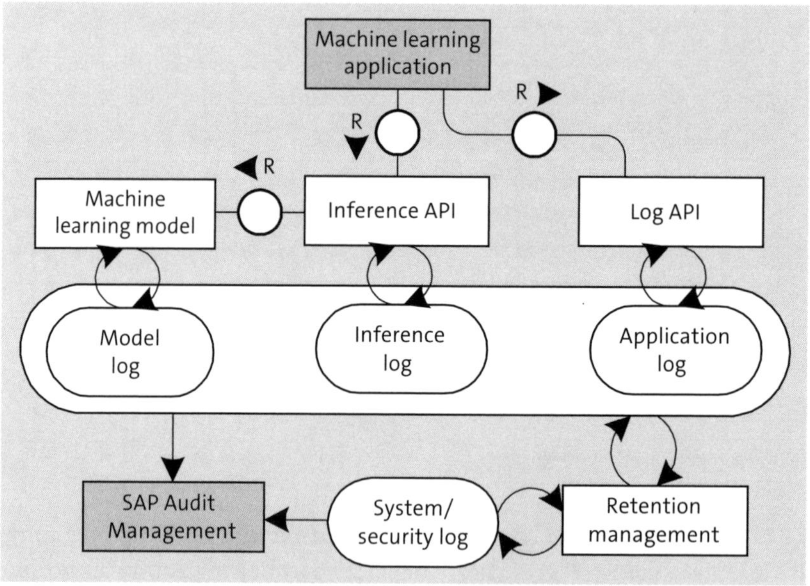

Figure 5.95 Machine Learning Logs Provisioning and Consumption

During the training process, the following logging entities can be captured by the training infrastructure automatically:

- Model algorithms/parameter
- Global explanation
- Training data

As already mentioned, storing snapshots of the training data depends on customers' configurations. The inference infrastructure can generically collect the following logging information:

- Request
- Response
- Local explanation
- Reason for action

Storage of the response can be avoided as it can be recalculated based on the machine learning model and the request data, which are logged anyway. Logging by the machine learning application is necessary for documenting the business process. This is specific for each scenario and must be considered by all machine learning applications. In this context, the concluding, final action will be logged, and references to the inference logs (e.g., local explanation) and model logs (e.g., global explanation) at least will be provided. Implicit and explicit feedback is covered with an experience management solution.

For system, security, and machine learning logs, integration into retention management is required. Thus, the logs can be periodically deleted/archived out of the system to reduce the memory footprint and be compliant with legal regulations.

The machine learning logs will be consumed by SAP Audit Management. This is a mobile-enabled, end-to-end audit management solution. The audit department can use it to build audit plans, prepare audits, analyze relevant information, document results, form an audit opinion, communicate results, and monitor progress. Key features of SAP Audit Management include the following:

- Multiple devices support
- Full coverage of the audit process
- Single source for audits
- Integration with third-party systems such as fraud or risk management
- Document management
- Global monitoring support
- Search functionality and intuitive UI

Figure 5.96 depicts the solution one level deeper. It illustrates details in context of provisioning the machine learning-specific entities for logging.

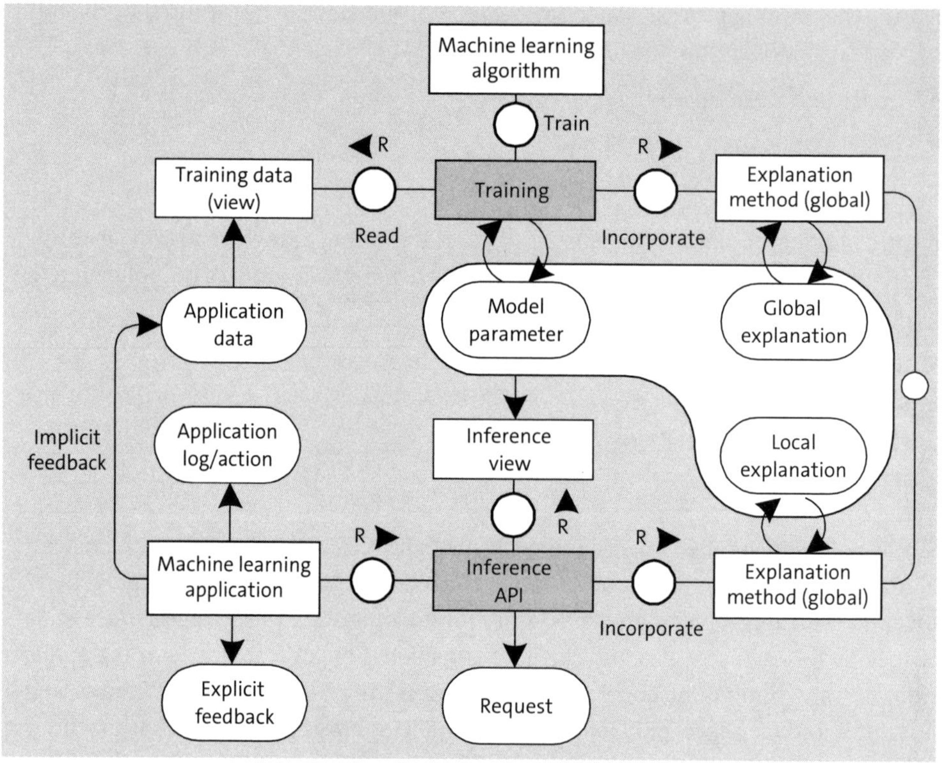

Figure 5.96 Machine Learning Logs Solution Details

The training job processes application data to train the selected machine learning algorithm and to store the trained model for consumption. During the training job, global explanations are computed by providing accuracy measures based on a specific explanation method. This is the basis for logging the model algorithm/parameter, global explanation, and training data entities.

The trained model is exposed to the machine learning applications by an inference API. The inference request is used to calculate a local explanation based on a selected explanation method. Local and global explanations are provided to the machine learning application as part of the response structure and can be displayed in the UI. This is the foundation for logging the request, response, local explanation, and reason for action entities. As part of the application log, the concluded and final action entities must be considered, which requires application-specific development and cannot be handled generically.

Figure 5.97 illustrates the interplay of the logging components in the context of the auditing process. While SAP Audit Management is in place, the machine learning-specific logging must be provided by the developer of the machine learning use case based on the described approach. The auditor initiates the process using the SAP Audit Management system, which supports the following steps:

1. **Planning**
   Audit planning is the initial phase of the auditing process. During this phase, the overall strategies and focus areas for the organization are defined, the audit plan for the upcoming audit period is prepared, and audit resources are arranged for the planned audits. Auditable items, audits, and audit plans are created in this phase.
2. **Preparation**
   In the audit preparation phase, the auditor develops and documents the audit work program that achieves the audit engagement objectives. The auditor sets up the structure of the work program, defines the detailed procedures for the audit, and obtains the approval from the audit manager before starting the audit. The audit manager receives the work program, reviews it, and decides to approve or reject it. If an audit announcement letter is required for the audit, the auditor can also prepare audit announcement letters in this phase. After the announcement letter is approved and distributed, the auditor continues with the preparation of the work program.

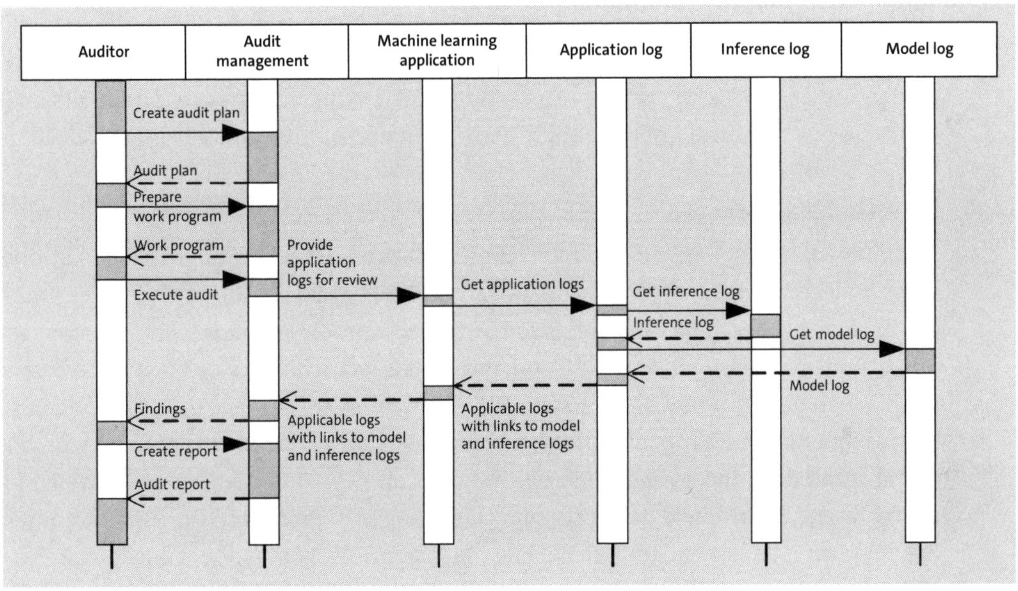

Figure 5.97 Integration of Machine Learning Logs in Auditing Process

3. **Execution**
   The execution phase is when the actual auditing activities take place. In this phase, auditors conduct interviews, gather information, record evidence, and prepare findings, conclusions, and recommendations. During the execution phase, the auditor

can also examine the machine learning applications. For this, the application logs are analyzed as they are the basis to understand and reconstruct the business processes into which machine learning capabilities are built. For displaying the application logs, a generic transaction with various filter criteria is provided. In addition, the machine learning applications can also contain UI sections to show the application logs. To support the drilldown from the application logs to the details, the inference and model logs are linked to the application logs.

4. **Reporting**
   Reporting is the final phase of the audit. In this phase, audit reports are created, reviewed, and communicated with the stakeholders. Auditors evaluate the adequacy, effectiveness, and timelines of actions taken by management on reported findings and recommendations. In reviewing the evaluation results, the auditor determines whether management has implemented the recommendations or accepted the risks of not implementing them.

5.5.10 Model Validations

Statistical methods exist to validate machine learning models with regard to the accuracy of predictions. However, for business applications this mathematical approach alone is not enough. Additional validation from functional and business process points of view are necessary. Therefore, this section explains the business requirements and the necessary technical implementation for validation of machine learning models.

Business Requirements

In context of intelligent ERP, machine learning discovers hidden insights and relationships by learning from application data, rather than programming explicit rules. Integrating machine learning capabilities across business processes helps identify untapped opportunities, expose hidden risks, and automate repetitive tasks or knowledge-based work. When designing intelligent applications based on machine learning algorithms, the machine learning model typically evolves over time. After being trained with an initial set of data, the model must be retrained on new data that becomes available during usage. This new data represents the changes in environment (e.g., customer behavior deviations) that are reflected by the machine learning model due to continuous retraining.

The process of retraining is one of the differences from classical systems, as we discussed in Section 5.5.6. It's a recurring process that results in new versions of the machine learning model. New model versions must be validated in terms of prediction power and robustness before they can be activated for usage. However, in the context of ERP systems, this is a very challenging task. This is because the classic validation approaches aren't sufficient for machine learning. Usually customers test business applications in quality systems and transport them to productive systems after successful validation. Quality systems typically contain only test data, which are adequate for testing functional correctness. However, for validation of machine learning models, the synthetic data in quality systems is not adequate. Thus, training machine learning models based on those synthetic data would result in incorrect machine learning models. Therefore, machine learning models must always be trained in productive systems in which live data is available. Only thus can it be ensured that the machine learning algorithms learn from the correct data foundation and identify the relevant patterns.

Live data usually can't be replicated from productive to quality systems due to GDPR compliance; developers and consultants working in the quality system, say, would then have access to live data which is forbidden from a legal point of view. Thus, in a quality system, only the functional correctness of machine learning applications can be tested; the quality of the machine learning model cannot be validated. This must happen in the productive system, which is the major difference from classical applications and brings specific requirements to the validation procedure, such as the involvement of new roles like the data scientist. However, testing in productive ERP systems is very problematic because each action affects business processes, change documents, and audit logs. This footprint cannot be reversed due to legal compliance reasons.

Thus, the key question is how to validate machine learning models from a business process point of view in productive ERP systems. The solution is the main objective of this section. Classic validation concepts and tools are taken as granted, such as testing the functional correctness of machine learning applications in quality systems or applying statistical techniques to determine the quality of machine learning models.

Technical Implementation

There are technologies that at first glance could be used as a basis for validation of a new trained machine learning model when another model is in use. However, taking a closer look, they are not adequate:

- **A/B testing**
  A/B testing is a method of comparing two versions of an application to determine which one performs better. A/B testing is essentially an experiment in which two or more variants of a page are shown to users at random, and statistical analysis is used to determine which variation performs better for a given conversion goal. In an A/B test, for example, an application screen could be modified to create a second version of the same page. This change can be as simple as a change to a single headline or button or could be a complete redesign of the page. Then, half of the traffic is shown the original version of the page (known as the control) and half the modified version of the page (the variation). As visitors are served either the control or variation, their engagement with each experience is measured and collected in an analytics

dashboard and analyzed through a statistical engine. Thus, whether changing the experience had a positive, negative, or no effect on visitor behavior can be determined.

The focus of A/B testing is on validating UIs and therefore it cannot be applied for machine learning models one to one. Machine learning models are deeply integrated into business processes. Thus, they impact the process flow and decision-making, which cannot simply be undone in ERP solutions due to legal compliance reasons; for example, bank transfers performed by a machine learning model cannot just be deleted. Therefore, having two applications with the same intent but with different behavior due to disparate machine learning models is not acceptable for ERP solutions. Hence, classic A/B testing cannot be applied for validation of machine learning models one to one.

- **Feature toggle**
  Feature toggle is a technique to selectively enable or disable a feature. A feature is a business functionality on the level of a user story. Feature toggles are used to prevent the release of incomplete or low-quality features to customers. Such features are delivered in an inactive state. A runtime switch (feature toggle) hides a new feature from customer access. The default status of the feature is off. In SAP development and test systems, it can be activated per user and per client. In customer systems, the feature cannot be switched on (hidden code delivery). The feature is a temporary switch to decouple the technical upgrade of the system from the functional update. After the feature is released, the feature toggles are removed. Feature toggles are not used as Customizing switches.

  The focus of feature toggles is on continuous delivery, beta shipment to selected customers, and phased rollout of functionality, and therefore feature toggles don't fit with validation of machine learning models by definition. Furthermore, feature toggles are temporary, so after final release of a feature to all customers, the feature toggle is removed and the corresponding source code is cleaned up.

- **Business feature**
  A business feature extends an SAP S/4HANA core business functionality with an additional opt-in feature. When a customer selects their scope, which references one or more business features, the corresponding Customizing entry is set in the central business feature Customizing table. Business features are defined by SAP and are delivered to customers as table content. By having a central Customizing table for all business features instead of multiple, heterogeneous implementations (per industry, per country, or per partner), transparency into the available business features is gained.

  From a development perspective, business features provide a fast and easy way to check if a customer has a certain functionality in scope. Typically, a business feature maps one-to-one to a business configuration content object (scope item, building block, or scope option), which represents the corresponding backend enablement. As for all business configuration content, the principle of incremental content design applies, and all dependencies between business features are managed on the business configuration content level.

The focus of business features is on checking whether a customer has specific Customizing settings in place for the given functionality. Thus, this technology doesn't help with validation of machine learning models because it's about proofing the quality, not about checking for Customizing settings.

- **Switch framework**

  The switch framework simplifies an ABAP-based system landscape by adopting one or more industry solutions in a standard system. The switch framework allows you to externally control the visibility of repository objects or their components by means of switches. By using the switch framework, all industry solutions and a restricted list of repository objects are delivered in an inactive state in the system. Usually an industry solution does not need to be installed but must be activated if required. The switch framework is integrated into the ABAP Workbench and works closely together with the enhancement framework. The overall goal of the enhancement framework is to provide a technology for creating modification-free enhancements and to unify all possible ways of modifying or enhancing repository objects. The core of the enhancement framework is a simple structure consisting of an enhancement option and an implementation element that can be attached there. The switch framework controls which enhancement implementations are to be performed.

  The primary focus of the switch framework is on industry verticalization and design-time artifacts. Thus, this technology doesn't fulfill the requirements for validation of machine learning models. Furthermore, new SAP S/4HANA-specific development won't be switched via the switch framework due to drawbacks tied to cloud qualities.

Because none of these technologies can be reused for validation of machine learning models from a business process point of view, we'll discuss a new approach next.

There are roughly two regimes for machine learning validation: offline and online. *Offline validation* happens during the prototyping phase, in which data scientists try out different features, models, and hyperparameters. It's an iterative process of many rounds of validation against a chosen baseline on a set of chosen evaluation metrics. Once there is a model that performs reasonably well, the next step is to deploy the model to production and validate its performance online, on live data. *Online validation* refers to techniques that are applied after the offline validation in order to continuously validate models and improve their performance with the arrival of new data.

In this section, we'll only discuss online validation. Figure 5.98 illustrates the state transitions of machine learning models covering the necessary validation steps.

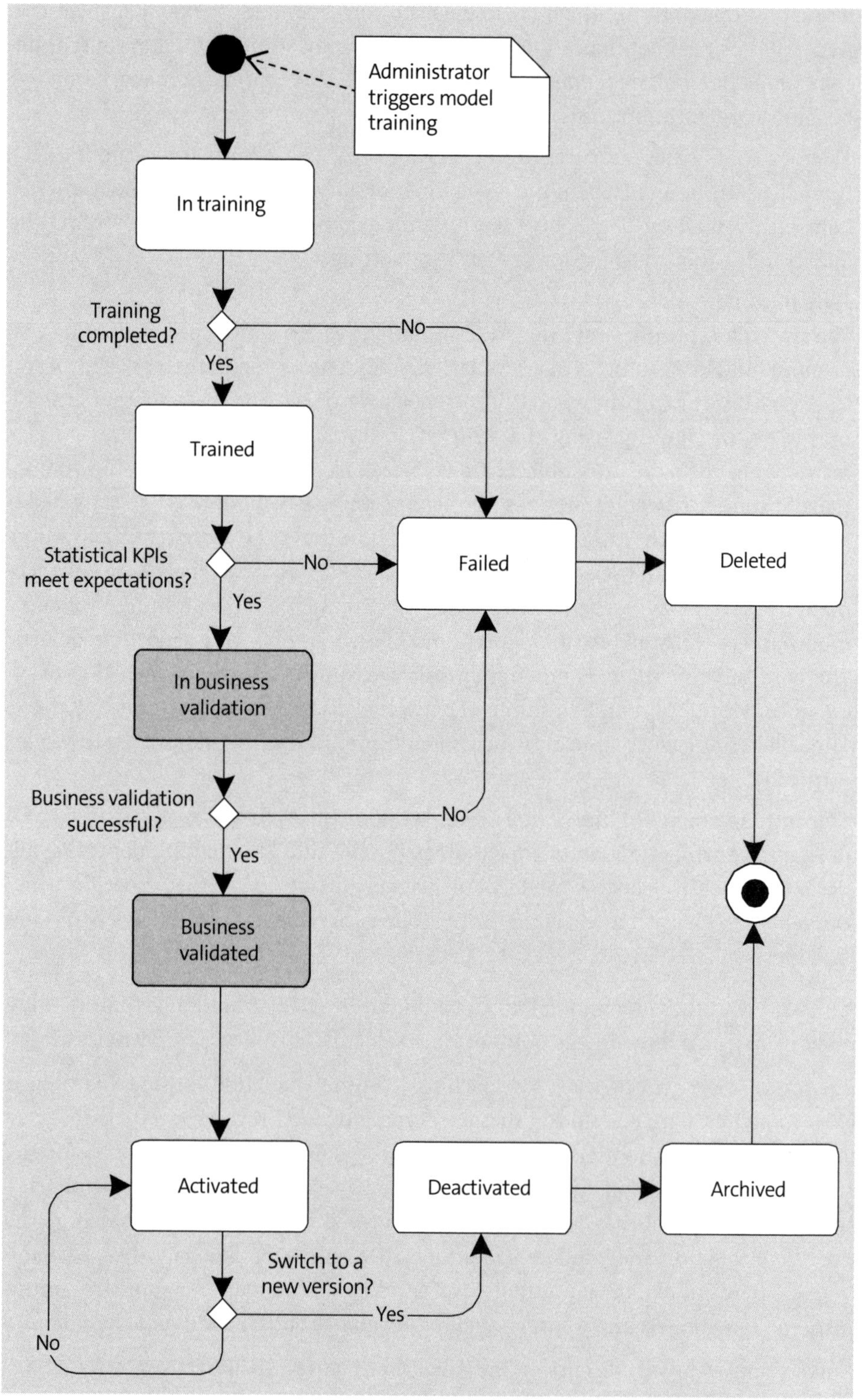

Figure 5.98 State Transition of Machine Learning Models

Let's walk through these steps:

1. **In training**
   The training of the machine learning model is triggered by the business administrator based on the Intelligent Scenario Management SAP Fiori app (refer back to Figure 5.2). Depending on the machine learning algorithm used and the underlying application data, the training can take anywhere from minutes to days. The training process can also fail—for example, if a specific configuration is missing or there isn't enough training data available.
2. **Trained**
   Once the model is trained, statistical KPIs are calculated for prediction power. The basic idea to determine the prediction power is to train a machine learning model on a given dataset and then use that underlying function on data points for which the value for the target variable is already known. If the model KPIs don't meet the expectations, then the model is useless and can be marked as failed.
3. **In business validation**
   The KPIs are just a static indicator for prediction power and not sufficient for productive usage of the machine learning model. Therefore, a business validation is required in which the machine learning model is evaluated in context of the underlying business processes and involved business users. The details of such business validations are described next. If the business validation isn't successful, then the machine learning model is marked as failed. Machine learning models with the failed status are usually deleted.
4. **Business validated**
   In the case of successful business validation, the machine learning model is marked as active for productive usage. From this point on, all inference calls are served by this machine learning model. Before a new model can be activated, the existing machine learning model must be deactivated; only one model can be active in the system. Deactivated machine learning models can't be deleted due to legal compliance reasons. They can be kept in the online database or archived to external systems for reduction of memory/storage.

As described previously, for business validation of machine learning models, classic concepts and technologies are not sufficient. Therefore, a specific solution is required, as shown in Figure 5.99.

The basic idea is to run one or more validation models in the background, parallel to the productively activated machine learning model. As illustrated in Figure 5.99, the end user consumes the machine learning application to perform predictions. Such requests are handled by the machine learning logic class, which wraps the consumption APIs of the trained model for integration into business processes. To be able to run validation models concurrently to the actively used model, an inference dispatcher is necessary. This component routes the inference requests to the validation and productively used machine learning models and receives the inference responses. While the responses from the active model are provided to the machine learning application, the responses from the validation models are stored locally with results of the active model. Thus, for each inference call, validation data is collected, which is exposed by the validation cockpit to the data scientist. Thus, data scientists can compare the prediction accuracy of the validation models with the productive model.

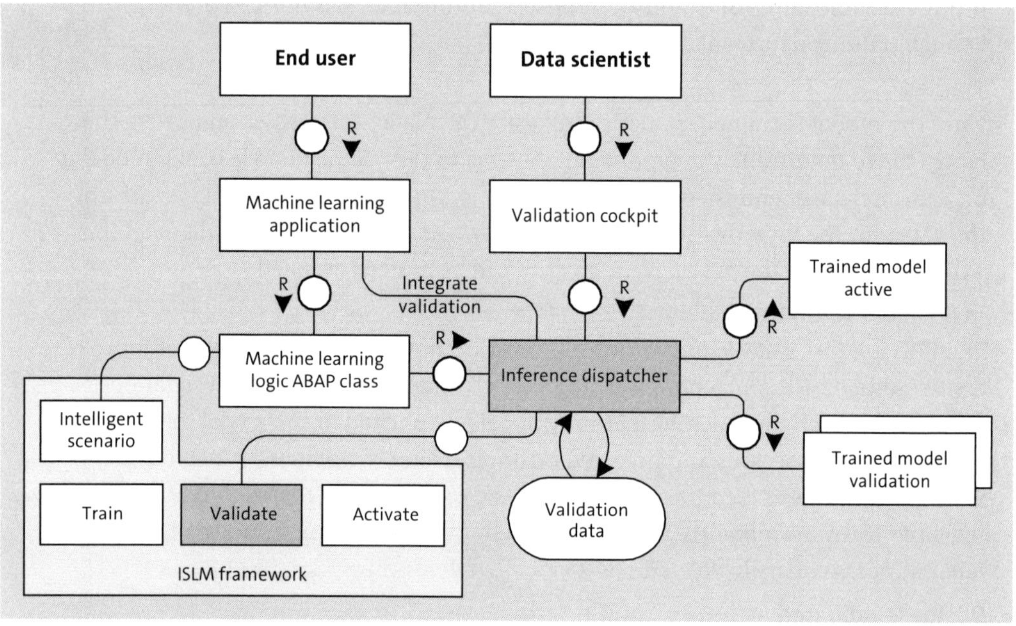

Figure 5.99 Machine Learning Model Validation Solution

For deciding which of the validation models will be activated productively, additional statistical methods might be applied by the data scientist. Optionally, the response of the validation models could be also exposed to the end user via the machine learning application. Hence, the end user could compare the results of the active and validation model and provide feedback. Such explicit feedback is transparent to the user but may impact the user experience, which can be a UI design challenge. On the other hand, it's a valuable input for validation of the machine learning model. If explicit feedback for model validation is integrated into the machine learning application, then the corresponding feedback data is also stored as validation data and displayed in the validation cockpit. Thus, the explicit feedback can also be considered by the data scientist as a basis for a decision.

The machine learning logic class is registered to the intelligent scenario and so associated with the ISLM framework. This framework handles the lifecycle management of machine learning applications from a consumer point of view (refer back to Section 5.5.1). The ISLM framework is the logical place to incorporate the validation functionality.

This means that validation of machine learning models must be explicitly triggered by the end user of the ISLM framework. This active decision is meaningful as the validation process consumes system resources in terms of storage of the validation data and CPU time for the training process. Figure 5.100 depicts the details of the validation steps.

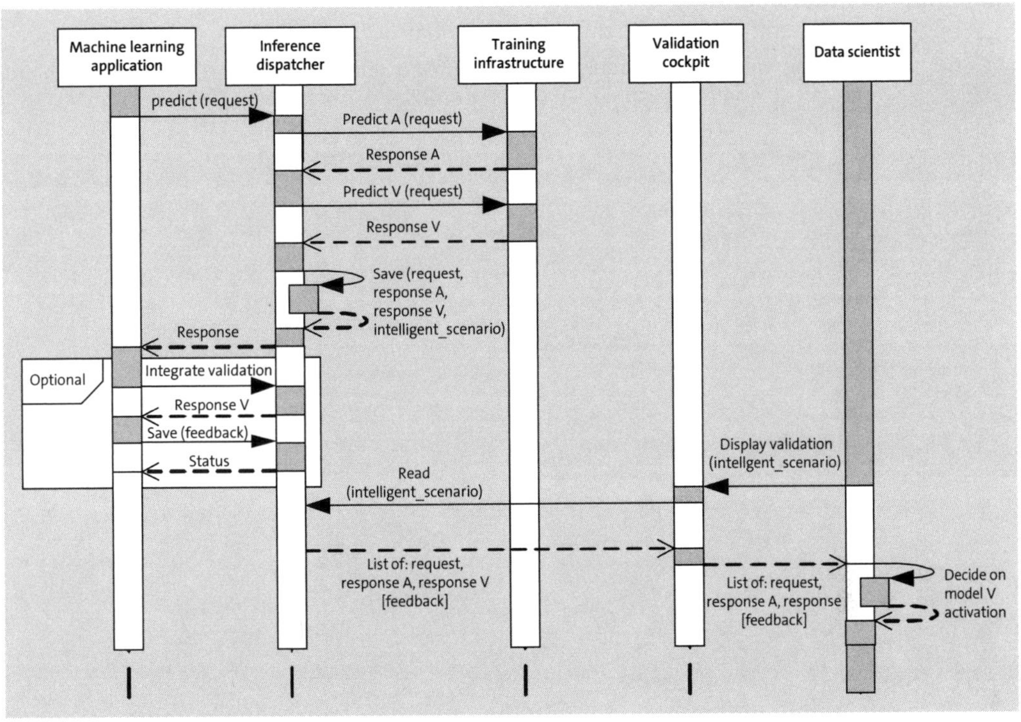

Figure 5.100 Machine Learning Validation Process

Figure 5.101 shows a UI prototype of the machine learning validation cockpit in order to explain its functional scope.

> **Note**
> A framework for this isn't yet available as of the time of writing (summer 2020), but the developer of the machine learning scenario could resolve this issue based on the described approach.

The UI consists of a search area, a list page, a detail view, and an action section. In the search area, the user can define filter criteria to search for available intelligent scenarios. The name of the intelligent scenario, the package, or the status are examples of such search criteria. The intelligent scenarios fulfilling the search criteria are listed on the left side of the cockpit. The user selects the relevant intelligent scenario and receives details for it on the right side of the cockpit. The table provides for each request the response of the activated and validated model. Optionally, explicit feedback of the consumer of the model is displayed. The user can forward-navigate to the request and compare the responses of both models. This comparison is supported by precalculated KPIs and time-dependent illustration of the responses. The user also can define their own KPIs or apply further data science techniques to the request and response data. Based on that analysis, the user can trigger different actions in terms of keeping the active model, shifting to the validation model if it has higher accuracy, or postponing the decision until additional data are available.

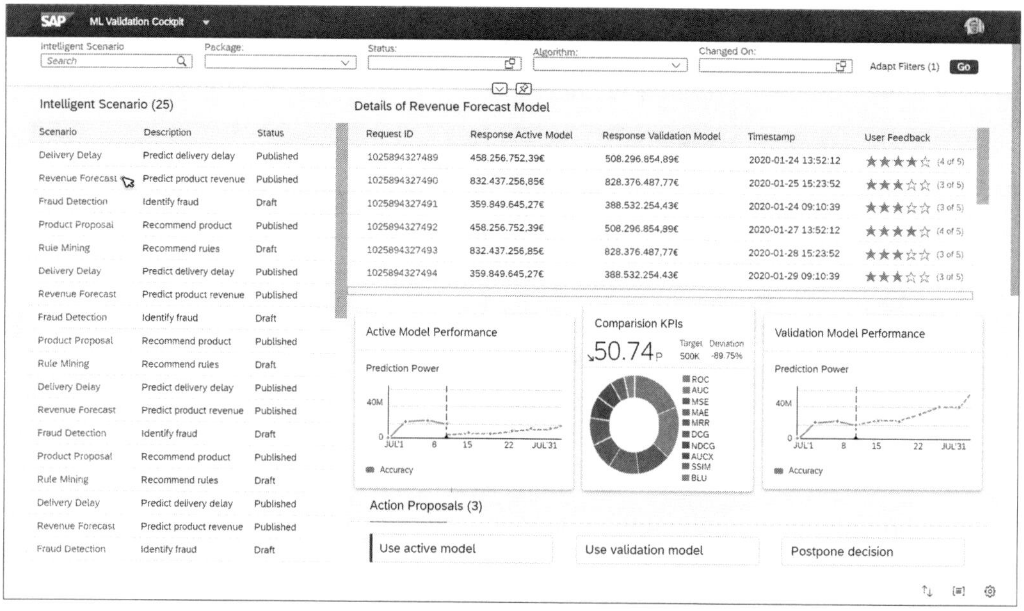

Figure 5.101 Machine Learning Model Validation Cockpit

5.5.11 User Interface Design

Incorporating machine learning capabilities into UIs needs a particular design and provisioning of new UI technologies. Only with proper integration into the UI is the instant value of machine learning exploited effectively. This section explains the business requirements and the necessary technical implementation for designing UIs for intelligent systems.

Business Requirements

A properly designed intelligent SAP system extends the cognitive capabilities of a human user. As with past generations of tools, our aim should be to empower users and improve the outcome of human work. To achieve this goal, we propose the following guiding principles for designing intelligent systems:

- **Human in control**
  In a business environment, actions triggered in a system have a tangible outcome in the real world that impacts the goals and profits of the company. Because the responsibility and accountability for these actions still lies with the human user, humans must always remain in control of the outcome.
- **Augment human capabilities**
  To gain the user's trust and foster successful adoption, an intelligent system should aim to upskill human experts, rather than replace them. Providing better transparency and efficient tools for decision-making processes, integrating user feedback, and presenting information in a way that makes it easier to understand are all measures that extend the power and reach of the individual. By contrast, hiding information, simplifying the truth, or reducing the number of options without sufficient transparency makes the user a slave to the system. The user must be able to understand and control the intelligent system.
- **Ethically aligned design**
  Machines do what they are told to do; there is no moral judgement in an algorithm. Designers and builders of advanced machine learning systems are stakeholders in the moral implications of their use, misuse, and actions, with a responsibility and opportunity to shape those implications. We need to establish clear and binding ethical rules that intelligent systems must obey.
- **Efficient automation**
  Intelligent systems should reduce the effort a user needs to invest to get something done. This means defining the right level of automation for each use case. Where full automation is not feasible, we should aim for greater efficiency. By combining automation with better use of existing information, transparency, and learning effects, intelligent systems can help users obtain the same results in fewer steps.

Technical Implementation

Depending on the use case and desired output, machine learning tasks can be grouped into several broad categories (such as classification, regression, or clustering tasks). In addition, the concrete approach for implementation can impact the UI in a very specific way. There are two important aspects to consider when designing intelligent systems that employ machine learning models:

1. **Explainable machine learning: How to build trust and empower the user**
   One of the key guiding design principles for intelligent systems is empowering the end user. This can be achieved by providing sufficient information about the underlying model and explaining the reasoning behind the results of an algorithm. Empowerment helps to build trust between human and machine. This topic was discussed in Section 5.5.7.
2. **Feedback loop: How to learn from the user and for the user**
   Different types of algorithms may require feedback from the end user to reinforce

the underlying data model. A carefully designed feedback loop is a challenging design task that can involve several new user roles (such as data scientists) and specialized UIs (e.g., for feedback monitoring and analysis). This topic was touched on in Section 5.5.6.

As described in Chapter 4, Section 4.2.1, several machine learning application patterns can be identified in context of SAP S/4HANA. In this section, we consider the general UI design for the machine learning patterns for matching, recommendation, and ranking. However, describing all the details would go beyond the scope of this book. Therefore, we focus on some key aspects just to explain the design proposals at a high level.

Intelligent Matching

Intelligent matching uses machine learning methods to identify related items and to generate matching groups. The system can present one or more matching strategies to link them. Users then only need to approve or reject the suggestions or adjust them to their needs. The different automation levels for matching are shown in Figure 5.102:

❶ **Select, validate, and rate**
This is the first level of automation for an intelligent system. Matching is still manual and user-driven, but the system verifies and rates the match quality on demand. The user can then decide to apply the match or refine it further.

❷ **Explore and adjust**
The system shows proposed matches based on different strategies. In this case, users pick one strategy, but often have to adjust.

❸ **Review and decide**
The system generates optimal matches and presents ranked recommendations. The user reviews the proposals and picks one option without further adjustment.

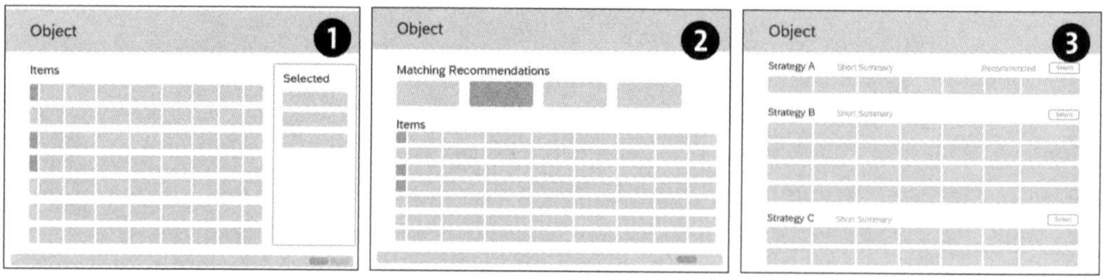

Figure 5.102 Matching Levels

Matching involves a group of items that are linked by certain criteria, such as similarity or the type of relationship. This is called a *matching group*. The following matching group characteristics are typical for the use cases we observed:

- **Group name**
  A meaningful label for the matching group. For example, this could be the name of the master record in a merging scenario or the name of the matching criterion.

- **Group summary**
  An optional description of the group contents. The summary can be presented as a text or using key/value parameters.
- **Group size**
  The number of items in the group.
- **Quality indicator**
  Rates the confidence of a matching proposal in relation to the defined goal. For example, this could be the similarity of business partner records when merging duplicates or the percentage covered when matching invoices.
- **Finalizing action**
  An action that can be applied to the whole group, such as approval of a matching proposal.

You can present the set of matching groups to users in different ways—for example, as a list, a grid, or even a graph. Users can then drill down into a specific group and analyze its content and match quality in detail. The presentation of a matching group depends on the use case and can vary from a simple list to a chart such as a network graph. Depending on the matching level and the match quality, the degree of interaction with the matched objects and groups varies. For example, if the matching proposal is not very accurate, then editing the matching group is an essential part of the interaction. Higher levels of automation already present good enough match proposals and differ only by strategy. Here, the following main or finalizing actions are possible:

- **Approve**
  The proposal is accepted as it is.
- **Reject**
  The proposal is fully or partially rejected. Partial rejection involves editing the matching group manually and then accepting it. Rejection can lead to a follow-up feedback mechanism (implicit or explicit).
- **Merge**
  Involves merging two or more groups into a *golden* one.
- **Split**
  Involves creating two or more groups out of one. This can be achieved by selecting individual items, which then are used to create new groups.

The concept itself does not have any limitations with regard to responsiveness and adaptiveness. It makes use of the standard SAP Fiori controls and floorplans and inherits their responsiveness characteristics and guidelines.

Recommendations

Intelligent systems can help users by recommending appropriate content or suggesting an action or input the user may prefer. In this case, we speak of a recommendation and its impact on the UI. As already mentioned, different recommendation patterns exist, such as input assistance or solution proposals. Input assistance recommends user input for a complete form or for a single element of a form. In both cases, we need to design the following microinteractions:

- **Detect**
  As a user, you need to see which values were suggested by the system.
- **Explain**
  As a user, you want to understand the reasoning behind the system recommendation.
- **Compare**
  As a user, you need to be able to compare existing human input with the system recommendation.
- **Act**
  As a user, you want to accept or reject system recommendations.

The guiding principle is that human input will always win against machine input. As illustrated in Figure 5.103, the system can prefill empty fields with recommendations. A user's input cannot be overwritten by a system recommendation without the user's approval. Active approval of the system recommendation by the user turns it into user input. Let's walk through the different input assistance options:

❶ Without human input and without system recommendation

❷ With human input but without system recommendation

❸ Without human input but with system recommendation

❹ With human input and with system recommendation

Users can accept the system recommendation implicitly, by submitting the whole form, or explicitly, by selecting an alternative in case of conflict. Users can explicitly reject a system recommendation by starting to edit it in the input field.

In most of the solution recommendation use cases, the solutions are presented as a list of recommendation items, which are grouped in a recommendation block. As a minimum, a recommendation item must have a meaningful title that links to an action. Depending on the use case, the presentation can be expanded by using a custom variant of the list item control. The example in Figure 5.104 shows the anatomy of the most complex case with all the optional elements:

- **Title and description**
  Give the recommendation a short and meaningful title that describes the proposed action. More information can be exposed by using an additional description. Longer texts can be truncated after the second line, but we recommend avoiding truncation altogether.

- **Action**
  To apply a recommendation, the user executes an associated action. Depending on the form of the recommendation item, the action can be triggered by clicking on the entire list item or by clicking on a dedicated button.
- **Selection**
  Depending on the use case, users might need to see a preview of the results first. In this case, the triggers for the preview and the final action must be separated.
- **Preferred proposal**
  Depending on the use case, you can highlight a recommendation item as a preferred proposal. For details, see ranked recommendations in the next section.

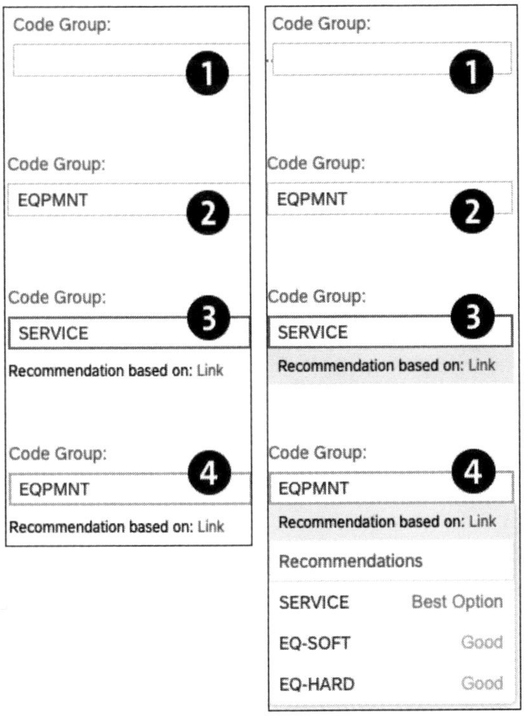

Figure 5.103 Input Assistance

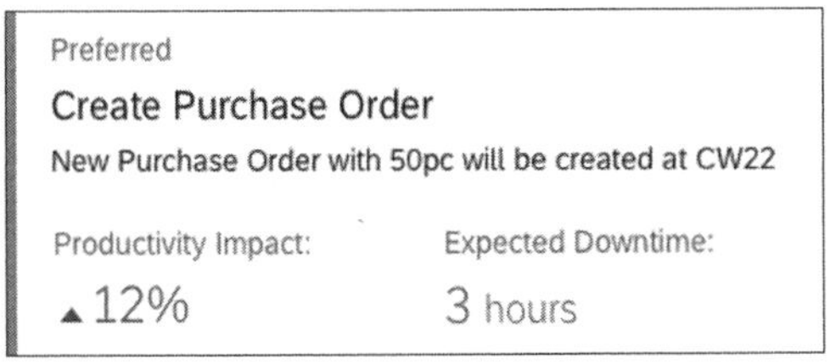

Figure 5.104 Recommendation Item

Several recommendation items form a recommendation block. The example in Figure 5.105 shows the anatomy of a horizontal recommendation block based on a grid list:

- **Recommendation block header**
  The block toolbar contains a meaningful and short block title ❶, which describes the set of proposed solutions ❷. It can also offer additional functionality ❸ that applies to the whole recommendation block, such as an additional explanation of the model behind the recommendations, or an option to provide explicit user feedback to reinforce the underlying model.
- **Recommendation block items**
  Recommendation items ❹ are the main content of the recommendation block. Depending on the use case and space constraints, they can be organized horizontally, vertically, or as a grid.
- **Ranked recommendations**
  You can highlight the first item of the recommendation block to indicate the preferred proposal. We recommend highlighting only one item (the preferred one) in the list and placing it at the top. You can use the semantic colors for highlighting if this is required by the use case.

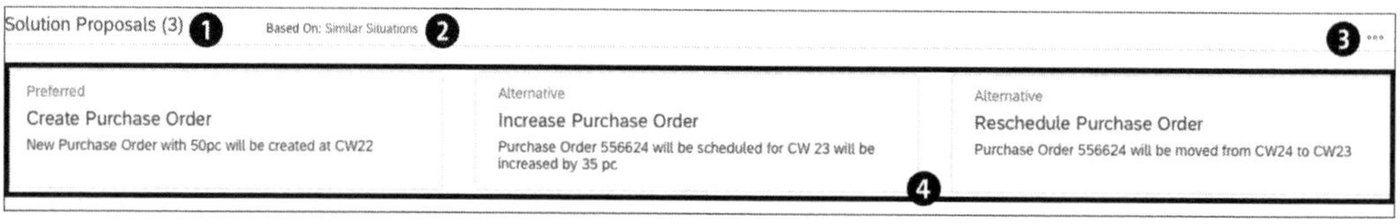

Figure 5.105 Recommendation Block

Ranking

Ranking simplifies complex decisions for business users by showing the best options first. Items in a group are ranked by comparing criteria that are relevant for the user's business context, such as an amount, the priority, or a score. In a ranked table or list, the results are always sorted to show the most highly ranked items on top. Ranking can incorporate intelligent algorithms, but they aren't required. Items can also be ranked using simple rules, thresholds, or heuristics. The concept for ranking in lists and tables comprises three basic elements grouped in a common reusable component, as illustrated in Figure 5.106:

❶ **Ranking value**
This is the common value on which the ranking is based. The value is mostly numeric and defines the rank of an object. It can be an implicit value, like price or delivery time, or a score. Depending on the use case, you can also show the score as a percentage or without a unit of measure. Another option is to use a microchart such as a radial microchart. Ranking based on a non-numeric value is uncommon but possible—for example, academic grading systems using letters (A+, A, A–; B+, B, B–; C+, C, C–; D+, D, D–). Clicking or tapping the ranking value opens a dialog with

ranking details for the corresponding item. If a list for some reason contains items that do not have a ranking value, show them at the bottom of the list. An example might be a new supplier that doesn't yet have enough data to be ranked by the system.

❷ **Ranking description**
The ranking description provides additional context for the ranking value. Terms like *best* or *alternative* can offer more specific guidance without being too complicated. It's important to understand that ranking descriptions are highly contextual. The wording and how they map to the ranking score depend on the content, use case, and process.

❸ **Change indicator**
The change indicator makes it easier to track the ranking value for fast-changing datasets, such as live streams for IoT scenarios.

For transparency and to support the user's decision-making process, be sure to provide additional ranking details whenever possible.

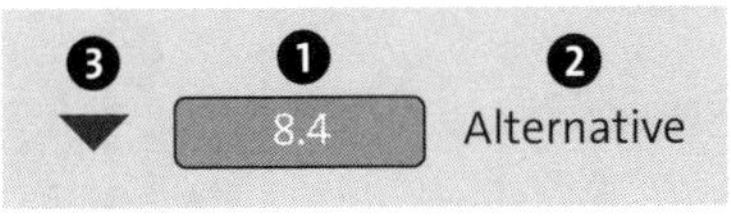

Figure 5.106 Ranking Elements

The ranking value is the minimum element needed to show a ranking for a list or table. The ranking description and change indicator support the basic value of the ranking and may not be required in all cases. In addition to the three basic elements of the ranking component, there are two other aspects to consider:

- **Semantic colors**
  Can be used to underline the ranking message. Use semantic coloring based on values when you want to show data points that are positive, neutral, or negative. Based on the defined threshold values, the color of each data point could be red, green, or orange.
- **Presentation variants**
  These variants help the ranking concept support different representations of the same data point. The appropriate representation depends on the role and use case, as well as the relative importance of ranking in the application. For example, if the ranking is part of the dataset but isn't relevant for the role or task, do not use a semantic color and display the ranking as for all other values in the table.

To gain additional insights into the ranking for a specific object, the user can view the ranking details as shown in Figure 5.107. The detail dialog is triggered by clicking or tapping the **Ranking** indicator ❶. In the ranking details, explain the position of the object in the overall ranking ❷ and the individual ranking components ❸. Also include a comparison ❹ between the current object and the average score for all other objects. Comparing individual values with an overall score helps users assess their importance.

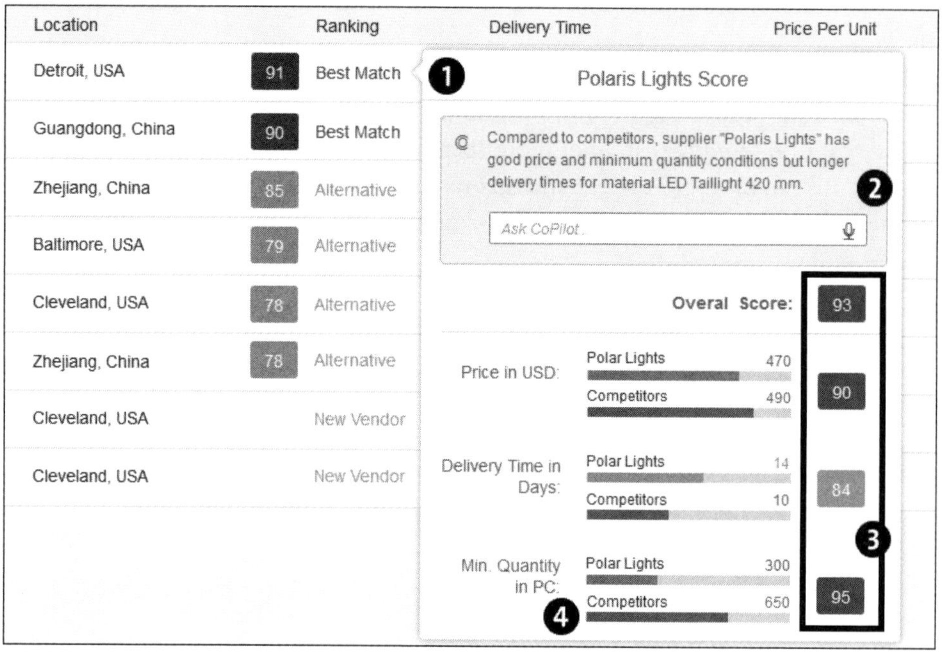

Figure 5.107 Ranking Details

5.6 Summary

In this chapter, we explained how you can implement embedded machine learning, side-by-side machine learning, and predictive analytics use cases based on SAP S/4HANA's infrastructure and programming model. For each of those approaches, we selected simple scenarios and guided you through the technical development steps. In particular, we provided decision criteria to determine when to apply which method. The machine learning applications you implement in the context of SAP S/4HANA must be enterprise-ready, which compromises qualities like compliance, lifecycle management, data and process integration, workload management, and performance. For all of those aspects we described the solution concepts and the implementation tasks so that you are able to ensure enterprise readiness for your machine learning applications.

In the next chapter, we'll describe the machine learning uses cases SAP S/4HANA delivers by applying the implementation approach explained in this chapter.

Chapter 6
Business Implementation

Now that you've learned how machine learning and predictive analytics are implemented technically, it's time to see how these technologies can be implemented in the business sphere. We'll discuss use cases from finance to sales to various other industries in this chapter.

SAP S/4HANA has been continuously augmented with machine learning capabilities. By adapting existing applications, rebuilding them, or even designing new ones from scratch, a whole portfolio of intelligent use cases has been created. Each application aims to improve a dedicated part of a business process by supporting users in the prediction of future outcomes, by augmenting their decision-making, or even by automating a process step.

In this chapter, we'll focus on the business implementation of predictive analytics and machine learning in SAP S/4HANA business processes. As discussed briefly in Chapter 1, we'll provide some end-to-end process examples for the following scenarios, then briefly explain the configuration basics. We won't go into step-by-step configuration detail but will show you how to access instructions specific to your use case via either the SAP Best Practices Explorer or the SAP Help Portal. We'll then dive into the various individual business use cases.

6.1 Overview of Intelligent Scenarios

Machine learning applications that are deeply integrated into business processes create a bigger impact, influencing across various technological, corporate, social, and community barriers. Let's now briefly look at a few examples of end-to-end business processes and see how they're implemented to address business problems.

The following are some of the scenarios we'll dive into to explain the end-to-end process in context:

- Intelligent procure-to-invoice scenario
- Intelligent source-to-contract scenario
- Intelligent invoice-to-cash scenario

6.1.1 Creating a Purchase Requisition as an Employee

In the context of the procure-to-invoice intelligent scenario, as shown in Figure 6.1, let's now briefly discuss an end-to-end process for an employee or a manager who creates a purchase requisition and see which intelligent technologies are leveraged.

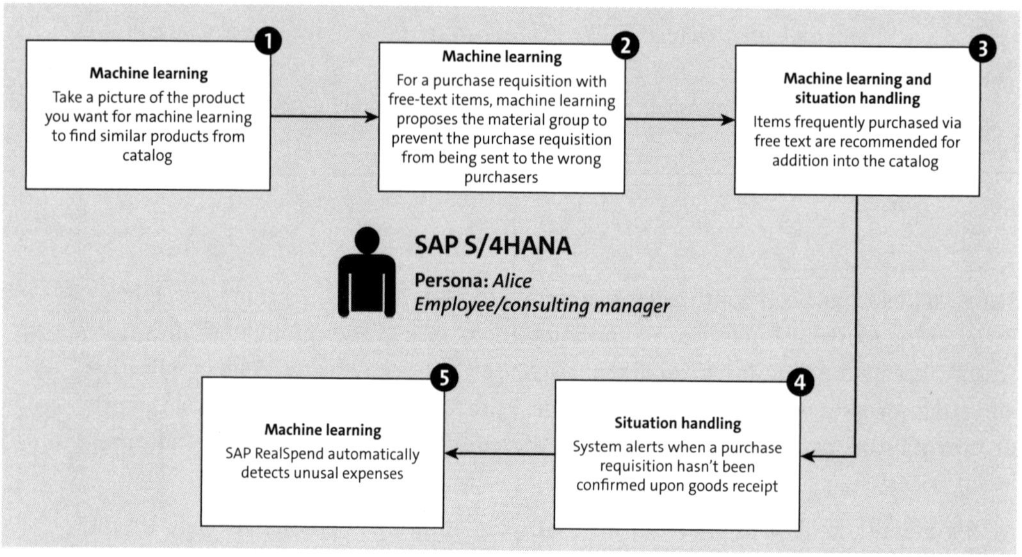

Figure 6.1 End-to-End Process Highlighting Purchase Requisition Creation for Employee

In the following steps, we dive into the details:

❶ In the first step, an employee takes a picture of the product needed and uploads it into the system. With the help of the machine learning algorithms, similar products are identified from the catalog and suggested for buying.

❷ In the second step, now that the purchase requisition is created with a free-text item, the machine learning algorithm can suggest a material group to be assigned based on historical information to prevent the purchase requisition from being sent to the wrong purchasers. This helps the employee save time and money for the company.

❸ In the third step, the items that are frequently bought via free text (purchased via searching for an item rather than selecting it from a catalog) are automatically recommended to be added into the catalog, leveraging the intelligent technologies of situation handling and the machine learning algorithms.

❹ In the fourth step, the system alerts the user if the purchase requisition hasn't been confirmed after the goods receipt, utilizing the situation handling framework.

❺ Finally, SAP RealSpend automatically detects unusual expenses using the machine learning algorithms.

6.1.2 Processing a Purchase Requisition as an Operational Purchaser

Now, in the context of the procure-to-invoice intelligent scenario, as shown in Figure 6.2, let's briefly discuss another end-to-end process for an operational purchaser processing a purchase requisition and discuss which intelligent technologies are leveraged.

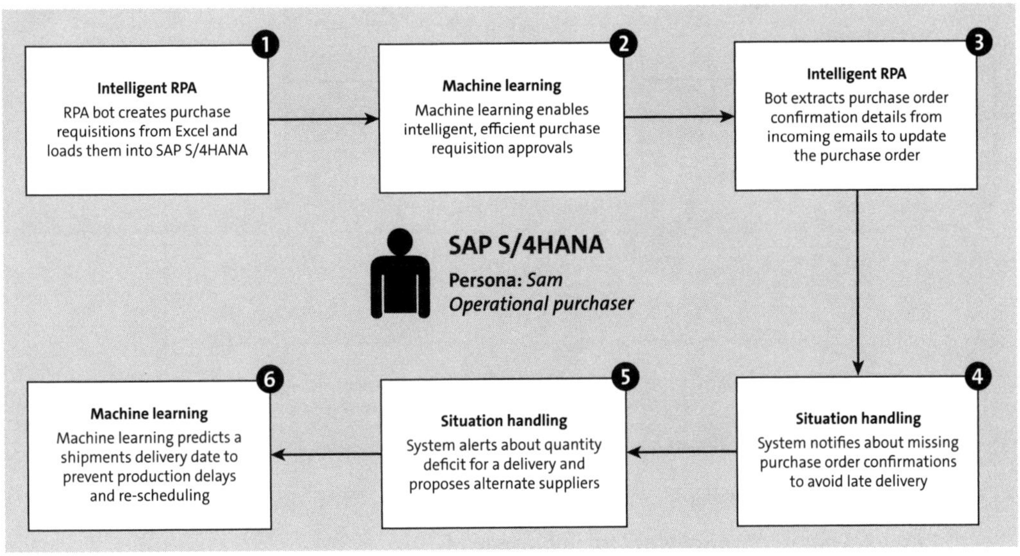

Figure 6.2 Operational Purchaser Processing Purchase Requisitions

In the following steps, we dive into the details:

❶ In the first step, a robotic process automation (RPA) bot automatically creates a purchase requisition from an Excel file and uploads it into the SAP S/4HANA system.

❷ In the second step, machine learning algorithms automatically approve regular purchase requisitions that contain all the relevant information.

❸ In the third step, an RPA bot now extracts the email for the purchase order confirmation and updates the purchase order in the SAP S/4HANA system.

❹ In the fourth step, system alerts are provided using the situation handling framework to identify purchase orders that have missing confirmation details and avoid delivery delays.

❺ In the fifth step, the system alerts the user of quantity deficits and proposes alternate suppliers, leveraging the situation handling framework.

❻ Finally, machine learning algorithms identify delivery delays ahead of time using machine learning models so that rescheduling of these items can be done to avoid potential production delays.

6.1.3 Monitoring the Spend as a Strategic Purchaser

In the context of the source-to-contract intelligent scenario, as shown in Figure 6.3, let's now briefly discuss an end-to-end process for the strategic purchaser monitoring the spend and discuss which intelligent technologies are leveraged.

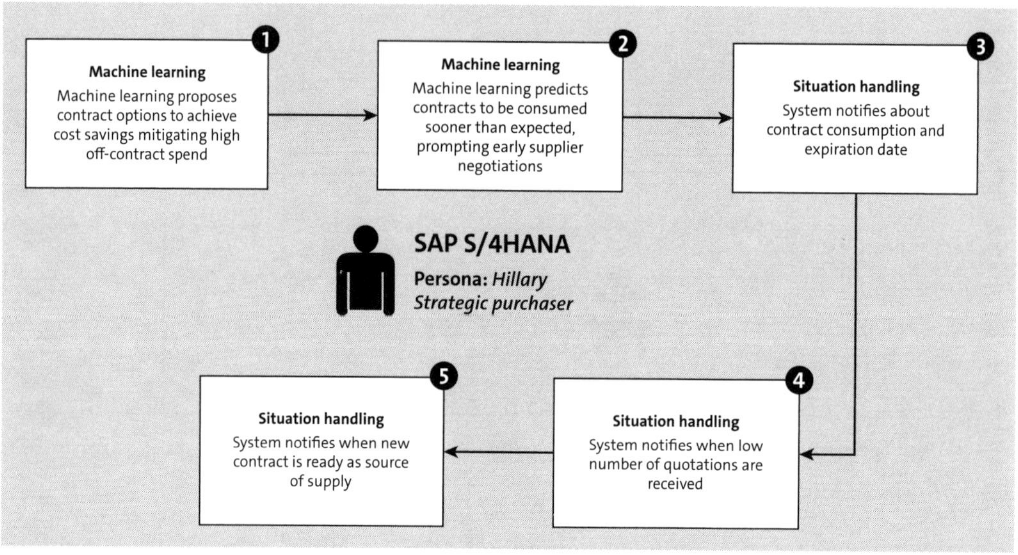

Figure 6.3 Strategic Purchaser Monitoring Spend

In the following steps, we dive into the details:

❶ In the first step, machine learning algorithms suggest alternate contract options to mitigate the high costs that are generated by off-contract spend.

❷ In the second step, machine learning algorithms predict the contracts that will soon be expired so that early renegotiations can take place with the suppliers.

❸ In the third step, the system notifies the user about the contract consumption and expiration dates, leveraging the situation handling framework.

❹ In the fourth step, the system alerts the user when a very low number of quotations are received so that the purchaser can be prepared ahead of time.

❺ Finally, leveraging the situation handling framework, the system notifies the user when a new contract is ready as a source of supply for the purchaser.

6.1.4 Creating Sales Inquiries as a Sales Manager

In the context of the invoice-to-cash intelligent scenario, as shown in Figure 6.4, let's now briefly discuss an end-to-end process for the sales manager creating sales inquiries.

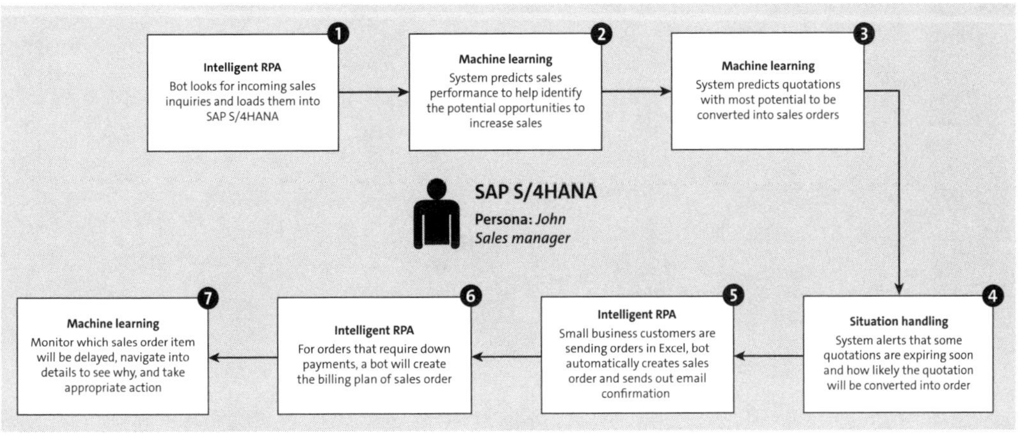

Figure 6.4 Sales Manager Creating Sales Inquiries

In the following steps, we dive into the details:

❶ In the first step, an RPA bot looks for incoming sales inquiries and loads them into the SAP S/4HANA system.

❷ In the second step, machine learning algorithms predict sales performance to help identify potential opportunities to eventually increase the sales for the sales manager's KPIs and the top-line revenue for the company.

❸ In the third step, a machine learning algorithm identifies the most sales quotations that could be converted into sales orders and provides these to the sales manager.

❹ In the fourth step, the system alerts the user that some quotations are expiring soon and how likely it is that they could be converted into sales orders, leveraging the situation handling framework.

❺ In the fifth step, the sales orders being received from Excel files are converted into sales orders in the SAP S/4HANA system and email confirmations are sent by RPA bots.

❻ In the sixth step, for the sales orders needing a down payment, the RPA bot creates a billing plan and uploads it into the SAP S/4HANA system.

❼ Finally, leveraging the machine learning algorithms, predictions are made about which sales orders are being delayed so that appropriate actions can be taken by the sales managers.

6.1.5 Recording Financial Transactions as an Accounts Receivable Manager

Now in the context of the invoice-to-cash intelligent scenario, let's briefly discuss another end-to-end process for an accounts receivable (AR) manager recording financial transactions and discuss which intelligent technologies are leveraged, as shown in Figure 6.5.

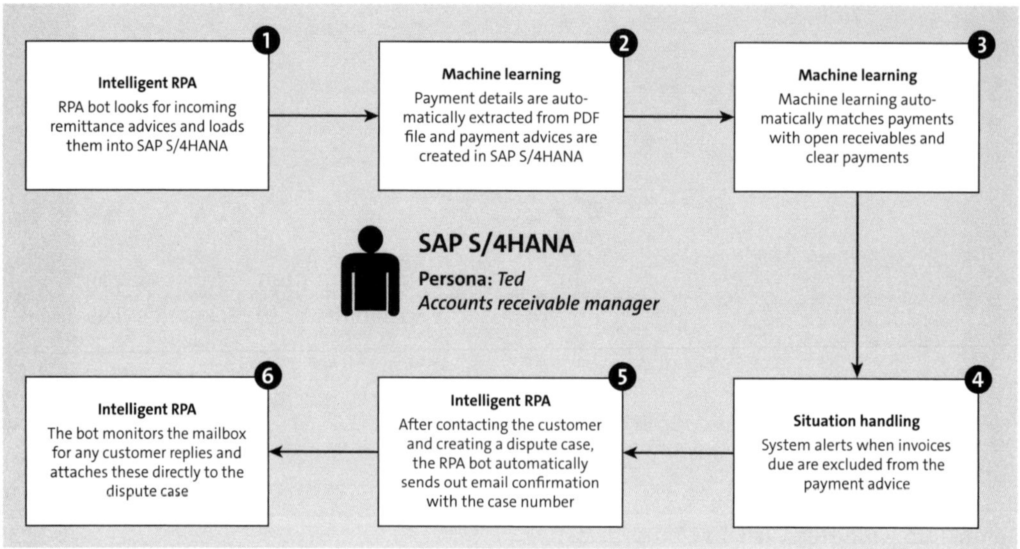

Figure 6.5 AR Manager Recording Financial Transactions

In the following steps, we dive into the details:

❶ In the first step, an RPA bot looks for incoming remittance advices and loads them into the SAP S/4HANA system.

❷ In the second step, machine learning algorithms extract the payment details from the PDF files, and payment advices are created in the SAP S/4HANA system.

❸ In the third step, a machine learning algorithm automatically matches payments with open receivables and clears payments.

❹ In the fourth step, system alerts are generated when invoices due are excluded from the payment advice, leveraging the situation handling framework.

❺ In the fifth step, after contacting the customer and creating a dispute case, an RPA bot automatically sends an email confirmation with the dispute case number.

❻ Finally, the RPA bot monitors the mailbox continuously for any customer responses to the dispute case and attaches them to the case on the SAP S/4HANA system.

6.2 Configuration Basics

In previous chapters, we focused on the approaches for performing machine learning with SAP S/4HANA, the architecture, and the technical implementation across these different tools and technologies. In the context of this chapter, we'll focus on discussing the configuration steps behind implementing machine learning and predictive analytics scenarios. In this section, we'll focus on an example that explains how to implement a particular machine learning scenario by following the configuration steps.

Predefined machine learning and predictive analytics scenarios are delivered with SAP S/4HANA. This means that for an intelligent application, all the needed development is provisioned by SAP: integrations to business processes and user interfaces (UIs), defining the data source for training, and selecting the appropriate machine learning algorithms. However, customers must adapt the delivered scenarios to their specific needs and activate them for usage. This procedure is known as configuration or Customizing and is performed at the customer site for utilizing the variability of the scenarios. The focus of this section is on machine learning- and predictive analytics-specific configuration; general concepts are taken as granted, such as configurations for user interfaces, business processes, or forms.

In general, for all the delivered intelligent applications based on machine learning and predictive analytics, corresponding configuration guidelines are provided by the SAP Best Practices Explorer and the SAP Help Portal. Those guides describe the necessary configuration tasks step by step. They also include generic tasks like model training using the Intelligent Scenario Management SAP Fiori app, as explained in Chapter 5. In this section, we explain how to deal with the tools containing those documents. Furthermore, for each of the described intelligent applications, we provide direct links to the configuration guides for ease of consumption and exploration. Consultants who are setting up systems for customers know that finding the appropriate configuration documents is very challenging and can be half of the job.

6.2.1 SAP Best Practices Explorer

Customers implementing an SAP solution can consult the SAP Best Practices Explorer as a knowledge base to find configuration content that's predefined by SAP for that solution. The key features are as follows:

- The catalog contains the whole portfolio of SAP Best Practices content. You can navigate through packages, click on an area of business, use the A–Z index, or search by inputting key words in the search field. The available content is free of charge.
- This knowledge base helps you understand the scope of the packaged solutions by navigating to the details. You can download the whole package or individual documents—for example, software and delivery requirements, configuration documents, and test scripts.

The necessary steps for using SAP Best Practice Explorer are as follows:

1. Open SAP Best Practices Explorer using the following link: *https://rapid.sap.com/bp*, as shown in Figure 6.6.
2. Select your solution, which in our case is **SAP S/4HANA**.

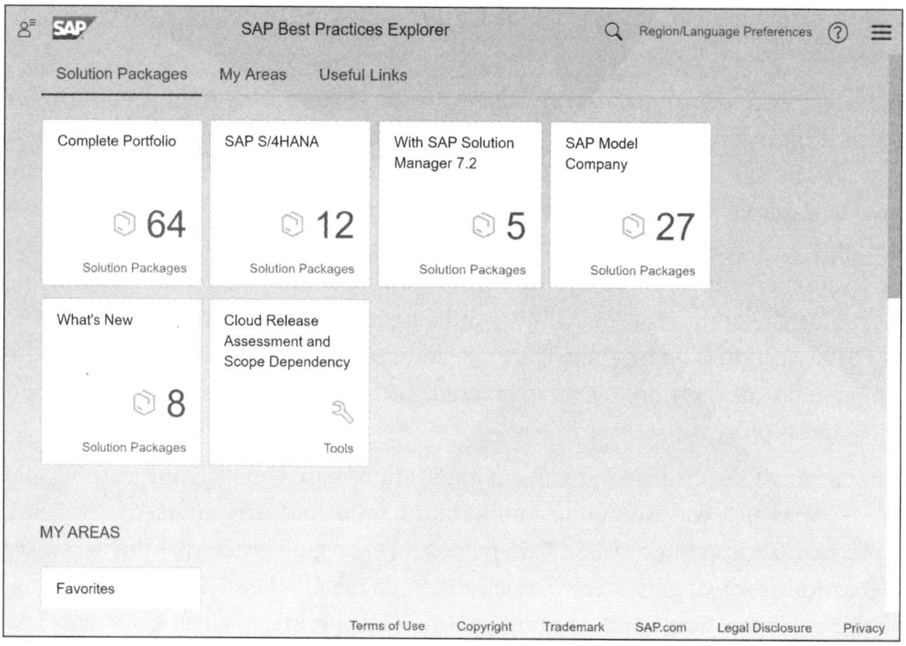

Figure 6.6 Launchpad for SAP Best Practices Explorer

3. Navigate to the required SAP Best Practices content by traversing the catalog tree or by entering search terms, as shown in Figure 6.7. You can access region-specific content by clicking on **Region/Language Preferences**.

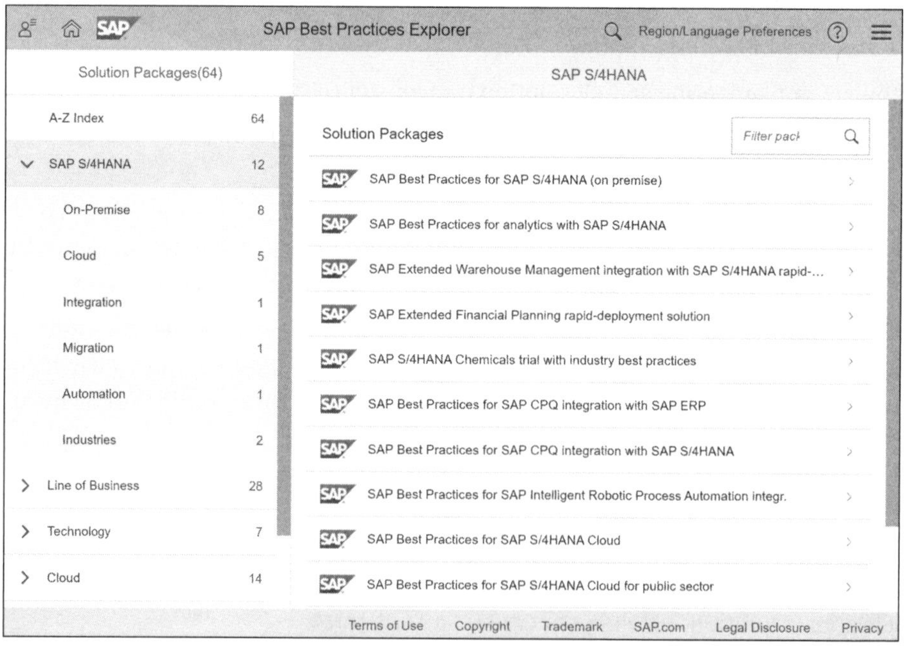

Figure 6.7 Navigation for Available SAP Best Practices Content

4. Consume the SAP Best Practices content for your selected scenario. Figure 6.8 illustrates the basic structure of the content for the SAP Cash Application scenario as an example.

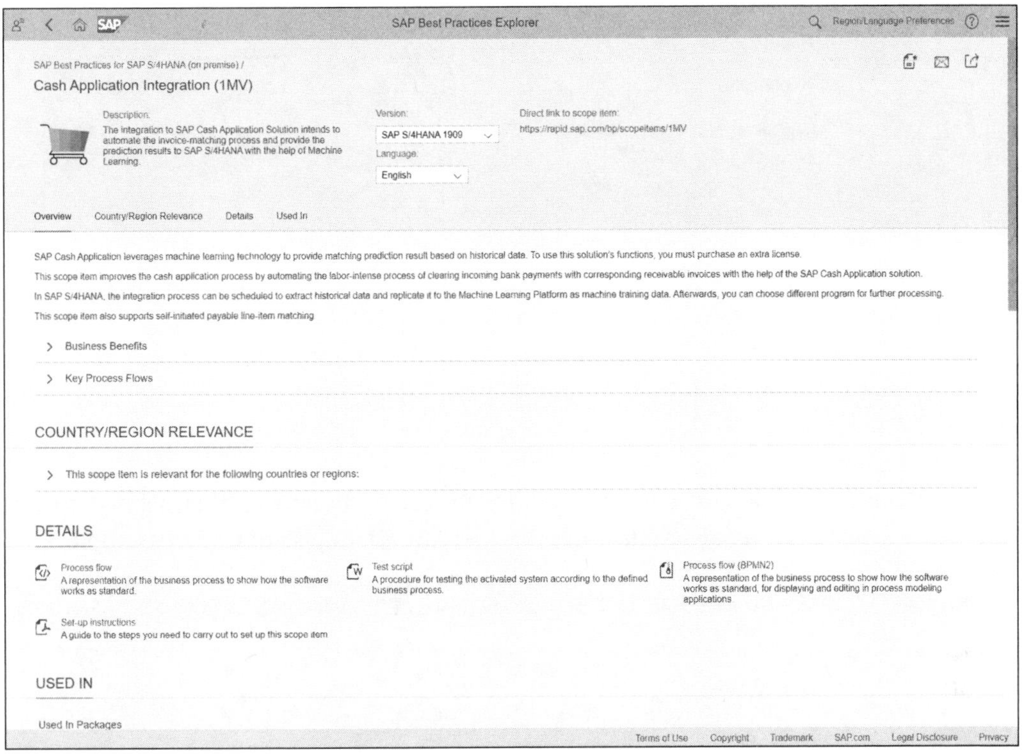

Figure 6.8 SAP Best Practices Content for SAP Cash Application

The header area contains a brief description of the scenario, including the underlying system release information. The **Overview** covers a more detailed explanation of the use case and lists the **Business Benefits** and **Key Process Flows** for the configuration. The countries and regions where the scope item is available are itemized in **COUNTRY/REGION RELEVANCE**. The **DETAILS** section is important because depending on the content, it covers the following information:

- **Process flow**
  This document describes the steps of the business process and their technical dependencies.
- **Set-up instructions**
  This document is actually the configuration guide and explains all the necessary steps for the configuration. Figure 6.9 shows this for SAP Cash Application as an example.

Content

1	**Purpose**	**3**
2	**Prerequisites**	**4**
3	**Configuration**	**5**
3.1	Setting Up SAP S/4HANA	5
3.2	App Configuration: Manage Payment Advices	6
3.3	Checking Scope-Dependent Appl. Job Catalog Entries	6
3.4	Scheduling Jobs – Automatic Bank Statement Reprocessing	7
4	**Appendix**	**8**
4.1	Ticket Component	8

Figure 6.9 Configuration Guide for SAP Cash Application

- **Test script**
  This section describes test procedures for each process step that belongs to this scope item. Figure 6.10 illustrates this for SAP Cash Application as an example.

Test Script
SAP S/4HANA - 07-05-20

PUBLIC

Cash Application Integration (1MV)

Table of Contents

1	Purpose	2
2	Prerequisites	3
2.1	System Access	3
2.2	Roles	3
2.3	Business Conditions	3
3	Overview Table	4
4	Test Procedures	5
4.1	Scheduling Accounts Receivable Jobs - Automatic Bank Statement Reprocessing – Receivables Clearing	5
4.2	Scheduling Accounts Receivable Jobs – Automatic Lockbox Processing (Optional)	7
4.3	Extracting Payment Advice Information	8
4.4	Scheduling Account Receivabes Jobs – Automatic Banks Statement Reprocessing – Payables Clearing	10

Figure 6.10 Test Script for SAP Cash Application

- **Process flow (BPMN2)**
  This section provides the file for the process flow in Business Process Modeling Notation (BPMN) format.

6.2.2 SAP Help Portal

For customers and partners, the SAP Help Portal is the main source containing the documentation for all SAP products, including machine learning and predictive analytics applications. The SAP Help Portal supports search-based interactions and provides navigation to product pages with logical groupings of topics. The necessary steps for using the SAP Help Portal are as follows:

1. Open SAP Help Portal using the following link: *https://help.sap.com*.
2. Navigate to or search for the solution for which you require the configuration steps. Figure 6.11 illustrates the start page of the SAP Help Portal, where we've entered "Machine Learning for SAP Utilities" in the search field as an example.

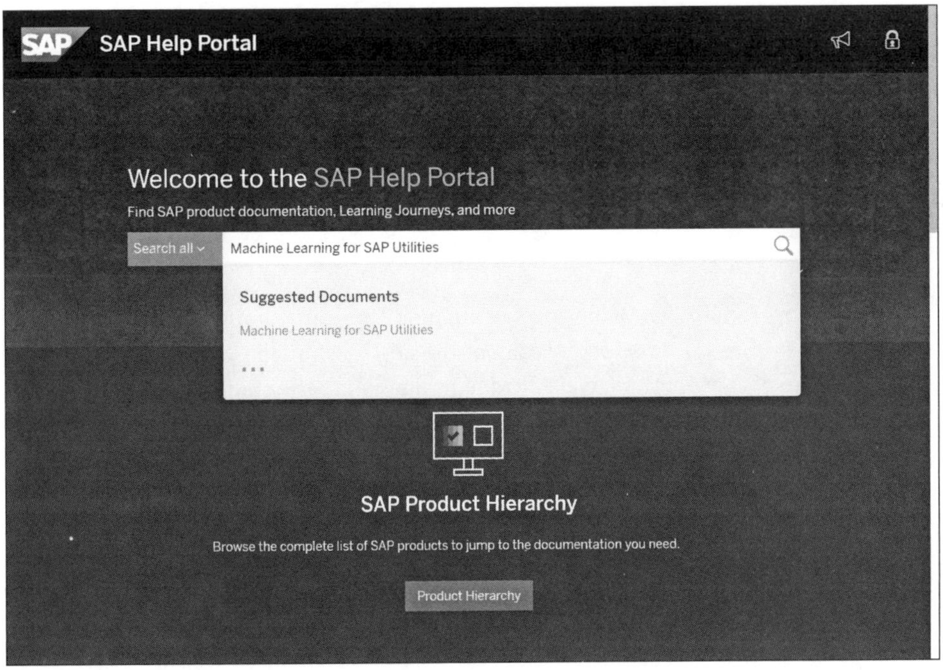

Figure 6.11 Launchpad for SAP Help Portal

3. Navigate to the configuration steps that are provided for your solution. Figure 6.12 shows the configuration steps for the SAP utilities machine learning use case.

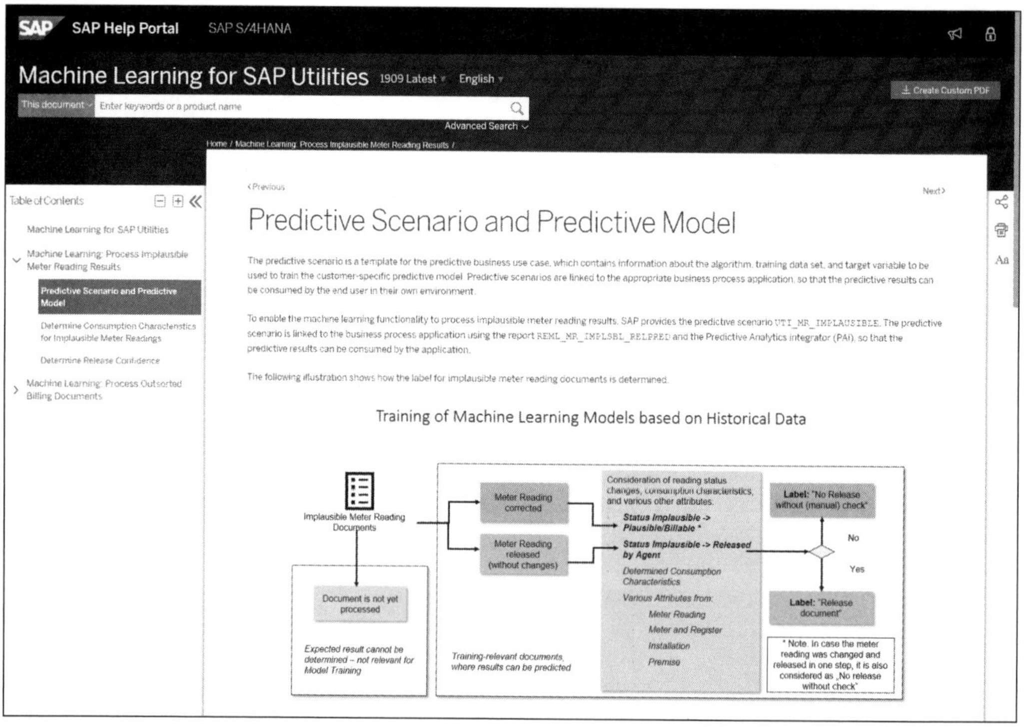

Figure 6.12 Configuration Steps for SAP Utilities Scenario

> **Note**
>
> As you're now familiar with the tools containing the configuration steps, we don't need to repeat this information for each of the described intelligent use cases. However, for ease of consumption, we'll provide navigation to the configuration guides in text boxes for all the described scenarios. Depending on the use case, we might provide links to SAP Best Practices Explorer and/or the SAP Help Portal.

6.3 Finance

In this section, we'll focus on use cases in the finance line of business (LoB) and how customers can benefit from this functionality. We'll give a brief overview of the use case requirements, the business processes involved, and how they're addressed. As you can imagine, the finance LoB is huge, with a lot of potential to automate and fine-tune the processes. This can help the AR/accounts payable (AP) clerks, the accountants and finance controllers, and others. We'll discuss topics centering on AR/AP, accounting and financial period-end close, financial planning and analysis, and governance, risk, and compliance (GRC) at length. These scenarios are discussed with a few ideas and use cases that are built with machine learning functionality.

6.3.1 SAP Cash Application

SAP Cash Application improves the cash application process by automating the labor-intense process of clearing incoming bank payments with corresponding receivable invoices. It's a cloud-based solution (via SAP Cloud Platform) that leverages machine learning technology to provide matching prediction results based on historical data. Therefore, the side-by-side machine learning approach is utilized with SAP Data Intelligence to build the machine learning services and eventually the machine learning application that is consumed by the business processes from SAP S/4HANA.

In SAP S/4HANA, the integration process can be scheduled to extract historical data and replicate it to SAP Data Intelligence as machine training data. You can then choose a different program for further processing. For example, you can schedule extraction of open receivable items and replication to SAP Data Intelligence for scoring. The prediction results from machine learning are proposed when the customer reprocesses the bank statement items.

Let's now look into the different functionality that is available as part of the SAP Cash Application. Figure 6.13 explains the process flow in the order-to-cash scenario for an AR manager/clerk. We'll explain how and where SAP Cash Application can help smooth the process.

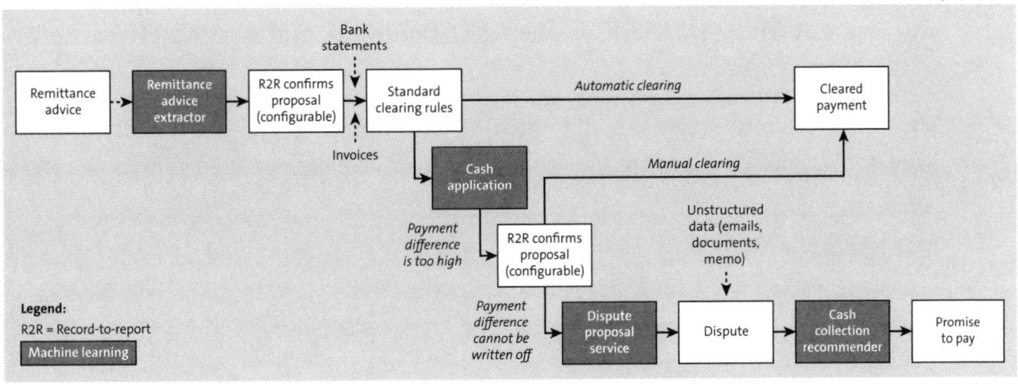

Figure 6.13 Order-to-Cash/Accounts Receivable End-to-End Process

In most organizations, a lot of manual effort is needed from the AR team or shared services team, from the moment they receive an order to the moment the payment is processed.

As companies grow their customer bases, AR needs to be able to process the increased volume of payments without adding more accountants. At the same time, it needs to keep the days sales outstanding (DSO) in check by ensuring that the cash doesn't sit too long in customers' accounts.

Existing tailored rules can achieve some level of automation but are hard and costly to implement or not comprehensive enough as many payments do not contain complete remittance information. A significant volume of payments still needs to be processed manually, such as if the payment contains no or incomplete invoice references, master data is not up to date, customers pay multiple invoices at once, and so on.

Machine learning applications seamlessly integrated into the existing business processes can be of great help, freeing up the AR teams from the daunting reconciliation efforts and letting them drive more strategic tasks.

The process flow in Figure 6.13 shows the end-to-end order-to-cash process enhanced with machine learning services. This is explained in the following steps:

1. The Remittance Advice Extractor application automatically extracts payment information from unstructured documents (PDFs, emails, papers, etc.) to be used in the payment clearing process.
2. The AR team confirms the proposal and moves to the payment clearing phase.
3. Standard clearing rules match the open receivables with the payments.
4. Payments that can't be processed by the standard rules are forwarded to SAP Cash Application, which infers matching proposals based on machine learning models applied to historical financial clearing information.
5. If the payment difference is too high, AR either clears the payment manually or rejects it, which triggers additional process steps that will be supported by additional machine learning SAP Fiori apps, like the Dispute Proposal Service and Cash Collection Recommender.

The machine learning services built into SAP Cash Application are further divided into a few topics because there are a lot of updates that have happened to help AR/AP staff.

Configuration Guide

Refer to *https://rapid.sap.com/bp/#/browse/scopeitems/1MV* for all the use cases related to SAP Cash Application in the SAP Best Practices Explorer.

Let's first look at receivables line item matching in the next section.

Receivables Line Item Matching

SAP Cash Application helps automatically match incoming payments with open receivables. It boosts the efficiency of the traditionally very labor-intensive clearing process, allowing it to be executed with nearly no user interaction. The application takes advantage of machine learning to enable an effortless setup and to increase automatic matching rates. Figure 6.14 explains the machine learning service built for the receivables line item matching functionality in SAP Cash Application, which we'll unpack in this section.

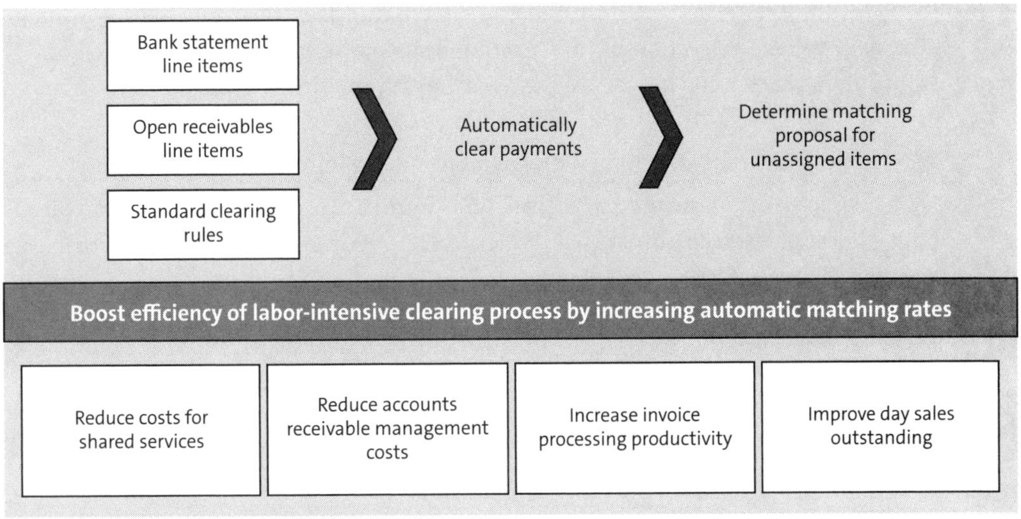

Figure 6.14 Receivables Line Item Matching

Note that the AR clerk needs to reconcile incoming bank payments with customer invoices and ensure proper payment is booked against the receivables. The following pain points are encountered by AR clerks when executing the regular job:

- Payment information may be incomplete (missing an invoice reference) or incorrect.
- Mismatched payments need to be processed manually, which is very labor-intensive and error-prone.
- Time dependent on reconciling the incoming bank payments.

Here are a few outcomes and benefits customers want to see to improve the business value:

- Improve DSO
- Reduce AR management costs
- Increase invoice processing productivity
- Reduce costs for shared services

The machine learning service helps the AR clerk by automatically matching the payments with open receivables without the need to manually match the data. In this scenario, algorithms such as random forest and XGBoost are used from the Scikit-Learn library. We won't be discussing the mechanics of these algorithms, only the business implementation—so that's where it ends for our purposes. These algorithms automatically learn the matching criteria from historical accountants' actions with the models running overnight and providing the needed results. This can help enable early detection of payment issues from customers who repeatedly send payments with the wrong invoice numbers. If this is customized, the system can clear the payments accordingly. The main value proposition of this machine learning service is to boost efficiency of labor-intensive clearing processes by increasing automatic matching rates.

Receivables Line Item Matching with Payment Advice Information Extraction

Let's now look into how the payment advice information extraction functionality is built and addresses customer needs. Figure 6.15 gives an overview of the machine learning service, which we'll explain in this section.

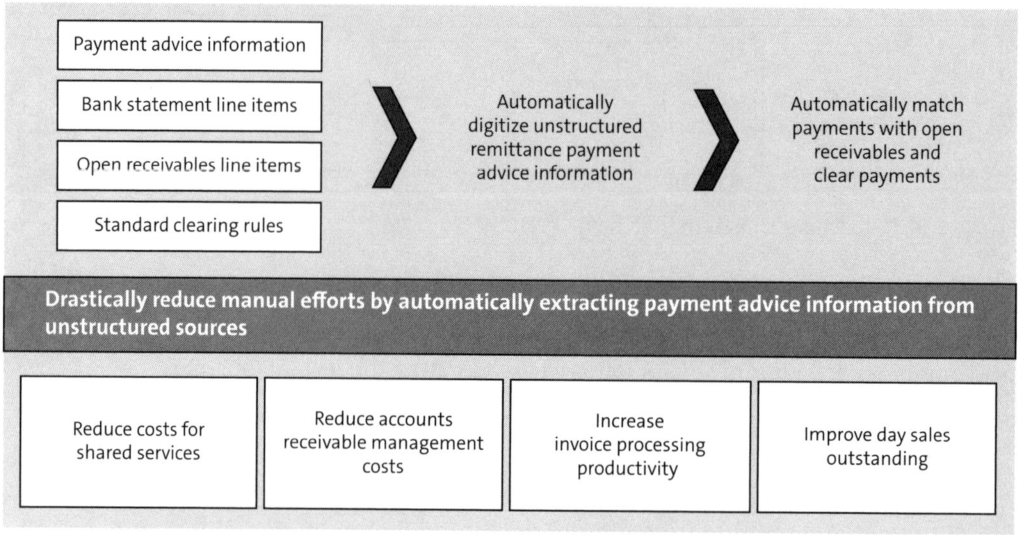

Figure 6.15 Payment Advice Information Extraction

How does the AR clerk and the shared services team leverage this machine learning service? Before going into the details, let's see what's needed. The AR clerk needs to record the payment advice notes received nonelectronically from customers. The payment advice note provides the information required for cleaning the customer accounts for invoice payments. Some of the pain points encountered by the AR clerk are as follows:

- Payment advice notes can be unclear or incorrect.
- Manual extraction from payment advice notes is time-consuming and error-prone.
- Manual reconciliation (matching) of payment advice notes and customer invoices can be difficult. Often payment advices are not entered into the system due to technical barriers and significant effort is needed to manually clear the payments.

Typically, there are a few outcomes and benefits the AR clerk and shared services can gain to improve the overall business value of their firm:

- Improved invoice processing full-time equivalent (FTE) efficiency
- Improved invoice receipt and processing FTE productivity

- Reduced payment processing and supplier inquiries with headcount costs
- Reduced external workforce invoice errors

Algorithms from the TensorFlow library are used in this scenario. The machine learning service automatically digitizes unstructured remittance advices/payment advices and extracts additional information about payments from unstructured advices (email, PDF, paper). This extracted information is used to prepare the automated clearing process via SAP Cash Application. The matching is done automatically and at scale by managing a wide array of customers' communication preferences. The main value proposition of the machine learning service is to reduce manual effort by automatically extracting payment advice information from the unstructured sources and using it to automate the clearing process.

Receivables Line Item Matching with Lockbox Information

Let's now look into how the lockbox information functionality is built and how it addresses customer needs. Figure 6.16 gives an overview of the machine learning service, which we'll unpack in this section.

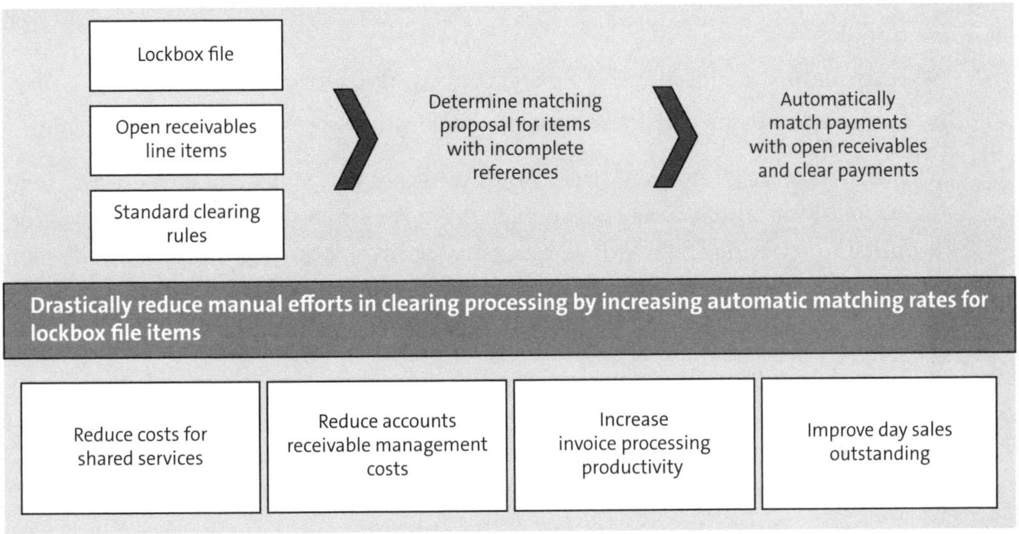

Figure 6.16 Receivables Line Item Matching with Lockbox Information

As the use of electronic payments increases, remittance information has become decoupled from payments, making it challenging to identify the payer or receivable in a timely manner. Remittance information and quality provided with checks will vary so that lockbox entries will also exhibit the same data quality issues, in addition to any manual entry errors introduced at the banking institution. The standard lockbox program has basic automation capability based on complete reference number matching, and additional manual effort or additional automation investment might be required.

Yes, there is an available solution! Lockbox entries can be sent to SAP Cash Application for enrichment through the machine learning program. The algorithms from the TensorFlow library are leveraged in this scenario. The machine learning model will identify the accounts and receivables based on the provided information and historical patterns. For example, the machine learning models can learn associations between the accounts and magnetic ink character recognition (MICR) or find receivables based on partial reference numbers. The enriched payment advice proposals are then returned to SAP S/4HANA, confirmed by an accountant, and cleared using the lockbox functionality provided in SAP S/4HANA. Auto clearing is also possible. The enriched fields include alternative customer, alternative company code, invoice number, and alternative account type.

Here are some of the business benefits of leveraging the lockbox functionality:

- Increases efficiency and reduced errors in the finance department
- Enables finance to focus on strategic tasks and scale with your business
- Processes incoming payments faster to reduce DSO
- Tightly integrated with SAP S/4HANA Cloud and SAP S/4HANA on-premise to extract the value from your historical data and maintain your current processing workflow
- Learns from accountant behavior and starts working for you immediately
- Continually adapts and doesn't require ongoing maintenance

The (shared service center) AR clerk needs to record the advice notices received non-electronically from customers. The payment advice note provides the information required for clearing the customer accounts for invoice payments. Some of the pain points encountered by the AR clerk are as follows:

- Payment advice notes can be unclear and/or incorrect.
- Manual extraction from payment advice notes is time-consuming and error-prone.
- Manual reconciliation (matching) of payment advice notes and customer invoices is tricky. Often payment advices are not entered into the system due to technical barriers and significant effort is needed to manually clear the payments.

By leveraging the lockbox functionality and the machine learning service, a lot of gains can be had:

- Improved invoice processing FTE efficiency.
- Improved invoice receipt and processing FTE productivity.
- Reduced payment processing and supplier inquiries headcount costs.
- Reduced external workforce invoice errors.

With the help of the lockbox functionality, you can automatically digitalize unstructured remittance advices/payment advices and extract any additional information about payments from unstructured advices. You can use the extracted additional information to prepare the automated clearing process via SAP Cash Application. The AR clerk would gain a lot by using this lockbox app because matching is done automatically and at scale. It allows for managing a wide array of customer communication preferences.

The value proposition is to drastically reduce manual efforts by automatically extracting payment advice information from unstructured sources such as emails, PDFs, paper, and so on, and use it to automate the clearing process.

Payables Line Item Matching

Let's now look at how the payables line item functionality is built and how it addresses customer needs. Figure 6.17 gives an overview of the machine learning service, which we'll unpack in this section.

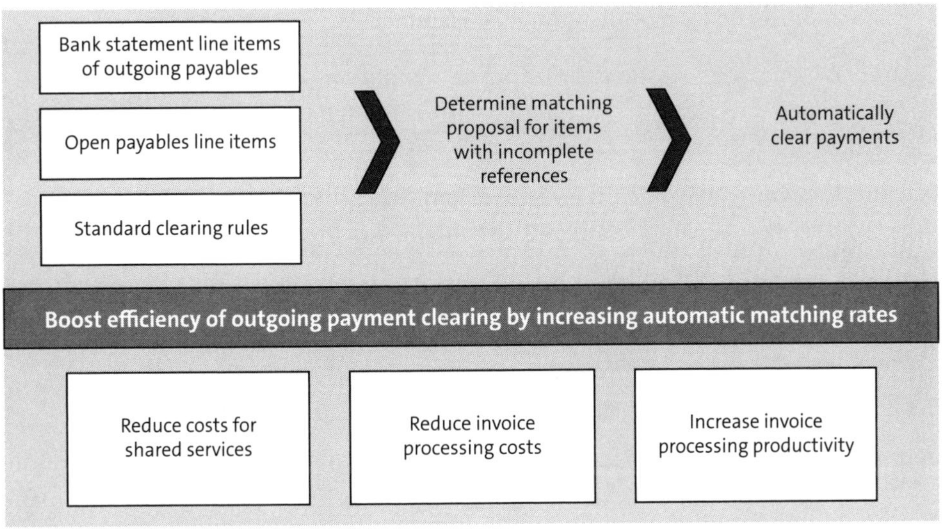

Figure 6.17 Payables Line Item Matching

Outgoing payments are payments customers make to their vendors; these are debit items in the bank statement. Vendor-initiated payments are initiated/triggered by vendors to pay a vendor invoice, which is deducted from the customer's bank directly. The system usually is unable to clear these payments automatically due to the minimum information stated on the bank statement.

The machine learning service will identify and propose the matching payables and automatically clear them with the following advantages:

- It learns from account behavior and starts working immediately.
- It continually adapts without requiring ongoing maintenance.

In this scenario, the random forest and XGBoost algorithms are used from the Scikit-Learn library. Basic data requirements include outgoing invoices, payments for vendors, bank statement items for outgoing payments, and so on.

The accountant needs to ensure the payment received is applied to the related payable item. The job is to match the incoming payments triggered by the vendor with the proper payable item.

This whole exercise is labor-intensive and requires investigative actions to match an amount to be paid, using the little information available in the bank statement. This whole process can reduce the labor required to match the payments with payables for machine learning-managed cases and enable a focused effort on the more challenging issues. This process also promotes the following:

- Increased staff productivity
- Reduced manual errors
- Faster month-end close with automated clearing

Customers can expect a faster clearing process combining standard rules and machine learning-generated clearing. The manual effort is limited to cases not yet managed by the model.

Machine learning enabled clearing versus manual clearing enables scalability and accuracy and leverages a customer's expertise contained in the system data.

The value proposition is to boost the efficiency of outgoing payment clearing by increasing automatic matching rates.

6.3.2 Accounting and Financial Close

In this section, we'll look into some of the machine learning services built for accounting and financial period-end close. Revenue accounting and reporting is used to manage the revenue recognition and revenue accounting process according to statutory regulations such as International Financial Reporting Standards (IFRS) and Accounting Standards Codification (ASC).

Inventory accounting is used to value and monitor your material and work-in-process inventories according to legal regulations and management accounting requirements. All goods movements are valued in the material ledger, which supports parallel, real-time valuation of inventories in multiple currencies. A special focus is on high throughput of logistics data that allows for managing massive data volumes. You can choose to value your material inventories at standard cost or moving average automatically. In addition, you may make manual adjustments to material costs and inventory values. You may also use periodic valuation of material inventories according to statutory requirements (such as lowest value, last-in-first-out [LIFO], or first-in-first-out [FIFO]) or product cost management requirements (such as actual costing or standard costing).

You can perform a local close for one or several companies of your corporate group monthly, quarterly, yearly, or at other intervals, with a freely definable range of closing operations. Local accounting principles and/or principles applied by the group can be taken into account. You use a consolidation foundation to create and monitor your financial data collection for corporate close, either for legal purposes or for the purposes of consolidated management reporting. This process offers a high degree of flexibility in the data collection process, as well as the configuration of different accounting principles. You can highly integrate with the accounting solution to automate the consolidation data collection process.

We'll briefly discuss the process and how machine learning services can improve it, beginning with the financial account reconciliation with goods receipt/invoice receipt (GR/IR).

Clear Goods Receipts and Invoice Accounts

Figure 6.18 explains how the machine learning service improves the process of clearing the goods receipt and invoice accounts in the context of the financial account reconciliation. We'll explain the process in this section.

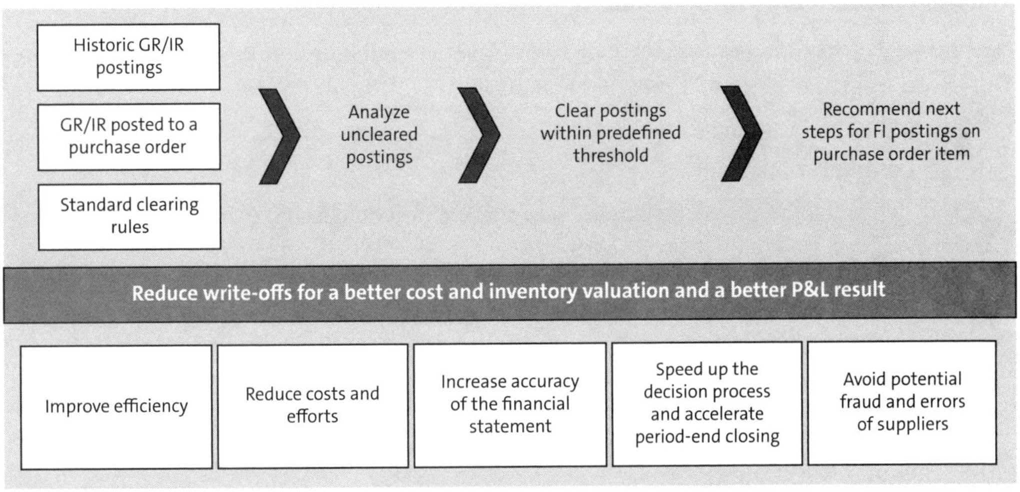

Figure 6.18 Clear Goods Receipts and Invoice Accounts

GR/IR account reconciliation is an exception-handling process for all purchase order items with differences between goods receipts and invoice receipts. The differences can be written off, or the root cause can be identified via the accounting, buying center, logistics, and supplier spheres.

There are a lot of pain points involved in the high manual effort required for creating reports, gathering all relevant information and investigating the root cause, and solving the issue by communicating with various departments. These efforts delay the period-end closing.

Customers can benefit from additional business value if this can be handled automatically. Some of the gains are explained here:

- Improve the accuracy of the financial statement.
- Reduction of write-offs leads to a higher accuracy for cost and inventory valuation, as well as a better profit and loss (P&L) result.
- Increased accuracy of supplier liabilities.
- Avoiding potential fraud and supplier errors.
- Compliance with corporate rules requires maintenance of the GR/IR accounts through Sarbanes-Oxley Act (SOX) compliance.

Pain points in this area can be countered through the powerful reporting of embedded predictive analytics algorithms in SAP S/4HANA Cloud. SAP S/4HANA Cloud provides real-time insights into all relevant information based on a simple and flexible SAP Fiori UI with integrated collaboration features. In addition, keeping track of process information enables process analysis and optimization.

Eventually, higher efficiency is attained through intelligent recommendations by the machine learning service that helps with faster period-end closing through less manual effort and real-time data analysis.

The value proposition is achieving better P&L results for companies; reducing write-offs improves cost and inventory valuation.

Configuration Guide

For configuration instructions, refer to the SAP Best Practices Explorer at *https://rapid.sap.com/bp/#/browse/scopeitems/2ZS*.

Accruals Management

Figure 6.19 depicts the machine learning service built to help understand the mechanics of the accrual management and accruals prediction processes. We'll unpack it further in this section.

The accruals prediction functionality estimates and predicts accruals in order to support cash flow planning. You can adjust the linear value proposed by the system based on your own knowledge of the purchase order, feedback of the business owner, and reported values to date.

Some of the pain points encountered include very hard to estimate accruals as projects rarely run synchronously, systems proposing linear values, adjustments always based on accountants' knowledge of the purchase order, reported values not up to date, and the feedback of the business owners being necessary. We can counter these issues by using existing data to group certain purchase orders, using the historical purchase order data to recommend how the purchase orders should be accrued, and so on. This data is used in creating the machine learning service for addressing these challenges.

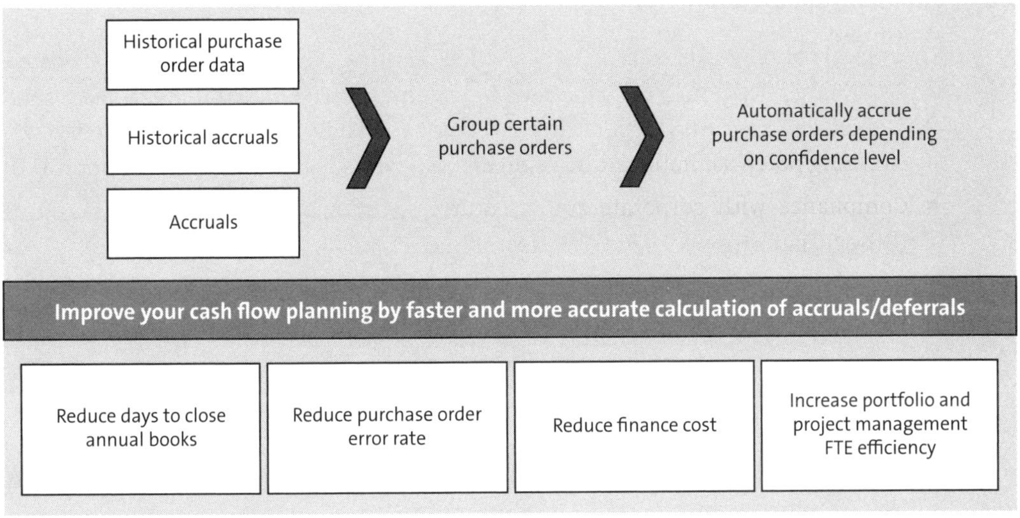

Figure 6.19 Accruals Prediction

Eventually leveraging the machine learning service, there will be gains pertaining to a reduced number of days to close the annual books, reduced purchase order error rates, increased portfolio and project management FTE efficiency, and reduced finance cost. The customer will benefit by automatically accruing the purchase orders, calculating the accurate total of the accruals, and so on. Algorithms such as the random forest and XGBoost are used from the Scikit-Learn library in the functioning of this side-by-side machine learning service consumed by the SAP S/4HANA processes.

The value proposition is to improve the cash flow planning via faster and more accurate calculation of accruals and deferrals.

Accruals are estimated amounts on a purchase order that have not been paid or billed yet because of the current status of a project. For example, assuming a project was estimated to cost 12,000 total over the duration of a year, the linear accrual would be 1,000 per month. However, projects rarely run that synchronously; perhaps in the first three months, nothing is paid instead of 3,000, but by month eight, the project is 80% finished, so 9,600 should be paid instead of 8,000. Currently accruals are estimated by an accountant who adjusts the linear value proposed by the system based on his knowledge of the purchase order, the feedback of the business owner, and reported values to date.

Leveraging the machine learning service can help solve this by basically using existing data to group certain purchase orders and, based on their past history, predict how they should accrue:

- Each purchase order that is accrued automatically reduces manual work.
- Accurate grand total accrual value helps avoid wrong cash flow planning and unnecessary capital costs.

The typical period-end closing activity also includes any accruals that need to be posted when liabilities occur for a third party with related costs. Flexible configuration options are provided by the SAP S/4HANA accrual engine to handle these kinds of accrual costs. The monthly accrual amounts can optionally be reviewed and accruals can be approved.

Configuration Guide

For configuration instructions, refer to the SAP Best Practices Explorer at *https://rapid.sap.com/bp/#/browse/scopeitems/3NF*.

Predictive Accounting

Figure 6.20 explains the machine learning service built to help understand the mechanics of the predictive accounting process, forecasting future revenues, costs, and liquidity. We'll unpack this service further in this section.

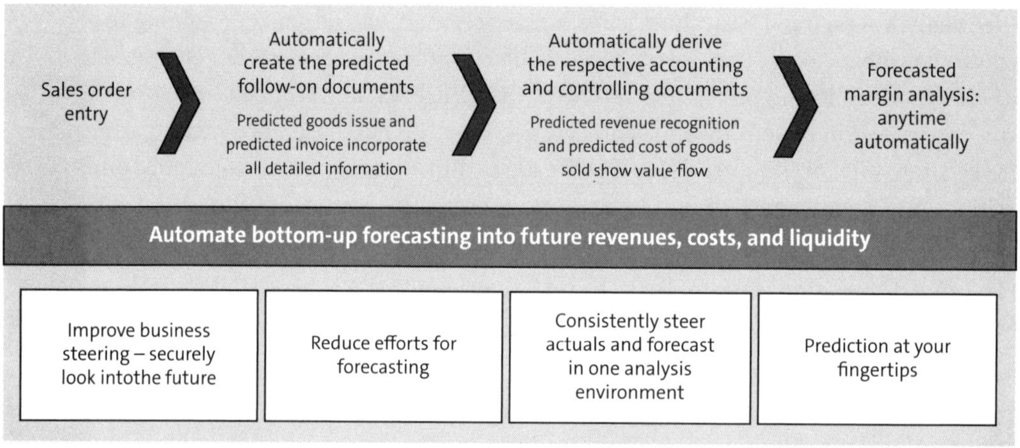

Figure 6.20 Predictive Accounting

In the traditional accounting approach, the account team had to perform a series of closing activities before the stakeholders could view how the company was performing. Businesses also wouldn't have the data readily available for making business decisions on the fly. This led to various innovations and changes in the SAP system for faster closing processes and reductions in the number of closing activities, which reduces the overall closing cycle. This is typically called *continuous accounting* because you can have most of the data in real time and not need to wait for the closing cycle to be completed. With the fast-changing business and industry processes, there was a need to have predicted accounting data for future periods as well as in a single source of truth table, like the Universal Journal, thus making the shift from continuous accounting to predictive accounting.

With the predictive accounting machine learning service in SAP S/4HANA, you can automatically extend the reach of accounting by consuming the machine learning service. This is the next step in continuous accounting, which helps organizations plan into the future for business operations and to make strategies. This functionality focuses on the revenue that organizations are making based on incoming sales order data, even if no goods receipt or invoice is booked. Basically, predictive accounting is an extension of your main reporting with forward-looking data. SAP provides SAP Fiori apps out of the box to see the impact of predictive accounting on gross margin and the like.

There are two steps involved here:

1. **Incoming sales orders translate into predicted revenue**
   Provides a comprehensive overview of all sales orders and their values for the time period regardless of the billing status. Predictive goods issued and billing documents are posted in the prediction ledger.
2. **Commitments translate into expected expenditures**
   This includes cost assignments to a work breakdown structure (WBS), order, cost center, and supplier. You also can view commitments for derived characteristics such as profit center, segment, and functional area.

The benefits of predictive accounting include the following:

- Accurately forecast upcoming revenue based on sales order predicted receivables and profit margins.
- Provide availability of reports for working capital as a future date based on the predicted revenue and expected expenditures.
- Identify exceptions and opportunities.
- Have an overview of future cash flow and other similar reports based on predictive data for future periods.

Configuration Guide

For configuration instructions, refer to the SAP Help Portal at *https://help.sap.com* and search for "Predictive Accounting".

6.3.3 Financial Planning and Analysis

In this section, we focus on the topics of financial planning and analysis and the machine learning services built to aid the customer during these processes. We will discuss cost management and profitability analysis, profitability and cost analysis, product cost management and overhead accounting, and more.

SAP Financial Statement Insights

Figure 6.21 explains the machine learning service built to detect the smart alerts for profit and loss leveraging SAP Financial Statement Insights, which we'll discuss in this section.

> **Note**
>
> SAP Financial Statement Insights is currently being retired as of the time of writing (summer 2020). This solution will continue to be available to existing customers, but there will be no new sales and marketing for this offering.

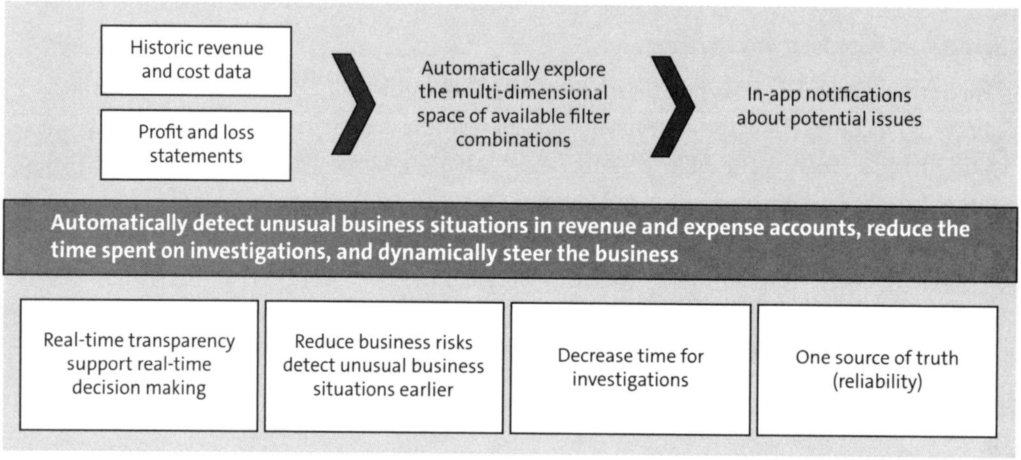

Figure 6.21 Smart Alerts for Profit and Loss

The controlling department needs to monitor, confirm, and report the financial condition of a company by analyzing accounting data and by P&L analysis. It also needs to provide an array of information needed by the business executives and to guide the financial business decisions for business executives. The challenges encountered by the controlling department are that its tasks are time-consuming and cumbersome and involve high effort. There are a lot of manual tasks with only a little system support for the controlling department, with complex UIs and no real-time reports.

The benefits and outcomes customers want are mainly real-time transparency, reduced business risks, support in real-time decision-making, and one reliable source of truth.

This is handled with in-app notifications about anomalies and automatically exploring the multidimensional space of available filter combinations to find attribute combinations that correlate to unusual behavior. Time for investigations is decreased by automatically detecting unusual business situations with unusual increases and decreases in single key figures and changes in trending behavior.

The controlling department could get additional help via the following actions:

- Showing detailed posting information for unusual business situations
- Providing smart alerts that will proactively inform it of potential issues
- Having lists of changes in trending behavior
- Viewing single line items for business exceptions found

The value proposition is to autonomously detect unusual business situations in revenue and cost accounts, which helps to decrease time for investigations and dynamically steer the business.

Overall, the smart alerts machine learning service aims to autonomously detect unusual business situations in revenue and cost accounts. This machine learning service will also proactively inform the group controller of potential issues and propose explanations for unusual business situations by analyzing the revenue and cost data. This is an embedded machine learning service leveraging the regression algorithms from the Automated Predictive Library (APL) on SAP Cloud Platform.

Smart alerts for P&L analysis allows your controller to do the following:

- Automatically explore the multidimensional space of available filter combinations
- Detect unusual increases and decreases in single key figures

The business benefits are decreased time for investigation and earlier detection of unusual business situations.

Configuration Guide

For configuration instructions, refer to the SAP Help Portal at *https://help.sap.com* and search for "SAP Financial Statement Insights for Profit and Loss".

SAP RealSpend

Figure 6.22 explains the machine learning service built to detect unusual expenses leveraging SAP RealSpend. We'll unpack the process shown here in this section.

Users with budget responsibilities perform spend analysis and get real-time aggregations on cost accounts and cost objects. In addition, the user can compare their cost situation against budgets from SAP S/4HANA or set budgets manually for any cost object/cost account. The challenges encountered by the users typically are complex UIs and reports, increasing complexity with higher volumes of expense bookings, time-consuming analysis, and highly cumbersome efforts needed.

Basically, customers want more real-time transparency, support for their real-time decision-making, a single source of truth for reliability, and the ability to report new dimensions and characteristics on the fly.

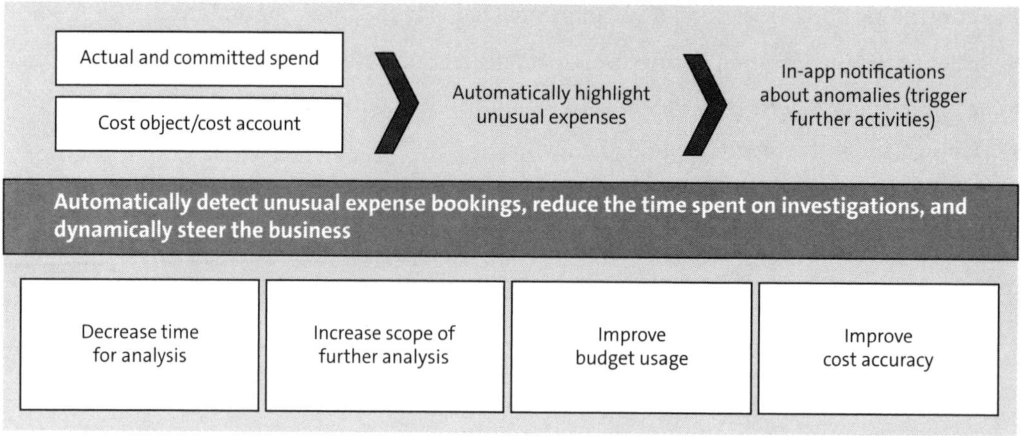

Figure 6.22 Detect Unusual Expenses

The users' pain points could be countered by automatically exploring the multidimensional space of available filter combinations to find attribute combinations that correlate with unusual behavior. Decreasing the time for investigations by automatically detecting unusual business situations would help as well.

The users could benefit from showing detailed posting information for unusual business situations, seeing in-app notifications about anomalies, and viewing single line items of identified business exceptions.

The value proposition is to autonomously detect unusual business situations in corporate spend, helping to decrease time for investigations and dynamically steer the business. Anomaly detection will proactively inform the controller of potential issues.

As a manager, staying on top of the budget requires frequently browsing cost center reports to find potential issues like wrong bookings, double bookings, amounts that are too high, and/or fraud.

Hence the SAP RealSpend machine learning solution helps with anomaly detection, reducing the need to browse randomly by automatically bringing noteworthy situations to the attention of the responsible manager. This uses the embedded regression algorithms from the APL for identifying and detecting the anomalies.

This offers great business value:

- Cost overruns are reduced by reducing the risk of issues being discovered too late to take action.
- The effort and cost of maintaining situational awareness are reduced.

Configuration Guide

For configuration instructions, refer to the SAP Best Practices Explorer at *https://rapid.sap.com/bp/#/browse/scopeitems/1KU*.

Invoice Payment Forecasting in Accounts Receivable

In this section, we focus on the invoice payment forecasting and the dashboards built for AR managers. AR is typically the largest asset for any organization's financial statements. Using the results from a prebuilt machine learning model, payment predictions are presented to enable the collections managers to prioritize their time on high-value invoices and customers with poor predicted payment behavior.

The idea is to analyze the current state of AR with a view of currently overdue invoices and those that will be due in the future. The machine learning model leverages the regression algorithms from APL via the ISLM framework. Here we're leveraging the concept of extending the predictive model functionality with SAP Analytics Cloud from the business processes in SAP S/4HANA. The regression model learns from historically cleared invoices and generates payment predictions for outstanding invoices. Predictions are made for whether invoices will be paid on time or late and, if late, by how many days. The predictive models are built in SAP Analytics Cloud by acquiring the data from a custom CDS view designed in the finance business process in SAP S/4HANA. The custom CDS view contains SAP S/4HANA Finance data from the general ledger (table ACDOCA).

The predictions dashboard in SAP Analytics Cloud is based on a model built from the custom CDS views in SAP S/4HANA. The dashboard in SAP Analytics Cloud helps to drill down into individual customers and understand their historical payment behavior, their current overdue invoices, and predictions of when they will pay outstanding invoices. With a view of historical and future behavior, you can take action to prioritize high-risk customers or renegotiate payment terms to minimize late payments and increase future cash flow visibility.

6.3.4 Governance, Risk, and Compliance

Let's now discuss the machine learning services built in the domain area of governance, risk, and compliance. Organizations are growing rapidly and changing frequently due to many situations and circumstances. Hence governance of these changes, managing risk for these changes, and adhering to compliance becomes even more critical. This is where SAP governance, risk, and compliance solutions help address organizations to manage the key operations and navigate through any risks with three main focal points: analyze risks, manage regulations, and monitor compliance. Some key features include the following:

- Identify risks and improve the risk-management activities
- Manage risk efficiently with little or no complexity
- Effectively prevent fraud in business processes and manage audits better
- Improve organizations' performance by protecting their core values

In the following sections, we dive into the details of how to leverage the machine learning service built around fraud identification/prevention and tax compliance.

Business Integrity Screening

Figure 6.23 explains the process involved in the machine learning service built around fraud identification and prevention in the SAP Business Integrity Screening application. We'll explain this process further in this section.

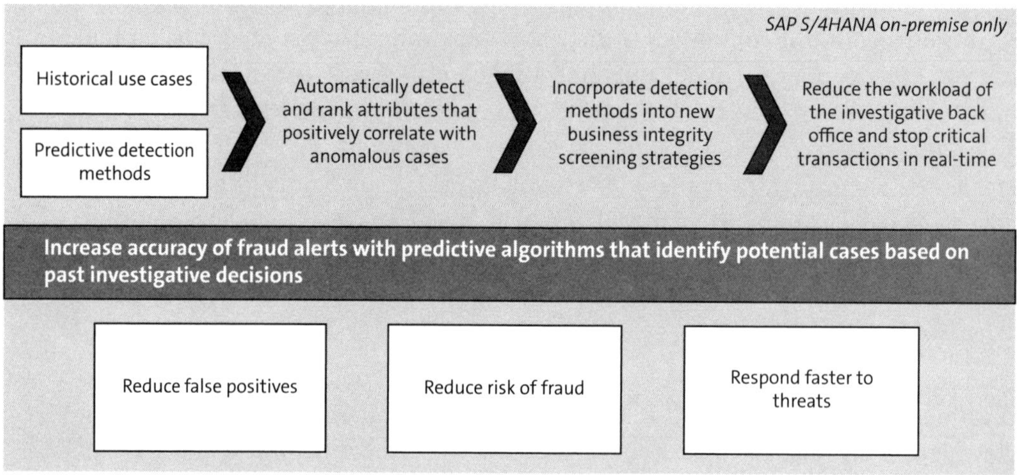

Figure 6.23 Fraud Identification and Prevention

Let's start our analysis with the problems faced by the customers, the pain points and challenges, and how the SAP Business Integrity Screening machine learning application can help address these issues.

Fraud investigators and screening specialists must handle an ever-increasing workload of suspicious or anomalous cases that have to be decided on and classified weekly, daily, and hourly. Some of the typical challenges faced by investigators pertain to time wasted on cases that are at the top of the queue but turn out to be of little or no risk or impact for the company, whereas some high-risk cases are never taken up. These investigators would benefit a lot from working on fewer cases that are more important to the existing workforce without losing the bulk of the most impactful cases.

So instead of working on a random queue of cases, predictive detection methods rank each new case according to the probability of it being impactful for the company's business. This would help the companies and greatly benefit the investigators with more productive work.

The investigators without deep knowledge of data science (citizen data scientists) leverage the threshold parameter of predictive detection methods to manage the workload and its prioritization, leading to important gains in efficiency and profitability.

As shown in Figure 6.23, by responding faster to threats and creating processes that deter fraud, organizations can take steps to prevent disruptions to critical business processes.

SAP Business Integrity Screening helps you reduce the risk of fraud by improving analysis capabilities, increasing visibility into possible fraud activities, and enabling better management decisions. It supports fraud scenarios relevant to your organization such as employee theft, corruption, and warranty fraud.

Powered by in-memory database technology and integrated into SAP S/4HANA and other third-party sources, this app allows investigators to collect, analyze, and act on very large amounts of data in real time.

SAP Business Integrity Screening now ships predictive detection methods that use predictive algorithms to identify potentially fraudulent use cases, learning from past investigative decisions garnered by SAP Business Integrity Screening. Algorithms can be updated regularly as fraud attack strategies change: by the day, hour, and minute. The classification and regression algorithms from the APL utilizing the ISLM framework are leveraged, and these algorithms are embedded into the SAP S/4HANA business processes.

The resulting real-time information gets to the right people to make the right decisions, yielding actions that effect the detection or prevention of a fraud attack.

To summarize, SAP Business Integrity Screening allows your business analysts and investigators to do the following:

- Automatically detect and rank attributes within classified data that positively correlate with anomalous cases
- Incorporate them with existing detection methods into new SAP Business Integrity Screening strategies
- Seamlessly integrate algorithms from APL
- Leverage robust, state-of-the-art machine learning algorithms

In a nutshell, this program offers the following benefits:

- Detection of new suspicious patterns
- Reduction of false positives
- Seamless integration of machine learning algorithms into the proven SAP Business Integrity Screening framework

Configuration Guide

For configuration instructions, refer to the SAP Help Portal at *https://help.sap.com* and search for "Business Integrity Screening".

Tax Compliance

Figure 6.24 gives an overview of the machine learning service built into the process of SAP Tax Compliance to address compliance issue processing, which we'll explain in this section.

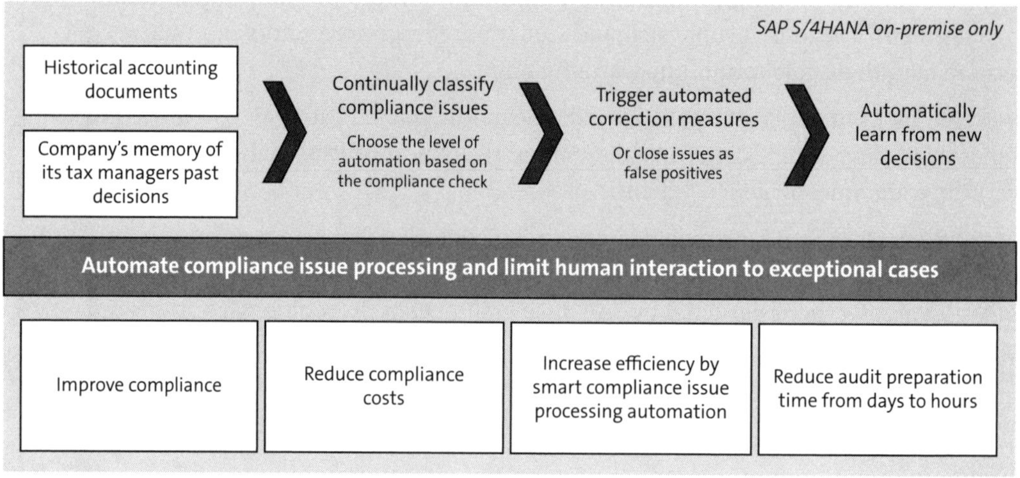

Figure 6.24 Compliance Issue Processing

With every new financial decision an organization makes, the company's tax knowledge base grows, and SAP Tax Compliance leverages this by continually classifying compliance issues and triggering automated correction measures or closing them as false positives. In rare cases, human interaction is requested for the decision-making.

Let's start our analysis with the problems faced by the customers, the pain points, and challenges, and how this SAP Tax Compliance machine learning solution can help address these issues.

The machine learning feature bridges the remaining gap in process automation between detection and issue remediation. The machine learning service enables automated decisions in which intensive manual interaction by bottleneck resources with domain expertise was irreplaceable. Some of the challenges encountered include releasing domain experts from routine tasks and putting their focus on critical issues. This leverages the regression algorithm from APL with the ISLM framework and is embedded into the SAP S/4HANA business processes.

Customers want outcomes and benefits with business value. This means that although SAP Tax Compliance enables the check of the entire data it doesn't necessarily mean that more domain experts are requested. The confidence level in compliance increases significantly with support dispatchers reducing the workload and concentrating on more rare, nonroutine cases. This also benefits the investigating clerks, who can avoid working on insignificant cases.

Learning from historical decisions and having the system propose the next steps helps to avoid triggering an unnecessary investigation. Also, this makes it easier to decide which hits require an immediate, thorough investigation and which can be not acted upon.

SAP Tax Compliance lets the customer gradually lower the dispatcher's workload in a controlled manner. First the dispatcher gets proposals for the dispatching decision, then the proposals can be calibrated/fine-tuned based on historical decisions to reduce the monetary risk; finally, the closing and a good part of the dispatching can be completely automated. Once trust in the system's decisions is established, investigating clerks can benefit automatically in a big way.

To summarize, the SAP Tax Compliance machine learning service allows your tax managers to do the following:

- Transparently apply SAP HANA's automated classification algorithms
- Benefit from the company's memory of its tax manager's past decisions
- Automate compliance issue processing and limit human interaction to exceptional cases
- Choose a level of automation on a compliance check basis

In a nutshell, the business benefits from leveraging the machine learning service include automatically learning from new decisions, increasing efficiency via smart compliance issue processing automation, and seamlessly integrating machine learning algorithms into SAP Tax Compliance.

The value proposition is provided by drastically reducing the manual efforts by automatically closing tax compliance cases or triggering a workflow to resolve them, leaving the dispatcher with a much-reduced workload for manual decisions and investigating clerks with mostly nonroutine cases. The path to automation is smooth and completely controllable by the customer on the individual check level. There are no special data requirements because the system rather rapidly (after one to two months) collects the necessary data to underpin the machine learning model. The regression algorithm yields quality indicators that inform the user when the model is worth considering for productive usage and ready to be trusted.

Configuration Guide

For configuration instructions, refer to the SAP Help Portal at *https://help.sap.com* and search for "Tax Compliance".

6.3.5 Detect Abnormal Liquidity Items

Figure 6.25 shows the machine learning service built around the process of cash and liquidity management, which we'll explain in this section.

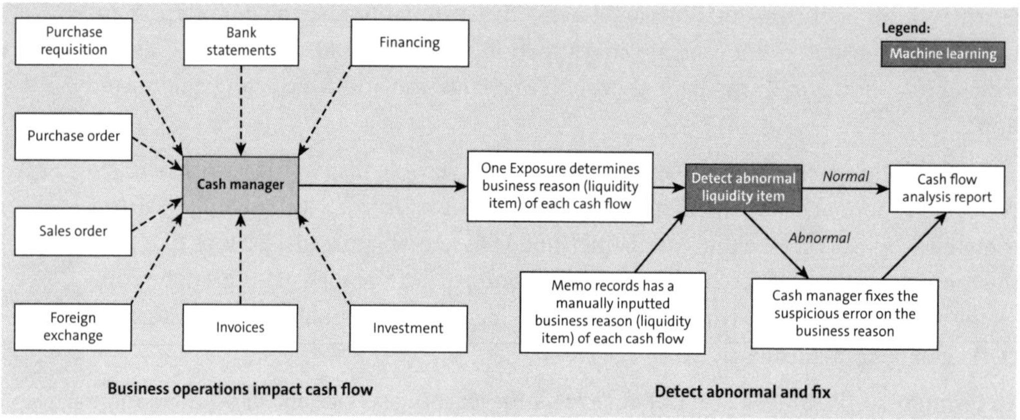

Figure 6.25 Cash and Liquidity Management

Customers have difficulty maintaining meaningful liquidity items considering the complexity of business processes, and it's also possible to define improper liquidity items when assigning them manually. Liquidity items are critical information for business users to understand the use and source of cash flows.

The business value for the customer can be measured in several ways. Liquidity items cannot be derived based on the configured rules because the business process isn't covered or because new business processes exist in which the liquidity item is wrong/improperly defined with manual intervention. More meaningful liquidity items cannot be derived because of improper setup. All these cases can be improved by leveraging machine learning functionality.

Cash managers and specialists want to analyze the business reason for the cash flow and detect abnormal liquidity items, eventually reducing errors. The machine learning service could discover potential errors and reduce the wrong liquidity items. More accurate liquidity assignment helps with more accurate analysis of the business operations. There is also a huge potential in discovering the percentage of errors.

The machine learning service would provide a liquidity forecast as a service to increase the accuracy. The idea is to forecast more accurately the future cash flow with machine learning capabilities and statistical models and show insights gathered by the machine learning algorithms as additional analytics reporting. Categorization and classification patterns are identified by leveraging the multiple decision tree and random forest algorithms of APL and embedding these into the SAP S/4HANA business processes with the ISLM framework.

The business benefits include the following:

- Detection of fraud in early stages
- Identification of correlations in the cash flow data

- Understanding the main influencers of the timing of cash flow and how they impact the difference between the actual date and planned date

Configuration Guide

For configuration instructions, refer to the SAP Best Practices Explorer at *https://rapid.sap.com/bp/#/browse/scopeitems/30K.*

6.4 Sourcing and Procurement

In this section, we'll focus on the machine learning services and embedded use cases built in the domain of souring and procurement. Figure 6.26 gives an overview of the procure-to-pay process in detail.

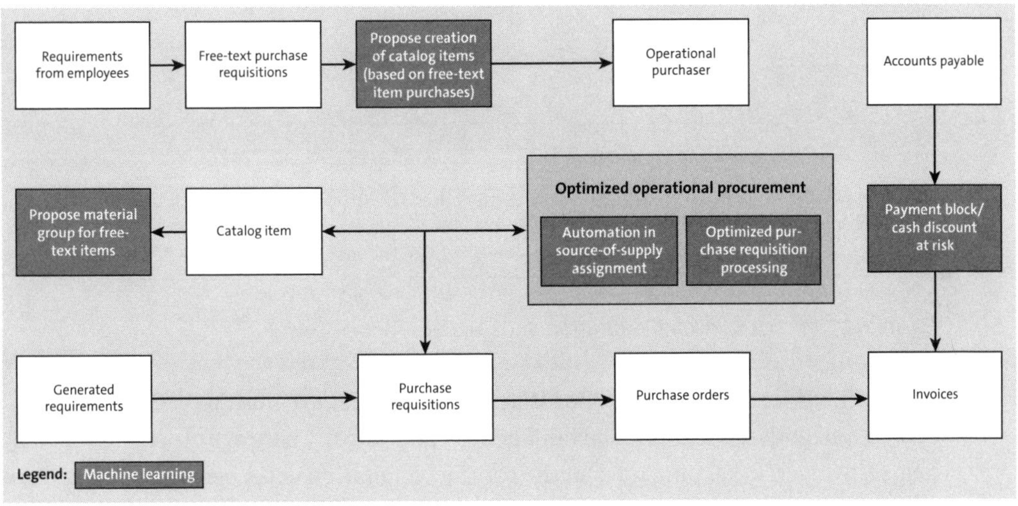

Figure 6.26 Procure-to-Pay Process with Machine Learning

Let's briefly review the procure-to-pay scenario and how the machine learning functionality is embedded into the process. Every organization that operates a business has to purchase materials, such as raw materials and services. The company can procure those materials via in-house production or external procurement.

At a high level, the procure-to-pay process consists of the following processes:

1. Material requirements planning
2. Supplier selection
3. Request for quotation
4. Purchase requisition
5. Purchase order
6. Goods receipt
7. Goods receipt invoice
8. Invoice verification
9. Payment to suppliers

In the subsequent sections, we'll look into some of these processes and how machine learning services are embedded or consumed with SAP S/4HANA.

6.4.1 Contract Consumption

We'll now focus on the machine learning service built around the quantity contract consumption functionality in SAP S/4HANA, which predicts the contract consumption and expiration ahead of time so that buyers can renegotiate contracts.

Figure 6.27 shows the machine learning service built for this process, which we'll explore in this section.

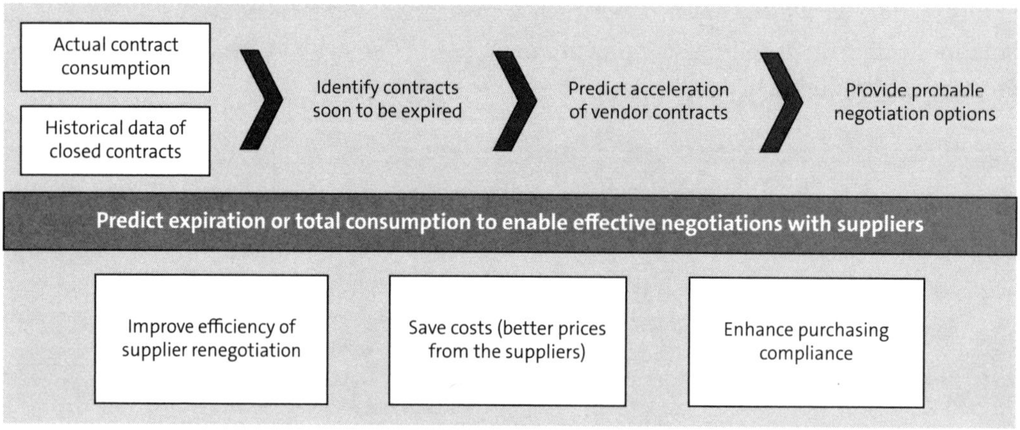

Figure 6.27 Predict Contract Expiration and Consumption

Procurement leaders need to renegotiate contracts that are soon to be expired (to be able to renegotiate them with advantageous conditions). There are quite a lot of challenges involved in keeping track of the contract expiration dates because contracts are often consumed sooner than expected.

The business benefits to the customers are early and efficient supplier renegotiation, better prices for goods from suppliers, predicted expiration of the contract ahead of time, and enhanced purchasing compliance. The machine learning algorithms identify the contracts that are at the end of the term and provide the probability for whether a contract is going to be consumed sooner than expected. A prebuilt set of KPIs allow robust analytics on SAP S/4HANA Cloud data with drilldown functionality, enhancing usability for the businesses by visualizing with predictions and providing probable negotiation options.

Basically, the SAP S/4HANA sourcing and procurement functionality enables companies to manage spend across every major category, reduce direct costs and administrative burden, and shorten the overall cycle time.

Together with predictive services from SAP Data Intelligence, the sourcing and procurement process is further enhanced with new insights into contract consumption dates.

Based on historical dates of closed contracts, the system predicts when the vendor contracts are going to be consumed at 100%, enabling the buyer to proactively plan contract renegotiations with suppliers right on time. SAP S/4HANA already analyzes contract consumption for quantity and value contracts by showing contract consumption per month. Now with the machine learning service for predicting contract consumption, companies can further improve contract negotiation, reduce overall costs, and enhance vendor management. In the current scenario, regression algorithms from APL utilizing the ISLM framework are leveraged and embedded into the SAP S/4HANA procurement business processes so that the predictions for contract expiry are provided ahead of time.

6

Configuration Guide

For configuration instructions, refer to the SAP Best Practices Explorer at *https://rapid.sap.com/bp/#/browse/scopeitems/1QR*.

6.4.2 Propose Resolution for Invoice Payment Block

We'll now focus on the machine learning service built around the payment blocks for invoices which risk the cash discounts that are available for the purchaser.

Figure 6.28 shows the machine learning service built for this process, which we'll explore in this section.

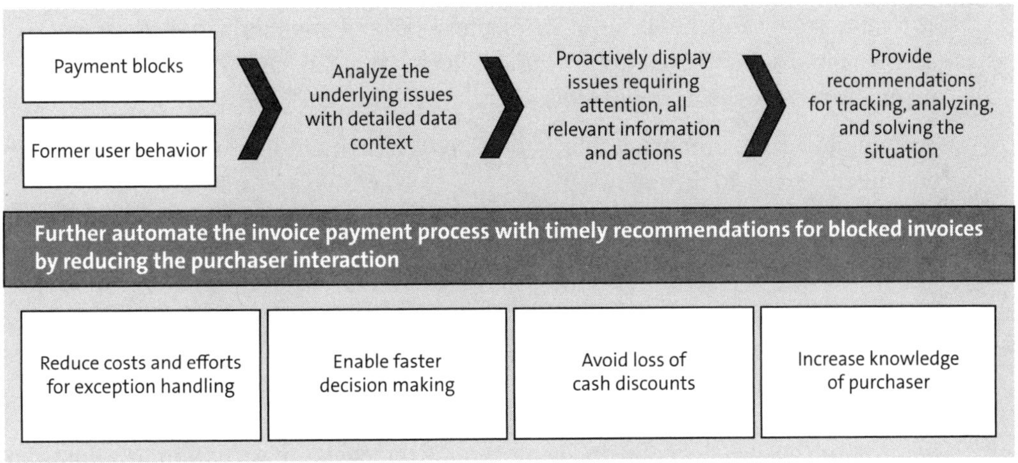

Figure 6.28 Resolve Invoice Payment Blocks

The invoicing process is already highly digitized and automated. Purchasers and accountants only need to manually intervene if there is a payment block on an invoice to be paid. The purchaser has to solve the underlying issue to remove the blocking reason. A few challenges are encountered by the purchaser during this process, such as clearing a blocking reason for a payment block, which is always time-consuming and very expensive. Most of the time, a manual process kicks in whenever there is an exception in a payment, which also results in buyers missing out on cash discounts and suppliers losing valuable business.

With the embedded machine learning service, algorithms check invoices against existing purchase orders and against procurement rules. Despite a highly automated invoice matching process, there are always exceptions that require manual interaction from purchasers or accounts payable experts, such as unplanned additional costs in the invoice or quantity/price deviations. These exceptions are time-consuming and expensive. Supplier invoices are blocked automatically if a blocking reason occurs, and the responsible purchaser has to solve the underlying issues. The process can be further automated by getting recommendations to track, analyze, and solve the situation. A faster processing time can also help buyers avoid losing out on cash discounts for early or timely payments.

This process will leverage the regression algorithms from APL, utilizing the ISLM framework, and is embedded into the SAP S/4HANA business processes.

Typically, businesses looking at benefits such as reduced costs and reduced efforts, faster decision-making, avoiding a loss of cash discounts, and enabling employees to focus on more important tasks than keeping track of these manual tasks. Suppliers would also benefit from faster invoice payments, cash discounts and cost savings, and so on.

The value proposition lies in automating the invoice payment process with timely recommendations for blocked invoices and reducing the purchaser's interaction time to manually remove the payment blocks.

To summarize, the propose resolution for invoice payment block allows accountants and purchasers to do the following:

- Proactively display issues requiring attention and all relevant information and actions in one place.
- Analyze the underlying issues with detailed data context.
- Receive recommendations based on former user behavior.
- Avoid a loss of cash discounts.

Some of the business benefits include reducing the effects of exception handling and recurring tasks, process automation enabling faster decision-making, avoid the loss of cash discounts, and enabling employees to focus on important tasks.

Configuration Guide

For configuration instructions, refer to the SAP Best Practices Explorer at *https://rapid.sap.com/bp/#/browse/scopeitems/2XX*.

6.4.3 Supplier Delivery Prediction

Let's now focus on the machine learning service built around purchase order processing to predict the delivery date for purchase order items.

Figure 6.29 shows the machine learning service built for this process, which we'll unpack in this section.

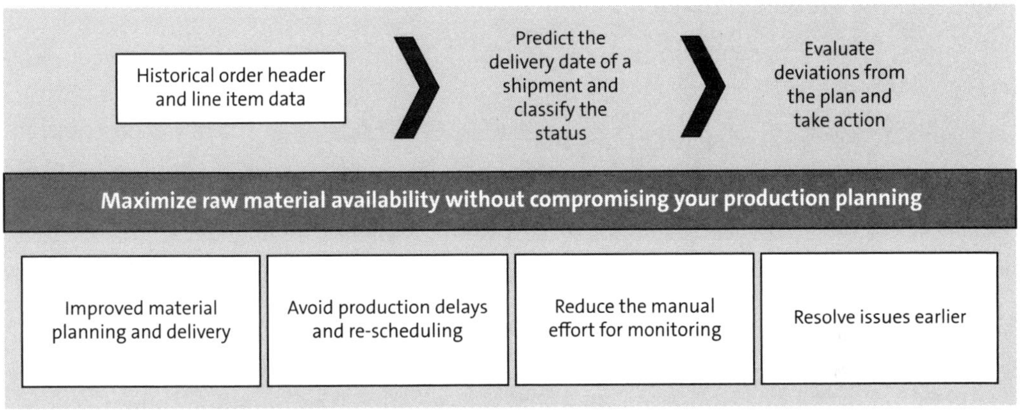

Figure 6.29 Predict Delivery Date for Purchase Order Items

During the purchasing process, often the materials required for production are supplied by multiple suppliers. Many times, the delay in delivery would impact on-time production at the plants and require rescheduling assembly lines, which becomes very costly. Indirect material delays also cause a huge delay in supporting employees with the required product/service. The machine learning algorithms identify the supplier delivery delays based on multiple situations and predict the chances of delays going forward. In this scenario, the regression and classification algorithms leverage APL, utilizing the ISLM framework to embed machine learning into the SAP S/4HANA business processes.

The machine learning algorithm behind the supplier delivery prediction allows purchasers to predict the arrival date of a shipment and classify its status. This allows the purchaser to convey the information to the product lifecycle management (PLM) department and eventually maintain the material requirements planning (MRP) and security optimization service (SoS) manually. In upcoming releases with updated machine learning functionality, there will be more to this process, making it so that the

MRP engineer and the design engineer won't need to manually update any records in the system.

Some of the business benefits include the following:

- Improved reliable planning and delivery of materials
- Avoidance of production delays and rescheduling
- Decreased manual efforts to monitor and resolve issues faster and earlier
- Up-to-date lead time info for goods and services at the master level for direct and indirect procurement

Configuration Guide

For configuration instructions, refer to the SAP Best Practices Explorer at *https://rapid.sap.com/bp/#/browse/scopeitems/3FY*.

6.4.4 Proposal of New Catalog Item

Let's now focus on the machine learning service built around the purchase order processing and the recommendations provided in the catalog items.

Figure 6.30 shows the machine learning service built for this process, which we'll discuss in this section.

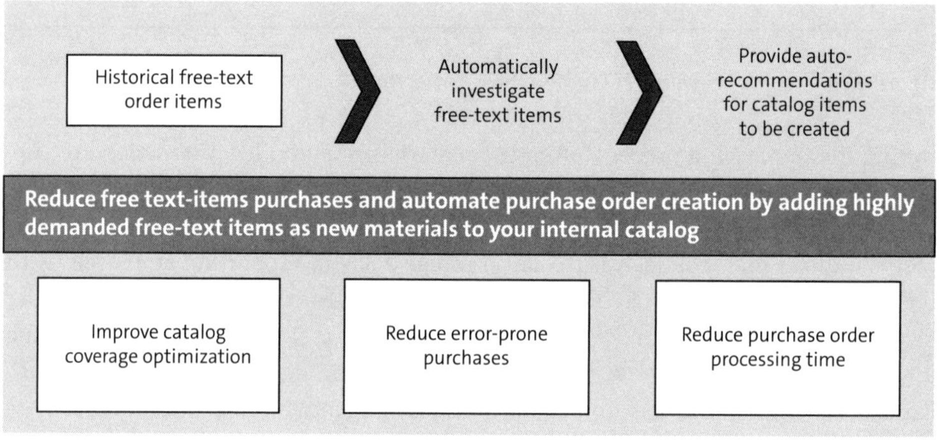

Figure 6.30 Convert Free-Text Order Items to Catalog Items

A purchasing manager manages the internal acquisition of goods and items and keeps the internal catalog up to date. There are quite a few challenges to be addressed by the purchasing manager. Huge amounts of free-text items are generated by users, and there is always a high amount of effort involved in investigating and analyzing these free-text items. Also, free-text items lead to error-prone purchases, and it's hard to always keep the internal catalog up to date. There would be real benefits in the purchasing department to improving the catalog coverage optimization and reducing the error-prone purchases, eventually reducing the total purchase order processing time.

With the propose creation of the catalog item machine learning service, the algorithm would take over the investigative actions and materials with high user demand and make them available in the catalog automatically. The purchase order creation would be automated, which ensures automated recommendations for items to be added into the catalog. Eventually this helps meet the purchasing manager's goal of reducing free-text items as much as possible.

When there are many free-text items generated by users, algorithms can help to decide when it makes sense to add a new item to the catalog. Existing materials will be made available in the catalog automatically if there is a high user demand. As discussed, via machine learning capabilities and statistical models, the machine learning service provides automated recommendations for catalog items to be created. These recommendations are enriched with insights gathered by the machine learning algorithm. The algorithms from the TensorFlow library are utilized and the machine learning service is built with SAP Data Intelligence and runs on SAP Cloud Platform. The respective SAP S/4HANA process consumes this machine learning service in a side-by-side model and delivers the required results.

The value proposition of this machine learning service highlights the learnings from the past to recommend or implement the automatic addition of new materials in your internal catalog, keeping it up to date and saving time.

Configuration Guide

For configuration instructions, refer to the SAP Best Practices Explorer at *https://rapid.sap.com/bp/#/browse/scopeitems/2XW*.

6.4.5 Proposal of Material Group

Let's now focus on the machine learning service built around the proposal of a material group for free-text items during the purchase requisition process.

Figure 6.31 shows the machine learning service built for this process, which we'll unpack in this section.

Operational purchasers need to identify wrongly assigned material groups and assign the correct material group for a free-text item, which reduces accidental purchases in the wrong category. Typically, the challenges encountered by the purchasing department include accidental purchases by the employees in the wrong category. During the creation of free-text items, employees often assign a material group that doesn't fit the required item. It's a process-intensive and resource-intensive manual task to then assign the correct material group for the free-text item. Purchasers need to manually change the material group so that the responsible colleague can take over. These manual efforts are expensive from process and resource perspectives.

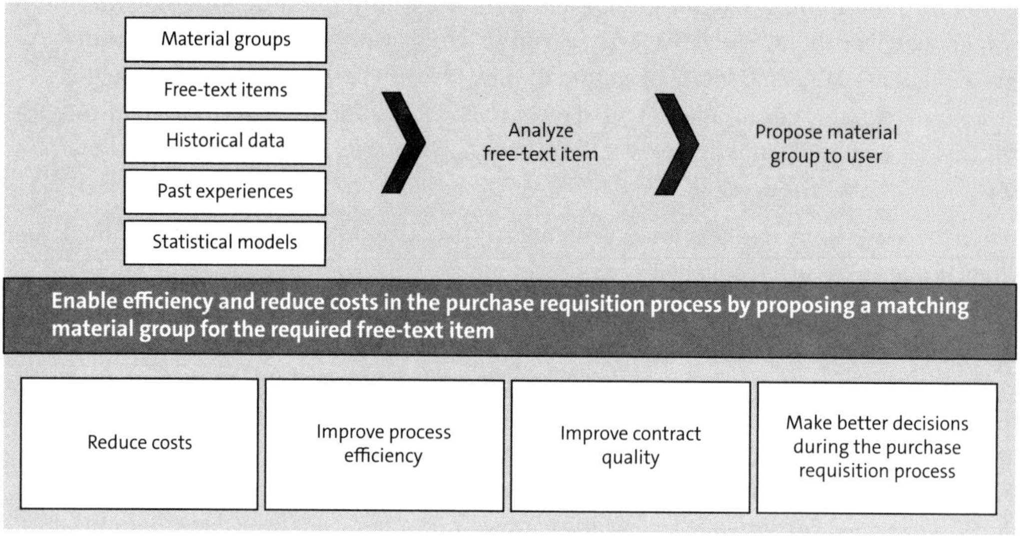

Figure 6.31 Propose Material Group for Free-Text Items

The typical benefits a customer would achieve by automating this process include reduced costs because of automatic assignment of a material group, improved process efficiency, reduced risk of error-prone purchases, and better decision-making in the purchase requisition process.

The propose material group for free-text items machine learning service will help to reduce the number of wrongly assigned material groups. The machine learning service proposes a matching material group for the required free-text item to the requester during the creation of the purchase requisition based on machine learning capabilities and statistical models. The algorithms from the TensorFlow library are utilized and the machine learning service is built with SAP Data Intelligence and runs on the SAP Cloud Platform. The respective SAP S/4HANA process consumes this machine learning service in a side-by-side model and delivers the required results.

The business benefits from the machine learning service include the following:

- Avoidance of manual activities
- Improved process efficiency and reduction in costs
- Reduced risk of error-prone purchases

Better classification of the free-text items under material groups helps control error-prone purchases. The proposal of a matching material group is based on past experiences and data. These machine learning models learn and provide suggestions based on historic data, leveraging the predictive algorithms with automatic material and category suggestions for the free-text items and enhanced usability for the purchasers.

The value proposition is to enable efficiency and reduce costs in the purchase requisition process.

Configuration Guide

For configuration instructions, refer to the SAP Best Practices Explorer at *https://rapid.sap.com/bp/#/browse/scopeitems/2XV*.

6.4.6 Proposal of Options for Materials without Purchase Contract

Let's now focus on the machine learning service built around reducing off-contract spend by prioritizing the materials without purchase contracts and suggesting options.

Figure 6.32 shows the machine learning service built for this process, which we'll unpack in this section.

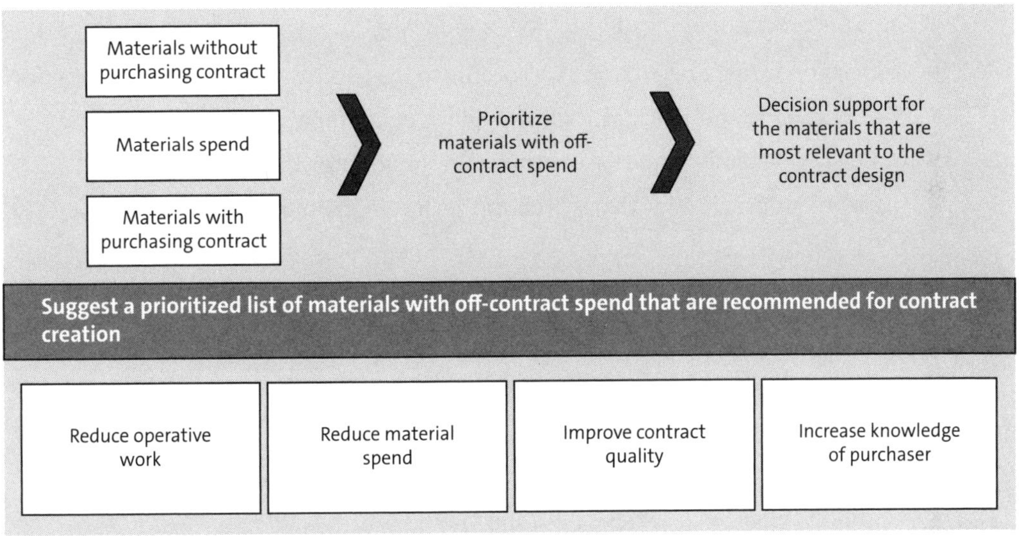

Figure 6.32 Reduce Off-Contract Spend

Purchasers need to identify materials with no contracts and create requests for quotations (RFQs) based on similar materials with existing contracts. Purchasers need to prioritize along the lines of volume, value, and so on to create an RFQ and then a contract for those materials.

Purchasers face quite a few challenges in the process to execute their corporate procurement tasks. A lot of off-contract ordering increases the overall off-contract spend because there is no attached purchase contract for a particular material. Creating an

RFQ from scratch is always a tedious process compared to creating an RFQ from a similar material. Prioritizing a list of off-contract materials also is time-consuming.

The purchasing department can benefit hugely if this off-contract spend can be handled intuitively. This would help reduce costs because RFQ proposals for materials with frequent off-contract ordering are created with the prioritized list of materials. Purchasers would be available ahead of time for early and efficient supplier renegotiations, leading to better purchasing prices for materials from the suppliers. The RFQ creation process thus is made more efficient.

To save money, buyers need to avoid off-contract ordering. In this sense, the machine learning algorithm will help the buyer identify the materials without a contract and relate them to similar existing ones with contracts based on specific factors, and it will propose these materials for an RFQ accordingly. The algorithms from the TensorFlow library are utilized and the machine learning service is built with SAP Data Intelligence and runs on the SAP Cloud Platform. The respective SAP S/4HANA process consumes this machine learning service in a side-by-side model and delivers the required results by providing recommendations for an RFQ creation.

It helps the procurement leaders to do the following:

- Anticipate contract negotiations according to the company guidelines and practices
- Negotiate early and efficiently with suppliers on materials often ordered
- Define the predictive models and train and run the scenarios
- Use a prebuilt set of KPIs that will allow the development of robust analytics on SAP S/4HANA

Some of the business benefits for the customers include the following:

- Enhanced usability for businesses to visualize the recommendation options with predictions
- Opportunities for purchasers to have early and efficient supplier negotiations
- Attaining better purchasing prices for materials from the suppliers
- Attaining enhanced purchasing compliance

The value propositions for the purchasers and buyers are reduced off-contract ordering and streamlined contract negotiations according to the company practices. The machine learning service benefits purchasers and buyers to negotiate early and efficiently with the suppliers while saving costs.

Configuration Guide

For configuration instructions, refer to the SAP Best Practices Explorer at *https://rapid.sap.com/bp/#/browse/scopeitems/30W*.

6.4.7 Image-Based Buying

Let's now discuss the details of the use case and the functionality around picture-based requisitioning in the purchase requisition process.

We already notice different human-to-system interactions in play nowadays, and we can only see that change growing as the technology unfolds. The procurement systems that have to deal with pictures and algorithms need to learn which internal material or catalog item a picture can best fit. That way, a user can just take a picture with his or her mobile device, and the cross-content search will start and create the respective purchasing demand automatically. This will greatly enhance the procurement experience for purchasers and can lead to huge cost savings. Many times, we see the description of an item provided by the user and the purchaser vary, and the process is delayed due to the wrong material procurement or unnecessary overhead costs in procuring unwanted material items. Addressing this issue increases user efficiency by leaps and bounds while generating purchasing demand that is validated ahead.

This machine learning functionality helps employees immensely when performing cross-catalog search while ordering any items. The algorithms from the TensorFlow library are utilized and the machine learning service is built with SAP Data Intelligence and runs on SAP Cloud Platform. The respective SAP S/4HANA process consumes this machine learning service in a side-by-side model and delivers the required results. By speeding up the process with the upload of an image and a cross-catalog search, a purchase requisition is created automatically. Some of the key aspects include the following:

- Compare images taken with catalog item images in cross-catalog search index to find matching patterns in pictures.
- Detect similar images by first normalizing the images—for example, flipping, rotating, adjusting brightness and colors, and so on—in order to increase the likelihood of similarities.
- Search results in the cross-catalog search displays the best matching items based on their similarity scoring (items are identified by internal catalog IDs).
- Images can either be added to the shopping cart from a local laptop or a dedicated image-based buying app that could prepare a draft shopping cart instead.

There are a lot of key benefits from using this machine learning service, which include the ease of use for the requester and avoiding expensive exception processing. It also ensures product compliance for the ordered products and avoids off-catalog spend, which results in huge cost savings; this is a big requirement in procurement departments.

Configuration Guide

For configuration instructions, refer to the SAP Best Practices Explorer at *https://rapid.sap.com/bp/#/browse/scopeitems/3UH*.

6.4.8 Intelligent Approval Workflow

The purchase requisition process is methodical and follows a few criteria. Let's look into the details of how to make the purchase requisition process semiautomated and intelligent based on history and past learning and by adding machine learning.

Figure 6.33 shows how the machine learning functionality is built in this context. Notice the **Top 5 Influencing Parameters** box, making it easier to approve a similar purchase requisition. The intelligent approval workflow machine learning service provides these suggestions for approval.

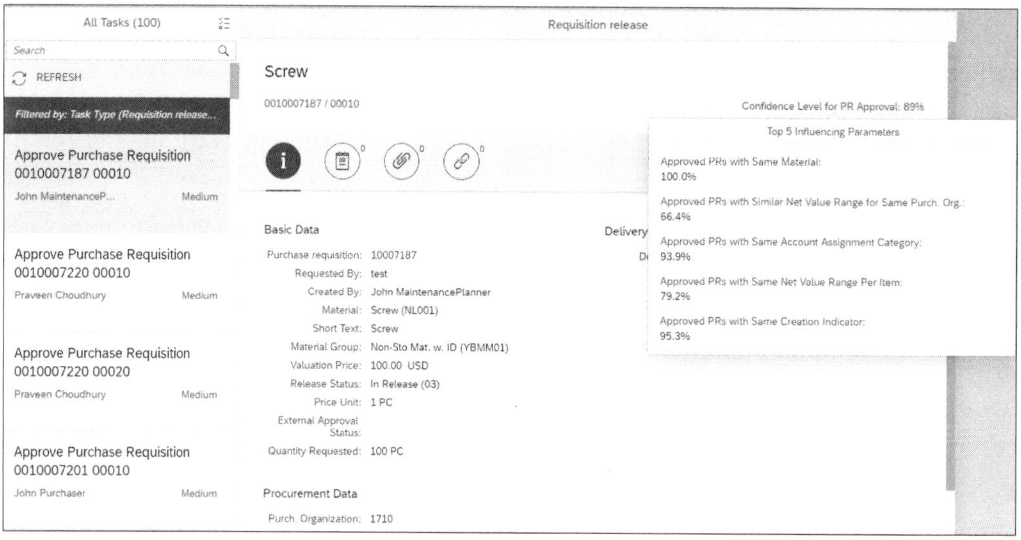

Figure 6.33 Intelligent Approval Workflow

The machine learning service analyzes the purchase approval processes and automatically approves requests that usually don't get rejected (those that have more than 90% approval based on historic transactions). The machine learning algorithms can identify the data patterns for automatic approval; these patterns could be cumbersome for purchasers to identify manually. This will help increase the efficiency of the purchasing department greatly and save costs. The algorithms from the TensorFlow library are utilized and the machine learning service is built with SAP Data Intelligence and runs on the SAP Cloud Platform. The respective SAP S/4HANA process consumes this machine learning service in a side-by-side model and delivers the required results.

Basically, functional and/or monetary approvers in the procurement department are usually flooded with numerous approval work items each day. To save time and costs and allow them to concentrate more on their strategic work, the intelligent approval workflow is aimed at classifying purchase requisition approvals into important and unimportant tasks by weighting them with a confidence factor in order to focus on the critical items. Some of the key aspects in rating the purchase requisition approvals with a confidence factor include the following:

- Easily distinguishes critical approvals from less important ones
- Focuses the approver's attention on critical work items only
- Groups all work items based on their confidence factor
- Prepares the stage for a potential mass approval of all grouped purchase requisitions in the approval stage

The business benefits are shorter approval times thanks to focusing on critical approvals only, more efficient usage of an approver's time and work effort, enhanced usability thanks to grouping approvals by their confidence factor, and process efficiency gained by automating the approval decisions.

Configuration Guide

For configuration instructions, refer to the SAP Best Practices Explorer at *https://rapid.sap.com/bp/#/browse/scopeitems/43E.*

6.4.9 Blockchain-Verified RFQ Processing

Let's now briefly discuss one of the blockchain use cases that have been delivered and released in the context of RFQ processing in procurement.

SAP strongly believes in the open-source ecosystem and in community efforts to advance blockchain technology and its ecosystem. This is why SAP participates in multiple collaborative activities. SAP is also a premium member of Hyperledger, an open-source consortium.

Figure 6.34 shows the steps involved in blockchain-verified RFQ processing, which we'll discuss in this section.

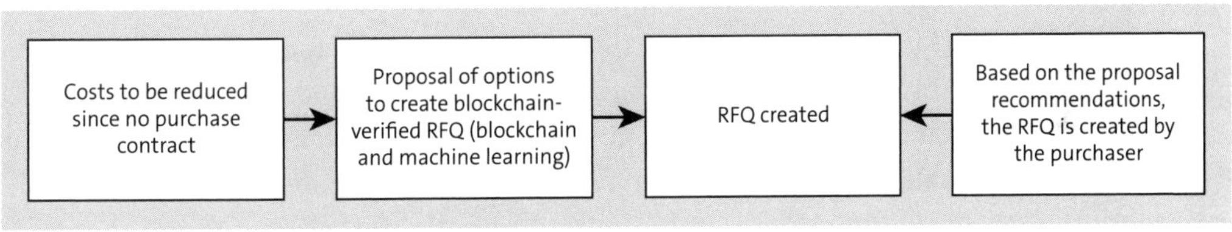

Figure 6.34 Blockchain-Verified RFQ Processing

There are typical challenges faced in a bidding process in which the suppliers and the purchasers collaborate. Confidentiality has to be maintained, and manipulation and fraud must be prevented. Collusion is possible between purchaser and supplier and therefore bids need to be disclosed immediately. Suppliers fear they are cut out of the process and submit bids as late as possible or invest in time-consuming lawsuits after a lost bid.

With blockchain-verified RFQ processing, all the documents involved in the process are fingerprinted in the blockchain. There are some additional benefits:

- Increased transparency and trust between business partners can prevent any kind of fraud.
- It becomes easier to avoid manipulation.
- In the case of an internal audit or complaints from suppliers, companies can prove that an identical RFQ was sent to all suppliers using technology with which manipulation is impossible.
- It helps reduce off-contract spend and overall costs.

Configuration Guide

For configuration instructions, refer to the SAP Best Practices Explorer at *https://rapid.sap.com/bp/#/browse/scopeitems/3DT*.

6.5 Inventory and Supply Chain

In this section, we'll focus on the machine learning services and embedded use cases built for the inventory and supply chain management area of business. We'll also briefly touch on the scenarios for warehouse management; most of that functionality will be linked to inventory management in the context of manufacturing.

Inventory management deals with the process of tracking the value and quantities of all the goods your company has in stock, whereas warehousing is the act of storing goods that will be sold or distributed later.

Typically, inventory management manages the products or materials a company sells to its customers in order to make profit. As part of the supply chain, inventory management includes several different aspects, such as controlling and overseeing purchases from suppliers and customers, maintaining the storage of stock, controlling the amount of product for sale, and order fulfillment. There are three core steps:

1. Purchasing inventory
2. Storing inventory
3. Profiting from inventory

In general, inventory types can be grouped into four classes: raw material, work in process, finished goods, and maintenance, repair, and operating supplies (MRO) goods.

The customer is king, and to keep customers loyal and happy, it's imperative to keep the promises that are made with unwavering quality. It's key for large companies with complex supply chains to maintain a healthy balance between manufacturing processes and inventory size. The supply chain managers typically handle the flow of inventory, considering the capacity and productivity issues along the way, while the inventory managers focus on local orders, tracking the suppliers for the deliveries, and so on.

Let us now start with the details of the stock-in-transit functionality.

6.5.1 Stock in Transit

Now, we'll focus on the machine learning service built around the stock-in-transit functionality to help predict the delay in stock that moves between different plants/storage locations.

Figure 6.35 shows the machine learning service built for this process, which we'll unpack in this section.

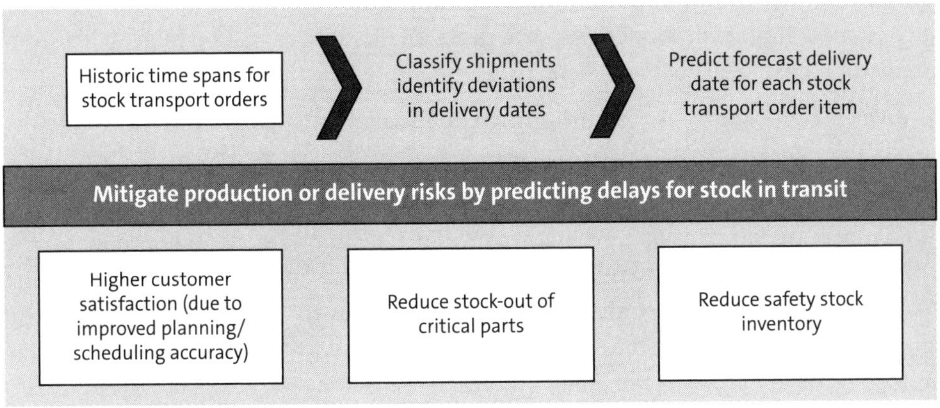

Figure 6.35 Predict Shipment Dates and Manage Stock in Transit

Most customers need to transfer stock between storage locations or plants, and therefore it's necessary to create stock transport orders. With the SAP S/4HANA Overdue Materials—Stock in Transit SAP Fiori app, it's possible for customers to track materials in transit/open stock transport orders for which no goods receipt has been posted by the receiving plant yet and which have, for example, already exceeded the expected time in transit. Customers usually have a schedule in place for how long the transfer should take between two plants.

The inventory manager ensures that the production plant/customer gets the material in time: the flow of materials needs to be managed among the different storage locations or plants. The stock transport orders are created to transfer stocks and track materials in transit to ensure on-time delivery.

The challenges encountered by inventory managers include lack of visibility into delayed deliveries between storage locations, hard to estimate stock transfer durations, and so on. With a negative time and money impact on downstream deliveries of finished goods, raw materials, or work in progress (WIP), there is a potential negative quality impact on downstream finished products, especially if goods are perishable or time-dependent.

Customers would gain immensely from a higher degree of improved planning and scheduling accuracy, which helps reduce critical stock-out scenarios and improve turnover. An overall reduction in safety stock inventory would thereby be achieved.

The predict arrival of stock in transit machine learning solution helps display items for which the planned delivery date significantly differs from the predicted delivery date. The list of materials and accompanying microcharts show the relation between the actual date, forecast delivery, and status of forecast, with early and efficient visibility for stock transport orders. The prediction of the arrival date (forecast delivery date) will be available during the creation of the stock transport order, with the embedded predictive modeling incorporated into the functionality. There will be huge gains from predicting the forecast delivery date of a shipment.

The value proposition lies in mitigating production or delivery risks by predicting delays for stock in transit.

Companies issuing and receiving goods from and to their plants need to track the status of the materials in transit. When there are delays, shortages, or other problems, the company needs to be able to respond and take action in a timely manner. The Overdue Materials—Stock in Transit app in SAP S/4HANA provides an overview of the open shipments so that the business user can take action. With the embedded predictive modeling, the application can predict shipment dates for each goods movement. Having this insight allows users to take action and manage delivery delays, production scheduling, and other downstream activities. The results will continue to become more accurate as users retrain the model with the latest behaviors and the most recent data stored within SAP S/4HANA. In the current scenario, regression algorithms from APL utilizing the ISLM framework are leveraged and embedded into the corresponding SAP S/4HANA business processes to predict the arrival of the stock that is in transit between locations.

To summarize, the predict arrival of stock in transit machine learning service will allow the warehouse and inventory managers to do the following:

- Predict the arrival date of a shipment and classify its status
- Define the predictive models and train and run the scenarios
- Use a prebuilt set of KPIs for robust analytics on SAP S/4HANA Cloud data with drill-down functionality

- Have early and efficient visibility into stock transport orders
- Use new integration capabilities with SAP S/4HANA Cloud to gain real-time insights into the production line of business scenarios with predictive analytics

Some of the business benefits include minimizing stock outages and stock overages, leading to eventually managing inventory costs in a much better way; optimizing and automating the business process of tracking stock in transit, helping to achieve more reliable planning and scheduling of goods in transit processes; and providing enhanced usability for the business by visualizing data with predictions.

With this new machine learning functionality, the warehouse managers and inventory managers can minimize cost outages and stock overages, better manage inventory costs, and predict the shipment duration in days.

Configuration Guide

For configuration instructions, refer to the SAP Best Practices Explorer at *https://rapid.sap.com/bp/#/browse/scopeitems/20N.*

6.5.2 Demand-Driven Replenishment

In this section, we'll focus on the machine learning service built around the demand-driven replenishment functionality to address the dynamic buffer level adjustment for the stock transfer materials.

Figure 6.36 shows the machine learning service built for this process, which we'll explore in this section.

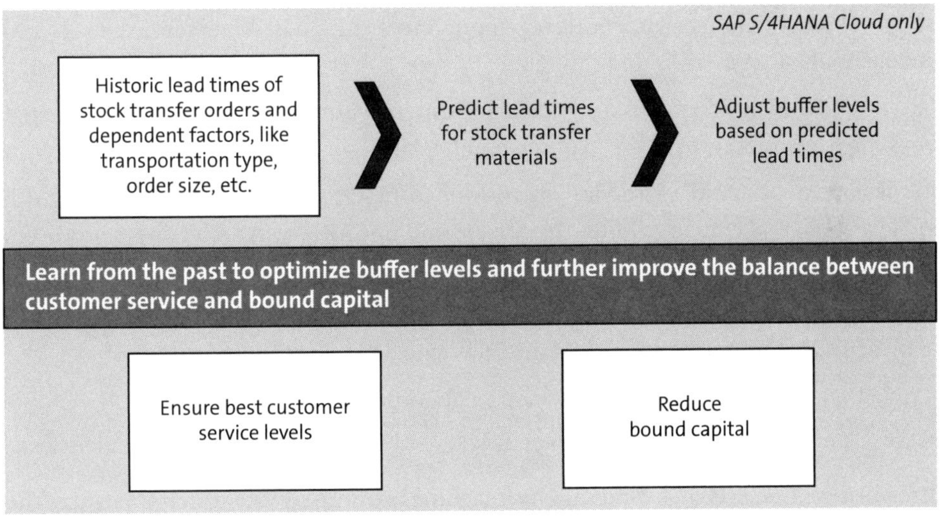

Figure 6.36 Demand-Driven Replenishment

The inventory manager ensures that the production plant/customer gets the materials in time and monitors replenishment planning, execution, and buffer levels. Inventory managers find it time-consuming and expensive to plan, track, and manage buffer levels because of ineffective and incomplete supply chain analytics and constantly changing demand signals.

The business value and benefits for customers would include reduced safety stock inventory, reduced inventory carrying cost, reduced days in inventory, and reduced inventory obsolescence. The system analyzes and classifies products that need buffer-level adjustments and selects products relevant to demand-driven replenishment (classification). Managing buffer levels for a stock-transfer product to adopt predicted lead time results in an optimal replenishment order generates buffer-level proposals. The value proposition lies in data-driven optimization of buffer levels to balance customer service levels and bound capital.

Predicting individual lead time for stock transport orders is the key for the inventory managers and the warehouse managers to better manage the stock at their plants or storage locations. Lead time is the latency between the initiation and execution of a process. For example, the lead time between the placement of an order and delivery of a new car from a manufacturer may be anywhere from two weeks to six months. Lead time calculations are predicted, and the jobs are then initiated to maintain the inventory buffer values. The new values can be checked into the application to manage the buffer levels with decoupled lead times.

In the planning method of demand-driven replenishment, one key element is the dynamic adjustment of buffer levels. A key input for the recalculation of buffer levels is the buffer replenishment lead time. The buffer levels depend proportionally on this lead time. The more accurately the lead time can be determined, the more appropriately the buffer can be sized: both big enough to ensure customer service levels and small enough to reduce bound capital.

Based on the historic lead times and their business context, buffer level adjustments are suggested to best serve the ever-changing reality.

In the current scenario, regression algorithms from APL utilizing the ISLM framework are leveraged and embedded into the corresponding SAP S/4HANA business processes to better calculate the lead time and buffer levels to be maintained.

To summarize, the demand-driven replenishment machine learning functionality allows the inventory manager to do the following:

- Evaluate historic lead times of stock transfer orders
- Consider contextual information of orders to identify patterns and dependencies

The business benefits of leveraging this machine learning service is that it ensures the best possible customer service levels at the lowest possible cost (i.e., bound capital through buffered stock).

Configuration Guide

For configuration instructions, refer to the SAP Best Practices Explorer at *https://rapid.sap.com/bp/#/browse/scopeitems/20N.*

6.5.3 Defect Code Proposal with Text Recognition

Before diving into the details of the defect code proposal scenario, let's walk through an overview of how defects are handled and how defect processing happens in quality management. Figure 6.37 explains the end-to-end process flow for defect processing and how and where machine learning can be leveraged, which we'll discuss in this section.

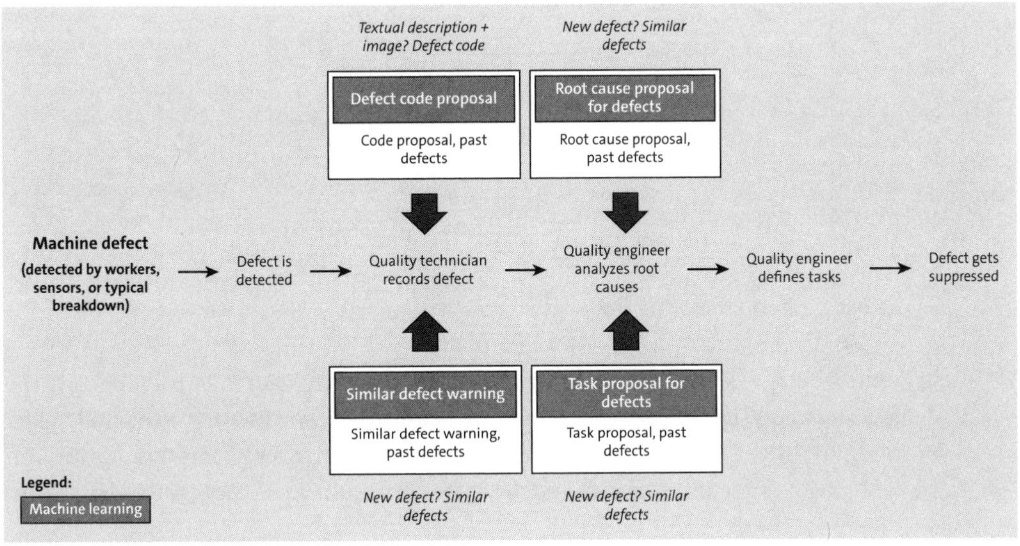

Figure 6.37 Defect Processing

As explained in the end-to-end process, we can categorize defect processing into the following groups (see Figure 6.37):

1. Defect code proposal with textual and image description
2. Similar defect warning
3. Root cause proposal for defects
4. Task proposal for defects
5. Clusters of similar defects

Figure 6.38 shows how the defects are processed and handled at the various stages by the quality technician and quality engineer.

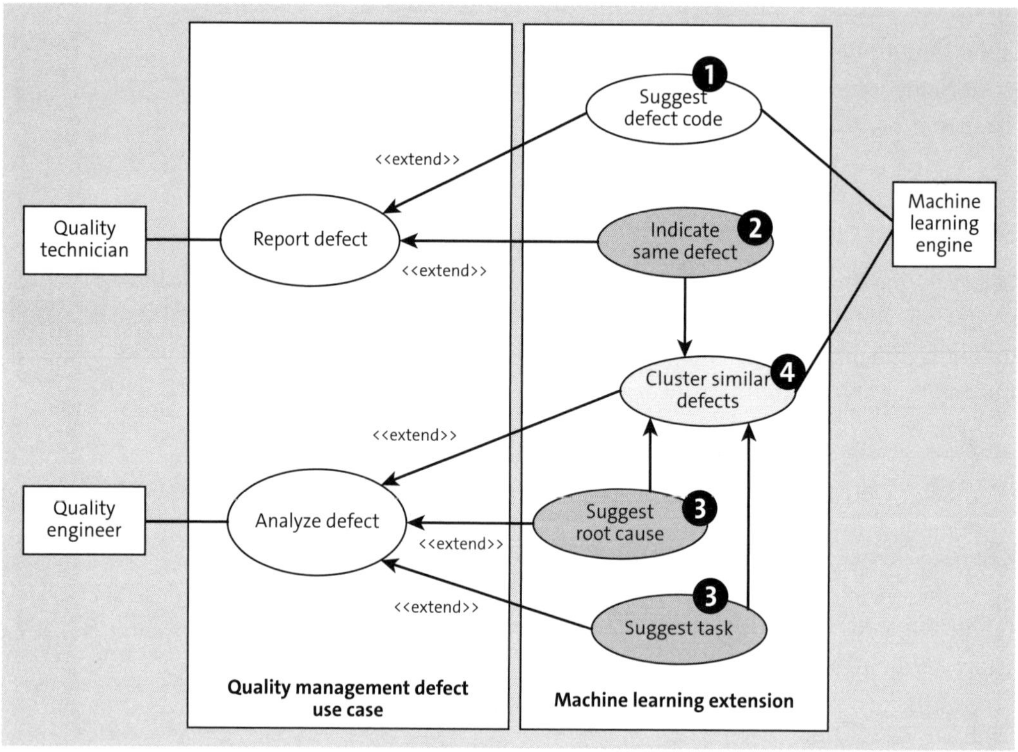

Figure 6.38 Defect Processing Flow

In Figure 6.38, notice the quality technician and quality engineer roles for reporting the defects and analyzing the defects, respectively, and how they fit into the overall defect processing context. The numbering indicates the various stages of defect reporting and analysis, from reporting steps (❶ and ❷) to understanding the root case ❸ to clustering after analysis ❹.

Let's now focus on the machine learning service built around the defect code proposal with textual and image descriptions in the quality management domain. Figure 6.39 shows the machine learning service built for this process.

Let's walk through the steps shown here. Say that a quality technician wants to get proposals for a particular defect code group and code based on a textual description and based on an image of the defect during the defect recording in order to find the code that fits the defect best. Based on the existing defects (with textual descriptions and images with a matching defect code group and code), the machine learning algorithm proposes a defect code group and a defect code for the new defect for which a textual description and an image already exists, while the quality technician records the defect. The machine learning algorithm is embedded into the SAP S/4HANA business process and helps the quality technician with recommendations and suggestions while working in the SAP Fiori launchpad. In the current scenario, classification and k-nearest neighbors algorithms from APL utilizing the ISLM framework are leveraged and embedded into the SAP S/4HANA quality management business processes.

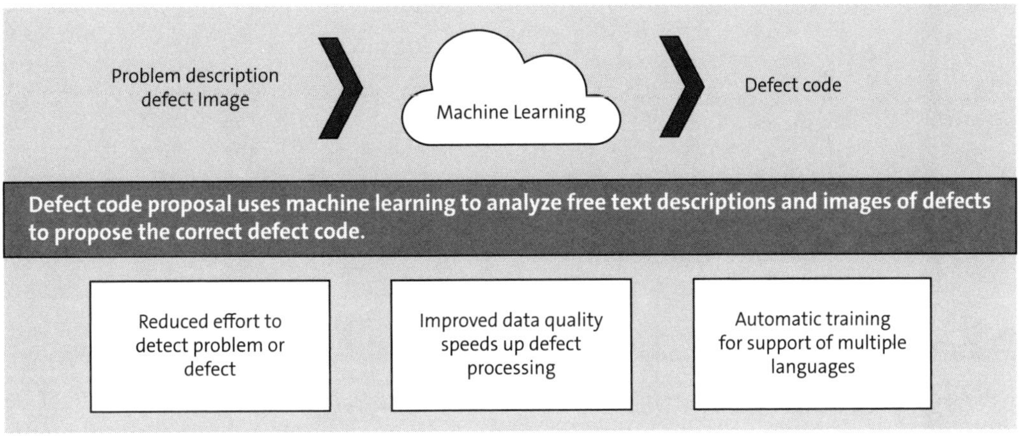

Figure 6.39 Defect Code Proposal

The machine learning model is built by considering various factors such as the defect category, the defect description (short and long description), and the defect picture/image, which eventually proposes the defect group and the code, leveraging the classification (neural networks) algorithms.

The value proposition includes streamlined defect code proposals on the historic trends so that the quality technicians can spend more quality time finding many other defects. The other use cases in the defect processing context are in the planning and evaluation phases and should be available soon.

Configuration Guide

For configuration instructions, refer to the SAP Best Practices Explorer at *https://rapid.sap.com/bp/#/browse/scopeitems/20N.*

6.5.4 Early Detection of Slow and Nonmoving Stock

In this section, we'll focus on the machine learning service built around the early detection of slow and nonmoving stock in inventory management.

Figure 6.40 shows the machine learning service built for this process, which we'll unpack in this section.

The production and inventory planner needs to plan the movement of the inventory between different locations—for example, if stock has to be moved to a nearby warehouse or a nearby dealer location at which the movement of the inventory is high. The planner needs to ensure the locations have the optimum quantity of stock for better movement of the inventory. With the help of the machine learning service, you can detect the slow and nonmoving stock in advance and plan/classify the product into different categories based on the historical data and information.

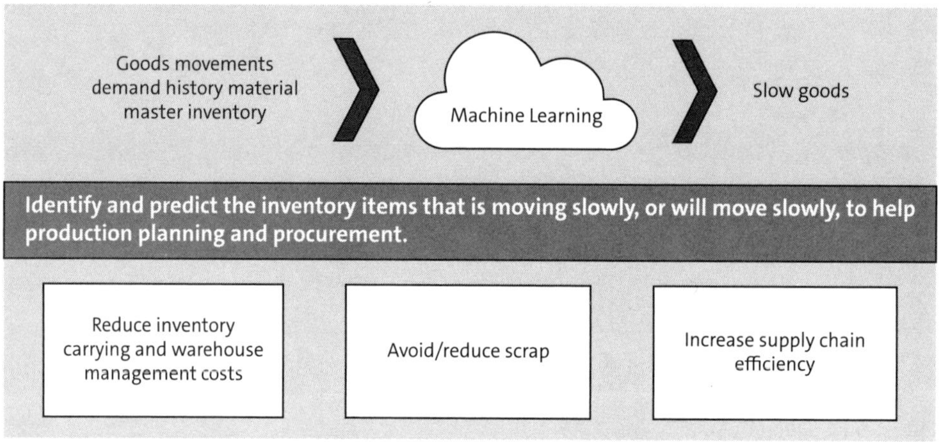

Figure 6.40 Early Detection of Slow and Nonmoving Stock

The machine learning algorithm checks the slow and nonmoving stock across various locations, such as plants, warehouses, and dealers, and identifies in advance via predictive capabilities the possible slow and nonmoving materials. The machine learning algorithms also check the new demands or requirements in the form of purchase requisitions, sales orders, or spares demand and even the expected orders and suggest material movements to avoid the slow and nonmoving stock situations. Historical sales, procurement, and manufacturing data information with a reasonable amount of data movement history would greatly help in designing and fine-tuning the machine learning algorithms. In the current scenario, regression and time series algorithms from APL utilizing the ISLM framework are leveraged and embedded into the corresponding SAP S/4HANA business processes for better prediction of the slow and nonmoving stock.

The typical business benefits for customers include the following:

- Reduction in inventory carrying costs
- Better planning of production, procurement, and related costs
- Increased supply chain efficiency

Configuration Guide

For configuration instructions, refer to the SAP Best Practices Explorer at *https://rapid.sap.com/bp/#/browse/scopeitems/20N*.

6.6 Sales

In this section, we'll focus on the machine learning services and embedded use cases built for the end-to-end process of optimized order-to-cash and the sales line of business. Figure 6.41 gives an overview of the order-to-cash process in detail.

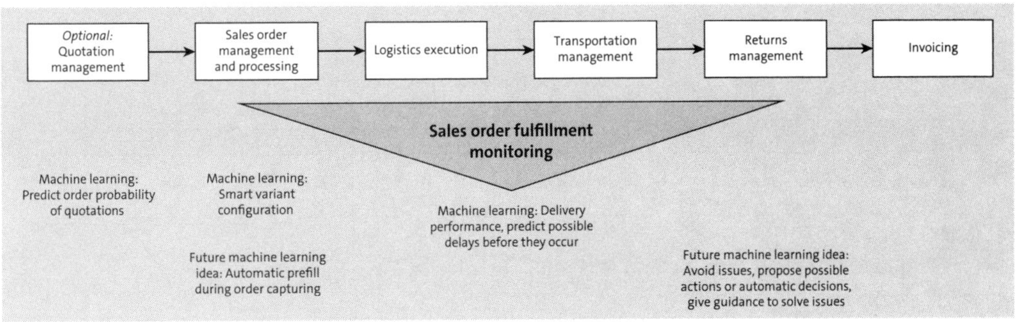

Figure 6.41 Order-to-Cash End-to-End Process with Machine Learning

As you can see from the end-to-end process shown in Figure 6.41, there are various steps involved, from quotation management to sales order creation to execution, transportation, returns, and invoicing. There are a few requirements from the sales manager and internal sales representatives to help them in the process by leveraging machine learning functionality. In this context, we are mainly focusing on sales order fulfillment and monitoring.

Here are some of the use cases that are already delivered as part of the SAP S/4HANA embedded business processes leveraging APL and PAL with the ISLM framework:

- **Sales quotation**
  Predict conversion of sales quotations to sales orders.
- **Sales performance**
  Predict sales forecasts.
- **Delivery performance**
  Predict delivery delays.

Before diving into the details of these use cases, let's also look into the order-to-cash process in the context of sales planning and analytics. We have a few machine learning use cases in that context as well, which are discussed in the next few sections. Figure 6.42 explains the process and the machine learning functionality at a high level.

As explained in the context of sales planning and analytics, the focus is mostly on the sales performance prediction with the structured data and providing sales forecasting details for the sales managers.

In this section, we'll dive into the details of these individual machine learning use cases that are embedded in the SAP S/4HANA business processes. We'll start with the process for predicting the conversion of sales quotations to sales orders in the following section.

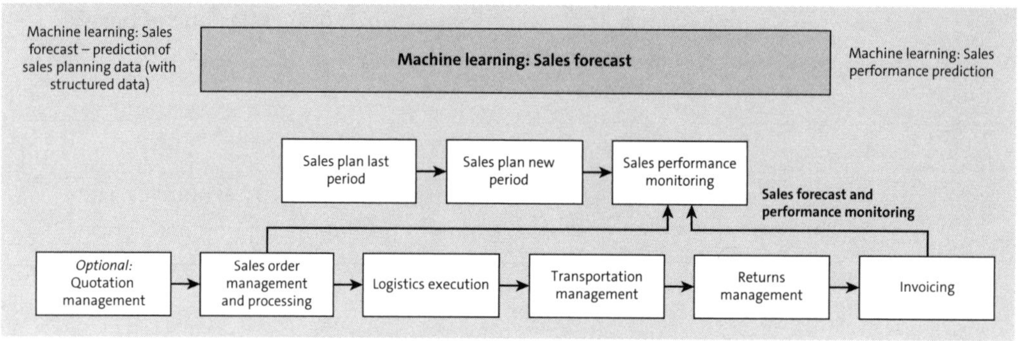

Figure 6.42 Sales Planning and Analytics in Order-to-Cash Process

6.6.1 Predict Conversion of Sales Quotations to Sales Orders

First, we'll focus on the machine learning service built around predicting the conversion of sales quotations to sales orders for a sales manager or an internal sales representative.

Figure 6.43 shows the machine learning service built for this process, which we'll unpack in this section.

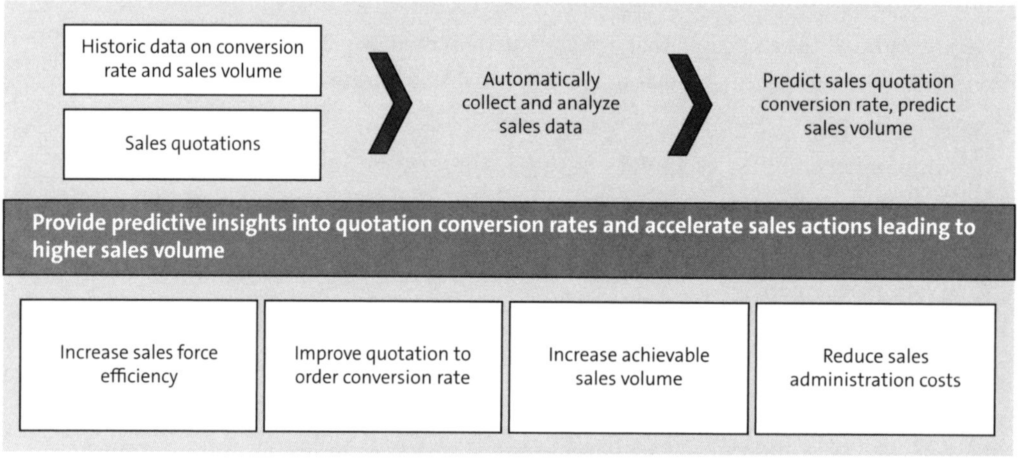

Figure 6.43 Predict Conversion of Sales Quotations to Sales Orders

Sales representatives need to track what percentage of sales quotations are being converted into sales orders, then review and follow up on available open quotations to increase sales volumes. The typical challenges for a sales representative include the difficulty of identifying quotations that are not being converted to sales orders as expected. Sales representatives put a lot of manual effort and all their experience and expertise into this highly time-consuming process. Sales managers expect increased sales force efficiency, improved quotation-to-order conversion rates, an increase in achievable sales volumes, and reduced sales administration costs to increase the top-line growth for their companies.

The machine learning algorithms can derive information from historical system data and provide accurate predictions of sales quotation conversion rates and the potential overall sales volumes. The system can manage complexity and automate the collection and analysis of the sales data with no more guesswork from the sales representatives. The sales representatives know which sales quotations might need more support and action to increase the sales volumes and continually monitor the progress. The underlying principle of the sales quotation conversion rate is to improve the overall sales volumes.

The sales quotation conversion rate (or sales order probability) is the probability that a sales quotation will be converted into a sales order item. The probability (expressed as a percentage) and net value of the sales quotation is used to calculate a total expected order value. Currently the sales order probability needs to be manually entered either in the customer master or document category. The manually entered probability relies on manual calculations and estimations or external (brought-in) data.

The quotation conversion rate machine learning functionality in SAP S/4HANA allows sales managers and internal sales representatives to get predictions for achievable sales volumes and quotation conversion rates in order to make decisions about which quotations have the most potential to be converted into sales orders and to increase the current sales volume. In the current scenario, regression algorithms from APL utilizing the ISLM framework are leveraged and embedded into the SAP S/4HANA sales quotation business processes for better predictions.

Some of the business benefits include the following:

- Receiving more accurate sales forecasts
- Less manual work and analysis
- Optimizing the win/loss ratio

The value proposition for the customers is getting predictive insights into quotation conversion rates and accelerating sales actions, leading to higher sales volumes.

Configuration Guide

For configuration instructions, refer to the SAP Best Practices Explorer at *https://rapid.sap.com/bp/#/browse/scopeitems/2YJ*.

6.6.2 Predict Sales Forecasts

Next, we'll focus on the machine learning service built around the sales performance prediction in the context of sales order fulfillment monitoring and sales planning.

Figure 6.44 shows the machine learning service built for this process, which we'll describe in this section.

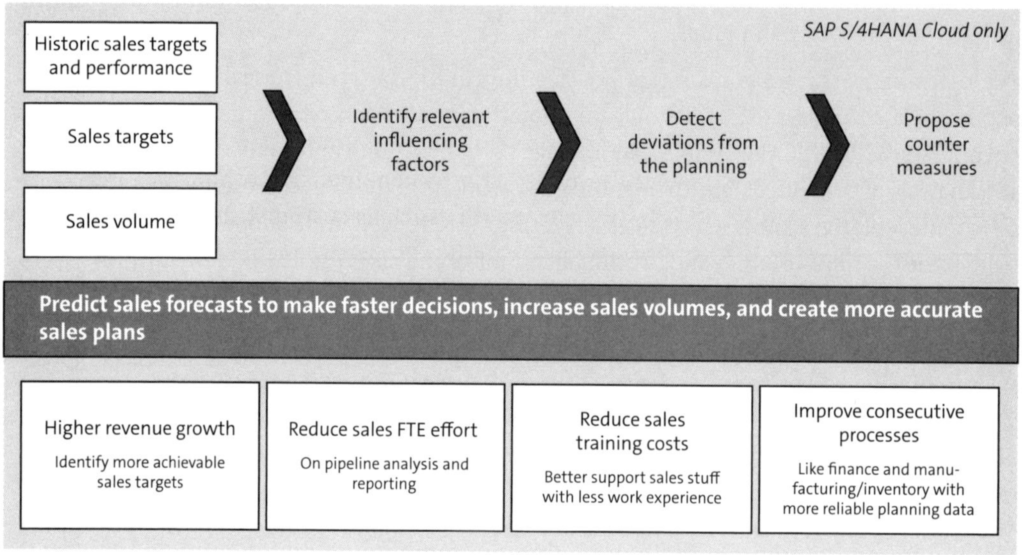

Figure 6.44 Predict Sales Forecasting

Typically, sales planning is a comprehensive analysis that requires high operational effort (collecting data from system and external data sources, formulating, planning, etc.), combined with human experience from a sales manager to make accurate predictions. To achieve greater sales volumes, sales managers need to analyze the sales performance, determine if sales targets can be met or exceeded, and propose proper follow-up actions, as shown in Figure 6.44. Currently, sales forecasting requires huge manual effort and is highly dependent on human experience to set historical data and other sources of information.

Sales managers manage the operational business by comparing actual sales with planned sales in order to increase the current sales volume. Sales managers also need to forecast the value of the sales of a company expects to make, set sales targets, and figure out ways to achieve them. The typical challenges encountered by a sales manager include high manual effort for collecting and analyzing data from systems and external inputs, repetitive and time-consuming tasks gathering the information, and more. There is high complexity in underlying data and uncertainty about how accurate the analysis and forecasting might be. There is an additional fear of not hitting quotas. For these reasons, sales planning and analysis requires sound knowledge and experience.

The benefits and business value expected by the sales managers are as follows:

- Higher revenue growth
- Identification of more achievable sales targets
- Reduced sales FTE effort for pipeline analysis and reporting

- More time to focus on selling
- Reduced sales training costs
- Better support for sales staff with less work experience
- More reliable planning data to improve consecutive processes in finance, manufacturing, and inventory

The machine learning service for sales performance prediction helps to automatically collect and analyze historical sales information. The system manages complexity and accuracy by identifying relevant influencing factors such as customer dependencies and product categories. The sales managers can easily compare the achieved sales volume with the predicted and planned sales volumes by leveraging the machine learning service. Predictions through sales performance analytics accelerate and simplify insights to action. With early detection of deviations from planning, it's possible to plan countermeasures to make informed, fact-based decisions from the data available in the system. In addition, developing a decision pool for sales analysis and KPIs with a focus on analysis can generate real value for the sales managers and the sales staff.

With the machine learning service for sales performance predictions, you can compare your achieved sales volume with predictions and analyze to what extent your sales targets are being achieved. You can gain predictive insights into your current sales performance and simplify your sales planning process. In the current scenario, time series algorithms from PAL utilizing the ISLM framework are leveraged and embedded into the SAP S/4HANA sales performance prediction business processes.

To summarize, sales performance predictions allow sales managers to do the following:

- Forecast the value of sales a company expects to make
- Compare actual, predicted, and planned sales volumes in different dimensions and visualize that comparison
- Increase planning accuracy with the help of the machine learning algorithms, delivering predictions/proposals and reducing manual efforts for sales planning and the like

The business benefits include increased actual sales volume thanks to providing better insights to take action where needed and providing predictions for achievable sales volumes, as well as overall decreased manual effort in sales forecasting and planning.

The value proposition is for the sales managers to help with the predictions of sales performance, resulting in faster decision-making, increased sales volumes, and building more accurate sales plans.

Configuration Guide

For configuration instructions, refer to the SAP Best Practices Explorer at *https://rapid.sap.com/bp/#/browse/scopeitems/2YJ*.

6.6.3 Predict Delivery Delay

In this section, we'll focus on the machine learning service built around the prediction of the delivery date functionality in the context of delivery performance of the sales orders.

Figure 6.45 shows the machine learning service built for this process, which we'll unpack in this section.

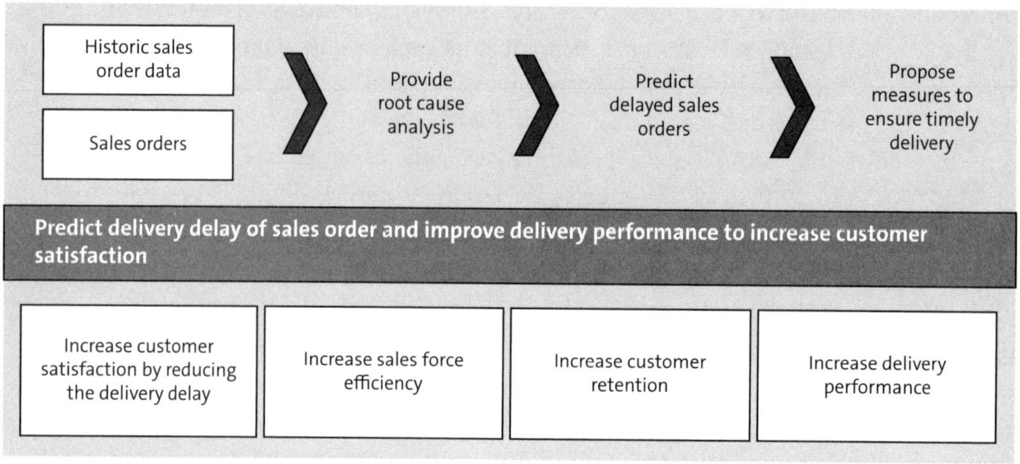

Figure 6.45 Predict Delivery Delay

Delivery performance is a broadly used standard KPI measurement in supply chains to measure the fulfillment of a customer's demand by the requested date.

As a sales representative, understanding which sales orders are delayed and finding the root cause for the delay and fixing it are of the highest priority. The challenges encountered are delays in delivery dates, lack of transparency into which orders will be delayed, lack of understanding of the causes of delay, customer escalations because of delivery delays, and added manual effort needed for repeated tasks tied to tracking the delivery performance.

Business value and customer benefits center on increased customer satisfaction thanks to reducing the delivery delay, increased sales force efficiency, increased customer retention, and increased delivery performance.

The machine learning service helps monitor how likely it is that a sales order item will be delayed and learns from past sales order data to predict delayed sales orders and provide root cause analysis. Potential delivery issues per sales order line item are shown for follow up by the sales representative, and the system proposes measures to ensure timely delivery.

With the Predicted Delivery Delay SAP Fiori app and its machine learning service, a sales manager is enabled to monitor the current delivery performance situation and instantly recognize the effect of delivered items (as a requested ratio of sales orders) to prevent a critical delay of delivered goods. This gives an idea of what ratio of sales order items are getting delayed, thus increasing customer satisfaction and retention in the future. Currently, the determination of the planned delivery date for a sales order item relies on the result of a planning or scheduling engine such as available to promise (ATP), so it can't consider probable deviations that arise in the future. In the current scenario, regression algorithms from APL utilizing the ISLM framework are leveraged and embedded into the SAP S/4HANA sales order delivery business processes.

To summarize, delivery performance/delivery in time allows internal sales representatives to do the following:

- Monitor how likely it is that a sales order item will be delayed
- Take appropriate actions to avoid the delay and to increase customer satisfaction, taking into consideration the influencing criteria of the probable delay

The business benefits include decreased manual efforts to monitor and resolve issues and achieving better delivery performance and hence increased customer satisfaction.

The value proposition for the sales representative is to predict the delivery delay of a sales order and improve delivery performance and increase customer satisfaction.

Configuration Guide

For configuration instructions, refer to the SAP Best Practices Explorer at *https://rapid.sap.com/bp/#/browse/scopeitems/2YJ*.

6.7 Research and Development/Engineering

In this section, we will focus on the machine learning services and embedded use cases built for the research and development (R&D)/engineering line of business. We'll be focusing on the *idea to design* concepts in the engineering and R&D domain. When we talk about idea to design, we're focusing on the project ideation, planning, preparation, and design phases before the projects are executed. In a nutshell, we'll be talking about the portfolio and project lifecycle management and related aspects.

Figure 6.46 gives an overview of the enterprise portfolio and project management in detail.

As you can see, project managers, project members, finance controllers, and accountants are involved in a lot of activities in the enterprise portfolio management processes, from project planning to project execution to period-end closing and more.

Let's now continue into the next section and discuss the details of the project cost forecasting functionality.

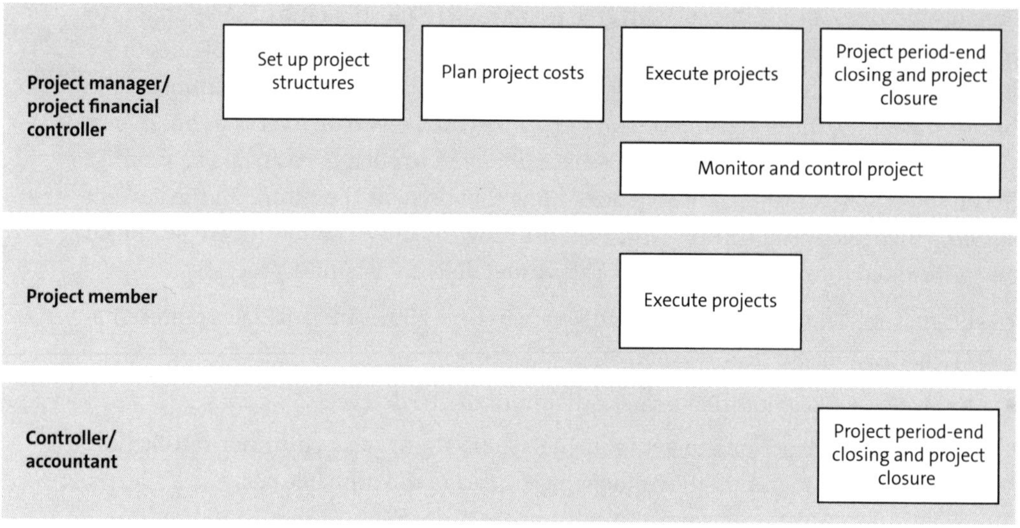

Figure 6.46 Enterprise Portfolio and Project Management

6.7.1 Project Cost Forecasting

First, we'll focus on the machine learning service built around predicting the project costs in enterprise projects while planning the portfolio for the upcoming months.

Figure 6.47 shows the machine learning service built for this process, which we'll describe further in this section.

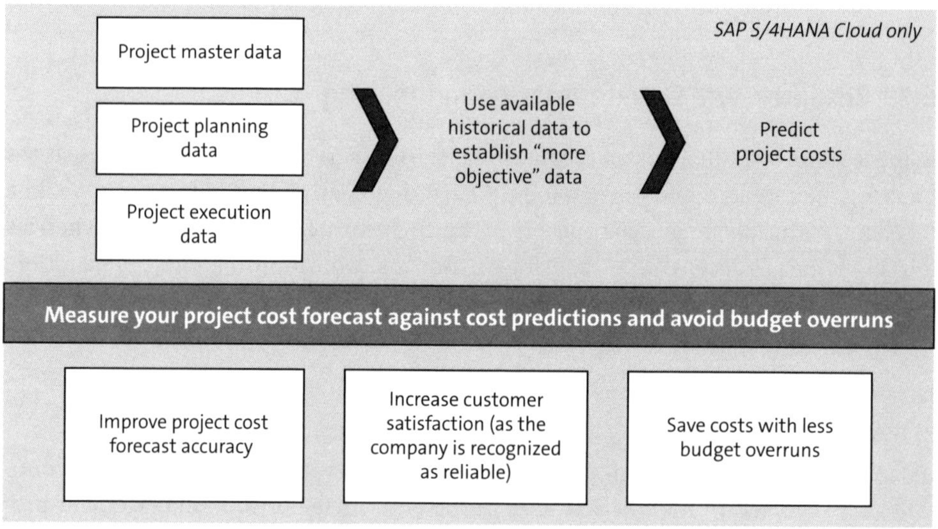

Figure 6.47 Project Cost Forecasting

For a project manager, keeping projects within the defined budget and avoiding budget overruns are the main criteria, assuming the given product scope, product quality, and delivery date are accurately maintained. To achieve this, you need to predict as accurately as possible the consumption of money and time related to the project.

The typical challenges encountered by project managers include the following:

- Inadequate or subjective data
- Difficulty finding the right information to calculate the estimated costs
- Time-consuming tasks to gather all the required information
- Lots of manual efforts and intense efforts needed
- Political and psychological factors
- Imperfect forecasting techniques

Project managers and finance controllers would benefit from improved project cost forecast accuracy and increased customer satisfaction: the company is recognized as reliable, saves costs with fewer budget overruns, and mitigates budget overrun risk.

The machine learning service leverages the available historical data by suggesting an ideal WBS. This is accomplished by establishing more objective and reference data gathered out of the project master data, project planning data and project execution data. The reference class forecasting algorithms helps the system learn from past projects and predict more reliable project forecasting. Using the machine learning service internally, the system will continuously monitor and control projects, notifying users of when overruns might appear by identifying available buffers and proposing possible reschedules and budget reallocations with a decision support mechanism. In the current scenario, classification and k-nearest neighbors algorithms from APL utilizing the ISLM framework are leveraged and embedded into the SAP S/4HANA enterprise portfolio and project management (EPPM) business processes.

There is a huge possible value in using reference class forecasting to get a coherent project plan and cost structure that enables more reliable project cost forecasting and reduces budget overruns.

Configuration Guide

For configuration instructions, refer to the SAP Best Practices Explorer at *https://rapid.sap.com/bp/#/browse/scopeitems/2Y7.*

6.7.2 Digital Content Processing

Next, we'll focus on the machine learning service built around document classification and on processing the digital content of these documents.

Figure 6.48 shows the machine learning service built for this process, which we'll unpack in this section.

Many paper- and document-based communications and processes reduce efficiency across organizations. Many organizations still rely on these processes; digitization would greatly improve efficiency. Digitization is all about finding valuable information in unstructured sources (text files, images, etc.).

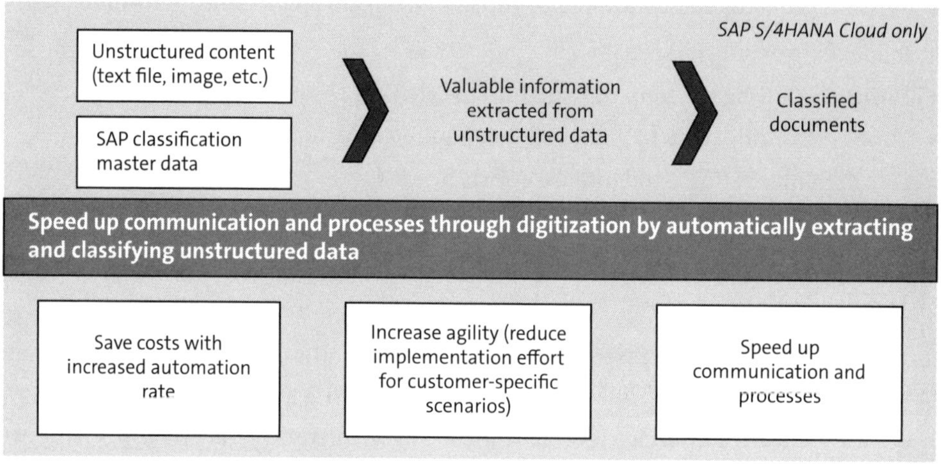

Figure 6.48 Digital Content Processing

Customers typically need to define various classification categories, identify them, and group documents according to those categories. The challenges encountered include a lot of manual work, which might be error-prone, and a manual paper- and document-based communication process that reduces efficiency. Customers benefit from increased business value related to cost savings and efficiency, leading to an increased automation rate. Reduced implementation efforts for customer-specific scenarios make them agile and speed up communications and processes through digitization.

You can produce savings in terms of time, money, and effort by relieving customer pain points via the following methods:

- By introducing new features, better performance, and enhanced quality, you can fix underperforming solutions.
- By making things easier or eliminating obstacles, you can remove negative social consequences faced by customers and eliminate their risks.
- By eliminating social, financial, or technical risks that could potentially cause problems by addressing significant issues, diminishing concerns, and eliminating worries, you could eradicate the most common mistakes that are committed during manual document handling and classification.
- By introducing lower or no up-front investment costs, you can eliminate barriers that are keeping customers away from adoption.

In this scenario, a few algorithms such as text analysis and classification are used from the Scikit-Learn library, and the machine learning service is made available on SAP Cloud Platform with SAP Data Intelligence. The respective SAP S/4HANA process consumes this machine learning service in a side-by-side model and delivers the required results.

After leveraging this machine learning service, digital content processing allows customers to do the following:

- It enables digitization of business processes by extracting structured and unstructured data from documents in the context of SAP Cloud Platform with SAP S/4HANA integration.
- The documents will be automatically classified using SAP classification classes.
- It allows for implementing specific use cases with minimum effort via this flexible and scalable digital content processing service.

The business benefits of leveraging the digital content processing machine learning service include the following:

- **Cost savings and efficiency**
  Replacement of manual work or increases in automation rate thanks to extracting structured content out of unstructured content and automatic document classification
- **Agility**
  Implementation of customer-specific scenarios with minimum effort by leveraging flexible and scalable document management services such as optical character recognition (OCR), rule-based content extraction, and machine learning-based file conversion.

Configuration Guide

For configuration instructions, refer to the SAP Best Practices Explorer at *https://rapid.sap.com/bp/#/browse/scopeitems/2YC*.

6.8 Industries

In this section, we'll focus on use cases across different industries and how customers can benefit from machine learning functionality.

Let's first look into the machine learning scenarios built for the professional services industry.

6.8.1 Professional Services

In the professional services industry, there are a lot of requirements for project management, resource management, service procurement, time recording, talent management, contingent workforce, billing, and other scenarios, which helps project managers, HR managers, accountants, and financial controllers solve their challenges.

Some of the scenarios include the following, which are subsequently planned for development and eventual delivery. We'll briefly look into these scenarios to understand how machine learning can improve and enhance processes for better business outcomes:

- Intelligent bid management
- Intelligent workforce planning
- Intelligent staffing and resource management
- Intelligent project profitability monitoring
- Intelligent WIP project tracking with analytics
- Intelligent bad debts risk analyzer for customer projects

Note

The use cases discussed here in the professional services section are in the planning and evaluation phases and are subject to change.

Intelligent Bid Management

In this section, we'll focus on the requirements for the intelligent bid management process so that professional services companies can better manage their top-line revenues.

Figure 6.49 shows how machine learning can improve the bid management process, which we'll unpack further in this section.

For many projects, potential profitability is determined at the bidding stage itself. It takes a very long time to create bids/proposals for acquiring clients, which jeopardizes the timeliness of the responses. It takes an intense manual process and weeks of research and effort to create and adapt a bid to meet the client's requirements. The scope inaccuracies from the bid stage are typically carried over and impact project delivery and profitability.

The solution is to build a machine learning service to successfully model projects in terms of what successful criteria are necessary when bidding for customer projects. Using an intelligently predicted successful criteria search on completed projects with similar criteria can deliver highly accurate bids and reduce the bid preparation time from weeks to days or even hours. The predictive modeling, with various key bid items such as project budget performance versus project plan and budget time performance versus project plan, can help determine key skill sets of the required resources to match the bid requirements to increase the overall profitability of the project. The idea is to leverage the algorithms from PAL/APL with SAP Analytics Cloud smart predict and visualize the results with the SAP Analytics Cloud explorative model by extending the custom CDS views from SAP S/4HANA.

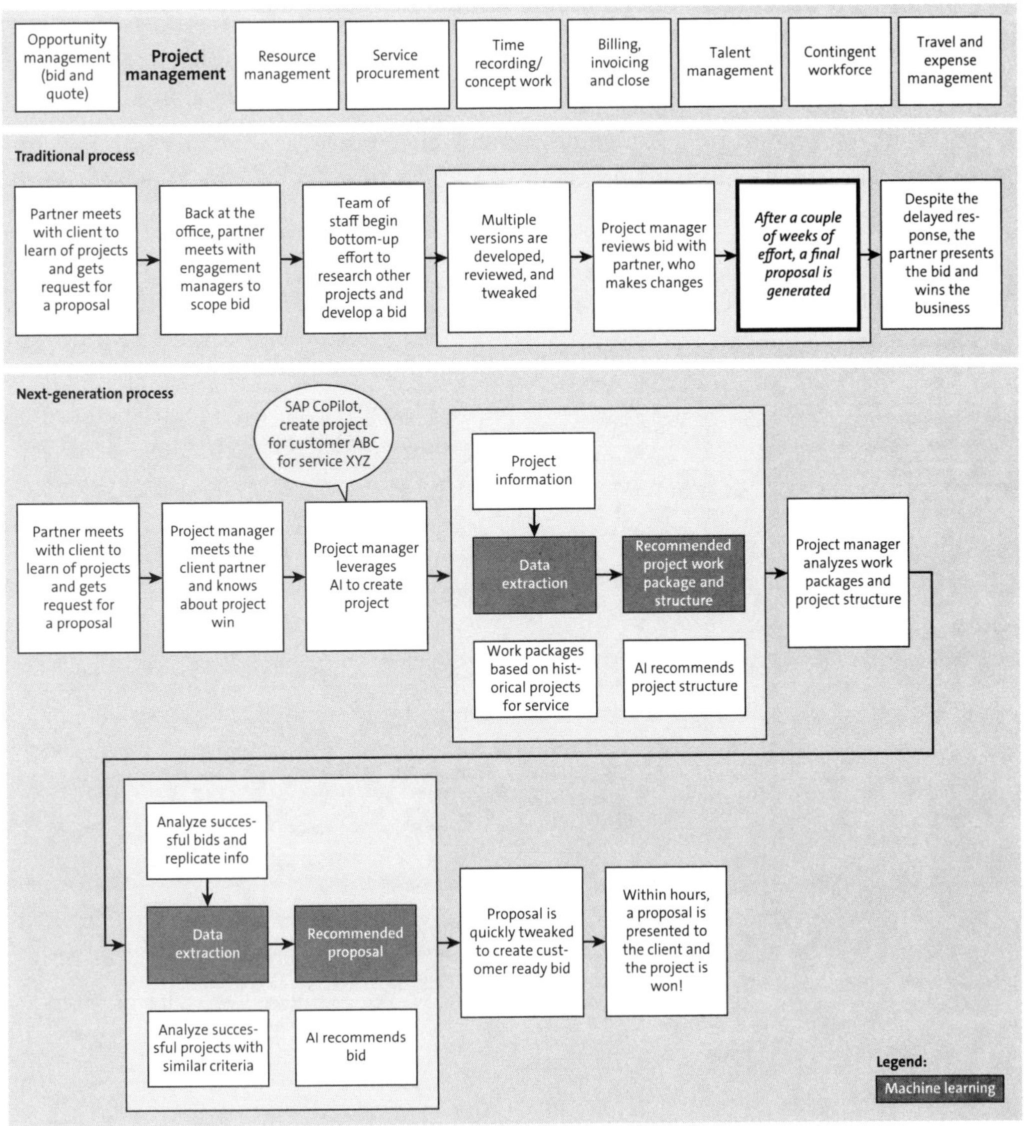

Figure 6.49 Intelligent Bid Management: Process Flow

The key business benefits include the following:

- Significant reduction in manual effort to respond to bids, from weeks to days or even hours
- Significant improvement in win percentage in sales pursuits by tapping into and leveraging similar previously delivered work
- Improved customer satisfaction by leveraging resources familiar to the customer
- Increased proposal accuracy and thus reduced risk related to delivery and profitability/margin erosion
- Efficiency in planned and forecasted resources

Intelligent Workforce Planning

Next, we'll focus on the requirements for the intelligent workforce planning process by which professional services companies can better manage their workforce.

Figure 6.50 shows the how machine learning can improve the workforce planning process, which we'll explain in this section.

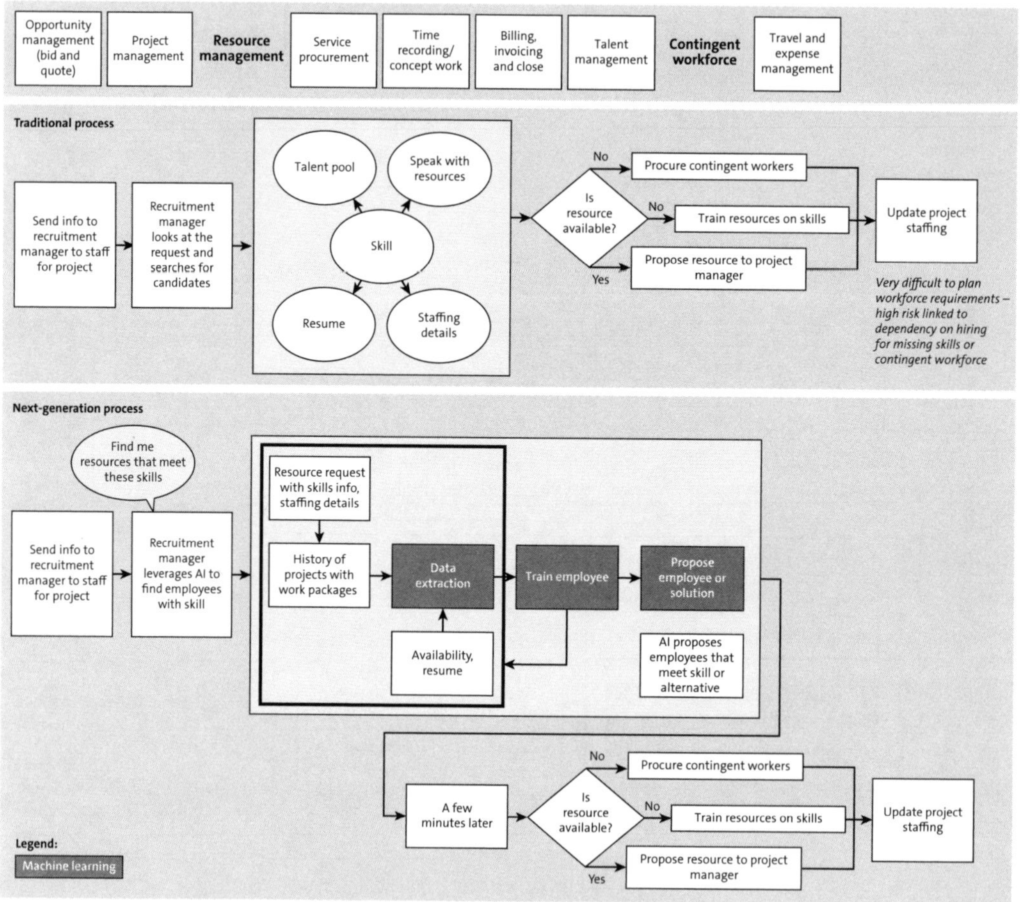

Figure 6.50 Intelligent Workforce Planning: Process Flow

Many projects have to be planned well to get the right number of resources available to complete the tasks within budget. It's very difficult to plan workforce requirements beyond the next 12 months in the pipeline. Capacity constraints and lack of visibility into the talent networks limit a professional firm's ability to bid with confidence that they have the resources to deliver the work. There is a high risk linked to a dependency on hiring for missing skills or the contingent workforce.

The solution is to build a machine learning service and hence leverage the total workforce management provided by SAP. Firms can then search talent across networks to determine if skilled resources are out there and available. By employing the predictive models, you can analyze existing projects and pipelines to determine what skills are increasing/decreasing in demand. Making educated decisions based on powerful predictive models, you can use contingent workforce versus potential skill acquisition versus training programs to meet the anticipated demand. The idea is to leverage the algorithms from PAL/APL with SAP Analytics Cloud smart predict and visualize the results with the SAP Analytics Cloud explorative model by extending the custom CDS views from SAP S/4HANA.

The key business benefits include the following:

- Increased revenue from adopting the best option to maximize margins and reduce risk
- Increased customer satisfaction
- Monitoring, engaging, and gaining insight from the most up-to-date information
- Rapidly developing or adjusting bid responses/proposals in response to customer reactions

Intelligent Staffing and Resource Management

In this section, we'll focus on the requirements for intelligent staffing and resource management so that the HR department can better plan its staffing and resource needs.

Figure 6.51 shows how machine learning can improve skills extraction from resumes and match the requirements accordingly, which we'll unpack in this section.

Often, hiring managers and HR departments have huge issues identifying the right candidates to be considered for the interview process. It can be difficult to quickly identify the right employees for the correct project engagements based on the available employees' skills, qualifications, and related data (structured and unstructured) because that info might be obsolete and scattered across various online and offline sources. The whole hiring process is time-consuming and has inherent human bias. According to a study done by hiring professionals, it takes about 52 days and $4,000 on average to fill an open position.

The solution is to embed machine learning into the intelligent resource matching service provided by SAP S/4HANA that automates the talent screening process and enables resource managers (and recruiters) to identify the best candidates with the best skills and education matches for a specific engagement role/job description. Historical resumes, collaboration and social media (SAP Jam, LinkedIn), emails, time entry data, and job descriptions are used to feed into a matching algorithm for predicting the best candidates. The idea is to leverage the algorithms from PAL/APL and build the machine learning service on SAP Data Intelligence using the side-by-side model for consumption by the SAP S/4HANA business processes.

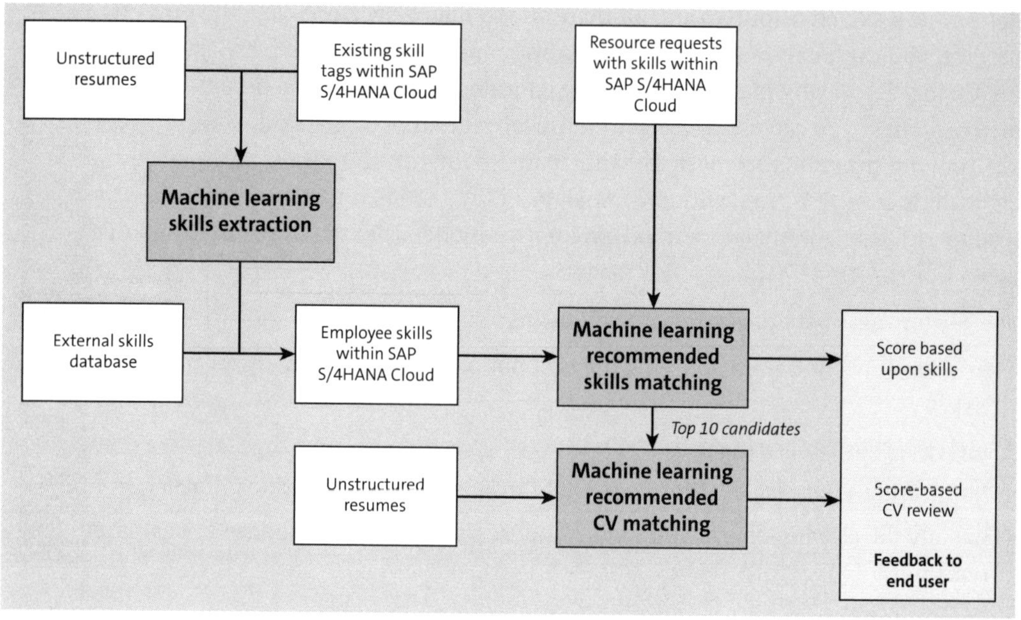

Figure 6.51 Intelligent Skills Extraction and Resource Matching

The business benefits include the following:

- Simplified and smart resource matching process with significantly less manual effort
- Leverage most up-to-date sources of information, removing limitations of obsolete skills and qualification catalogs
- Reduced costs and shorter staff hiring cycles
- Identify candidates with the best skills and education with reduced bias
- Increased quality of candidates, reduced possibility of missing potential candidates, and fewer false positives
- Increased organizational success

Intelligent Project Profitability Monitoring

Now, we'll focus on the requirements for intelligent project profitability monitoring, by which project managers can better understand what projects are more or less profitable than others.

Figure 6.52 shows how machine learning can improve the project profitability by monitoring for any known issues and cost trends, which we'll explain further in this section.

Project managers often find it difficult to manage a large number of projects. However, many large firms manage thousands of projects at any one time across the globe. There is a lack of transparency into eroding project profit margins or projects that are described as on track to underperform. Providing early warning signs of troubled projects would be helpful.

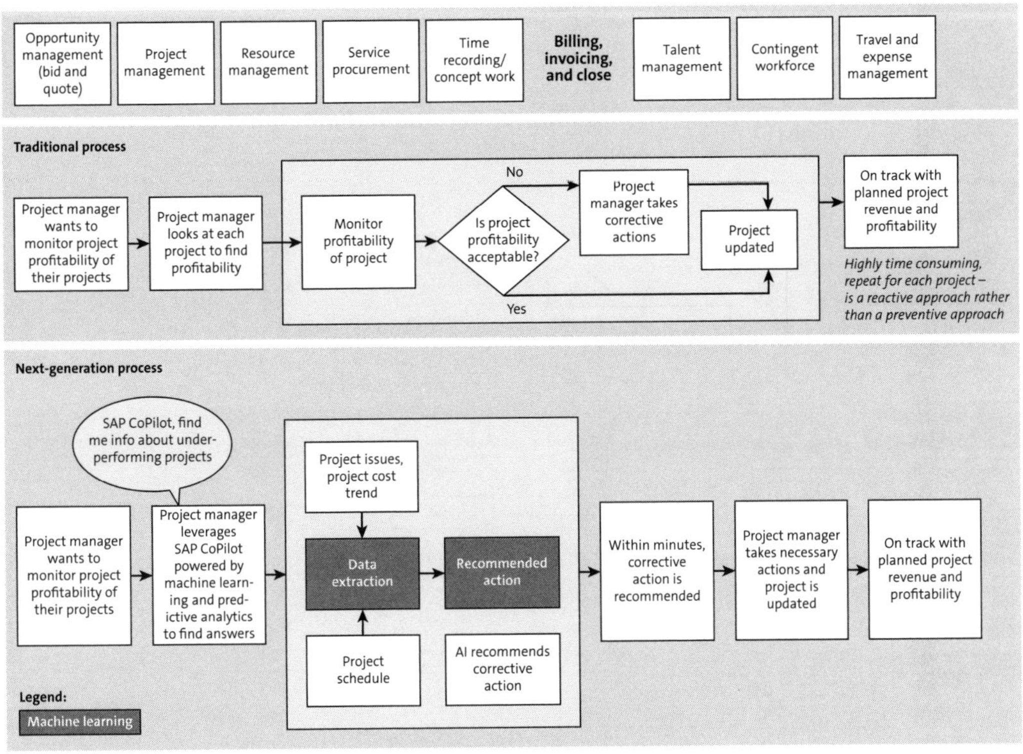

Figure 6.52 Intelligent Project Profitability Monitoring: Process Flow

The solution is to embed machine learning into intelligent project profitability monitoring to create models that can help identify project trends that result in suboptimal outcomes. Machine learning algorithms can also help predict projects that are behind schedule, projects trending with numerous issues, or planned costs being exceeded. The idea is to leverage the algorithms from PAL/APL and build the machine learning service on SAP Data Intelligence using the side-by-side model for consumption by the SAP S/4HANA business processes.

The business benefits include the following:

- Significantly reduce and eliminate issues by predicting and identifying risks early and taking immediate actions to mitigate them.
- Deliver the planned project revenue and profitability.
- An automated process can profile early warning signs of projects in trouble or underperforming situations and offer alerts with appropriate actions.

Intelligent Work-In-Progress Project Tracking

In this section, we'll focus on the requirements for intelligent WIP project tracking, by which project managers can better understand which projects are progressing on track.

Figure 6.53 shows how machine learning can improve the project tracking and improve the project delivery, which we'll discuss in this section.

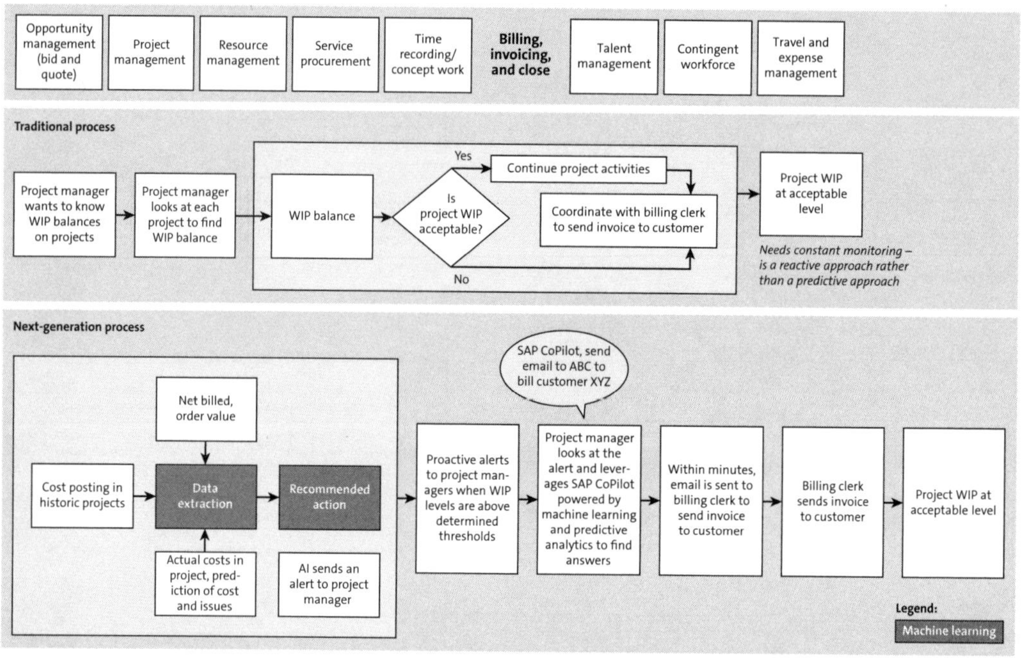

Figure 6.53 Intelligent WIP Project Tracking

Project managers often find it difficult to keep up with growing WIP balances for the vast majority of firms—that is, work that was completed but not yet billed. WIP has a direct impact on DSO and can result from bills not being generated in a timely manner or from other project issues that cause a delay in billing. Delayed billing reduces the timeliness of payment and often causes disputes for charges that were weeks or longer from the actual service delivery.

The solution is to leverage machine learning services to issue proactive alerts to project managers when WIP levels are above determined thresholds. The machine learning algorithms help to 1) determine timeframes that typically have an impact on billing "collectability" factors and 2) provide a proactive mechanism to ensure timely and accurate billing of WIP maintained across the firm, ensuring DSO levels can be properly controlled. The idea is to leverage the algorithms from PAL/APL with SAP Analytics Cloud smart predict and visualize the results with the SAP Analytics Cloud explorative model by extending the custom CDS views from SAP S/4HANA.

The business benefits include the following:

- Timely and accurate billing
- Reduced and eliminated revenue leakage
- Increased customer satisfaction

Intelligent Bad Debts Risk Analyzer for Customer Projects

Now, we'll focus on the requirements for the intelligent bad debts risk analyzer for customer projects, by which project managers and project finance controllers can do better forecasting of project cash collection and the like.

Figure 6.54 shows how machine learning can predict bad debts that are at risk and reduce the budget overruns, which we'll explore in this section.

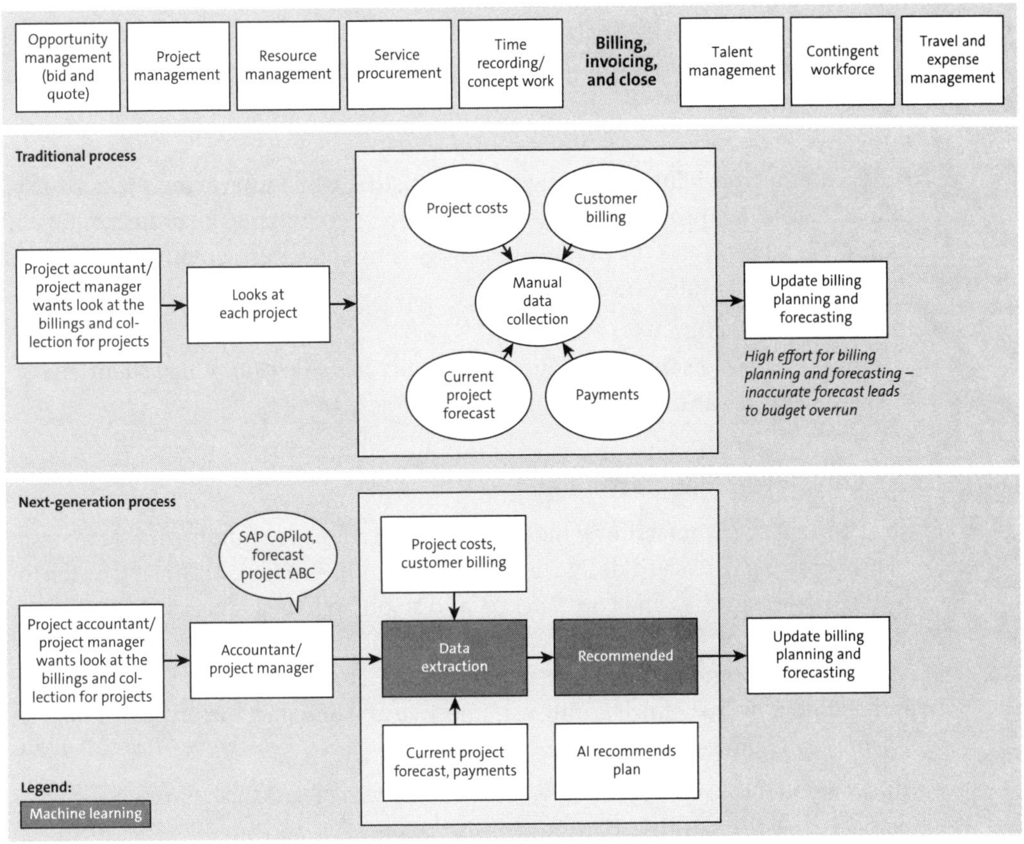

Figure 6.54 Intelligent Bad Debts Risk Analyzer for Customer Projects

Project managers and project finance controllers often find it difficult to find bad debts that are at risk for customer projects. Reducing the manual effort involved with an intelligent invoice tracking system and automation for clear and detailed invoices and reminding clients of the deadlines is the need of the hour. Predicting bad debt risks when planning or creating new customer projects is very helpful for project managers and finance controllers.

The solution is to embed machine learning services into the planning and creation of customer projects, which can help identify customers with past due accounts, bad debt data, and write-off accounts. The machine learning algorithms can be designed with the threshold definition of bad debt percentage and provide recommendations and alerts accordingly. This can greatly help reduce the risk of budget overruns because the models would predict ahead of time the projects running into bad debt risks. The idea is to leverage the algorithms from PAL/APL and build the machine learning service on SAP Data Intelligence using the side-by-side model for consumption by the SAP S/4HANA business processes.

The business benefits include the following:

- **Cost savings**
  Fewer budget overruns and better project investment decisions based on more realistic estimations of customers with overdue payments, and improved cash flow with intelligent collection.
- **Efficiency**
  Reduced effort for billing planning and forecasting while improving the accuracy. Efficient customer project billing collection processes covering the project billing of customers for project services rendered, processing collection activities, and recording the receipt of money owed.
- **Customer satisfaction**
  Improved customer satisfaction due to the ability to deliver simplified client invoicing and measure the profitability of contract project services.

6.8.2 Component Manufacturing

In the component manufacturing industry, we have a lot of requirements for streamlined procure-to-pay, idea-to-design, accelerated plan-to-product, optimized order-to-cash, HR, project services and core finance, and so on, to help address and solve a few challenges. Some of the scenarios include the following, which are subsequently planned for development and eventual delivery. We'll briefly look into these scenarios to understand how machine learning can improve and enhance the processes for better business outcomes:

- Automate root cause analysis
- Optimize inspection plans
- Defect recording
- Analysis of backflush issues
- Predict output batch

Note

The use cases discussed here for component manufacturing are in the planning and evaluation phases and subject to change.

Let's now get into the details of these use cases.

Automate Root Cause Analysis

First, we'll focus on the requirements for automating the root cause analysis functionality for component manufacturing companies so they can better manage the time of their quality engineers and production specialists.

Figure 6.55 shows how machine learning can improve and automate root cause analysis, which we'll unpack in this section.

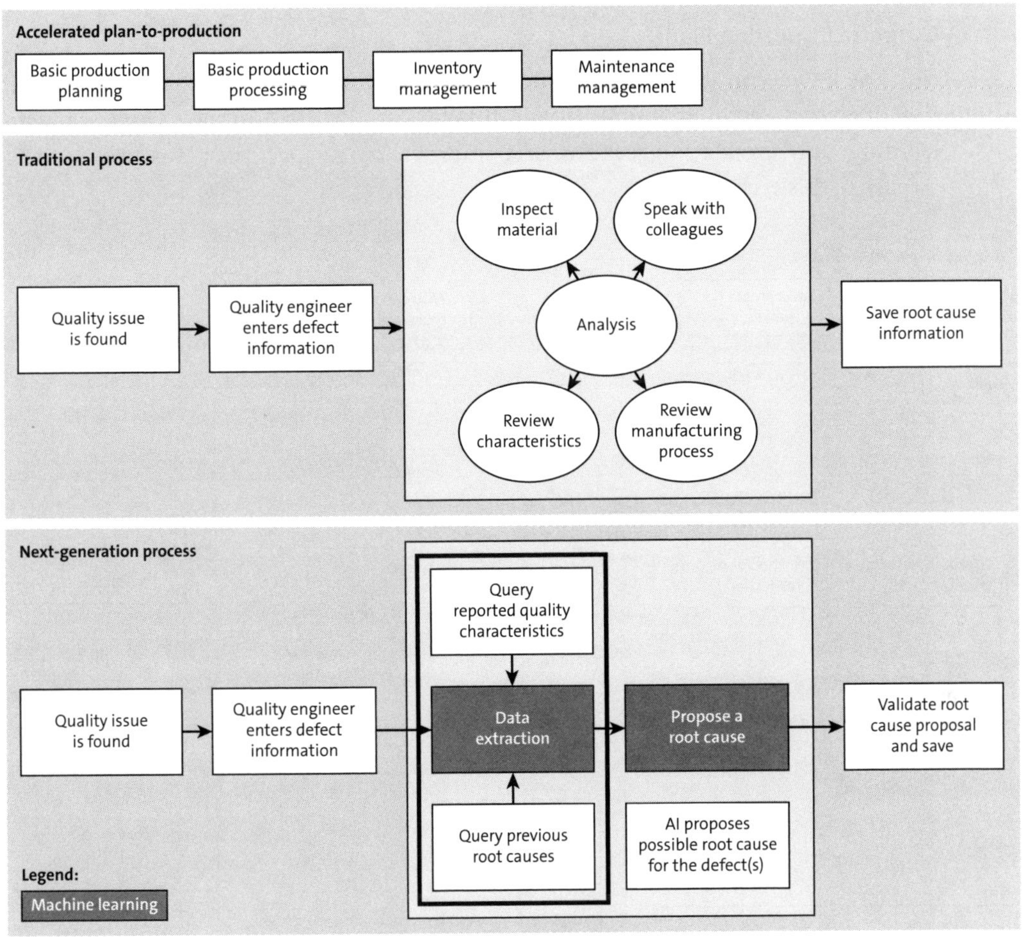

Figure 6.55 Automate Root Cause Analysis

In a traditional process, when the quality engineer detects problems, there's a lot of analysis done by inspecting the materials, speaking with various colleagues and people, reviewing the different characteristics of the issues found in the manufacturing process, and documenting them. Later, detailed root cause analysis is performed based on these various sources and snippets of information.

The proposed solution leverages a machine learning service that proposes possible root causes for the various defects based on the historic information of the root causes and the different quality characteristics. The machine learning algorithms run through the different information provided and come up with various root causes, which can help the quality engineer to provide a fix, reducing the downtime for the equipment and assets. Because this functionality is in the evaluation phase, the architecture decisions are not yet finalized in terms of which approach would be considered: embedded, side-by-side, hybrid, or explorative analytics.

Optimize Inspection Plans

Next, we'll focus on the requirements for optimizing inspection plans so that production operators and quality inspectors can save costs. Figure 6.56 shows how machine learning can provide suggestions and improve the inspection plans, which we'll unpack in this section.

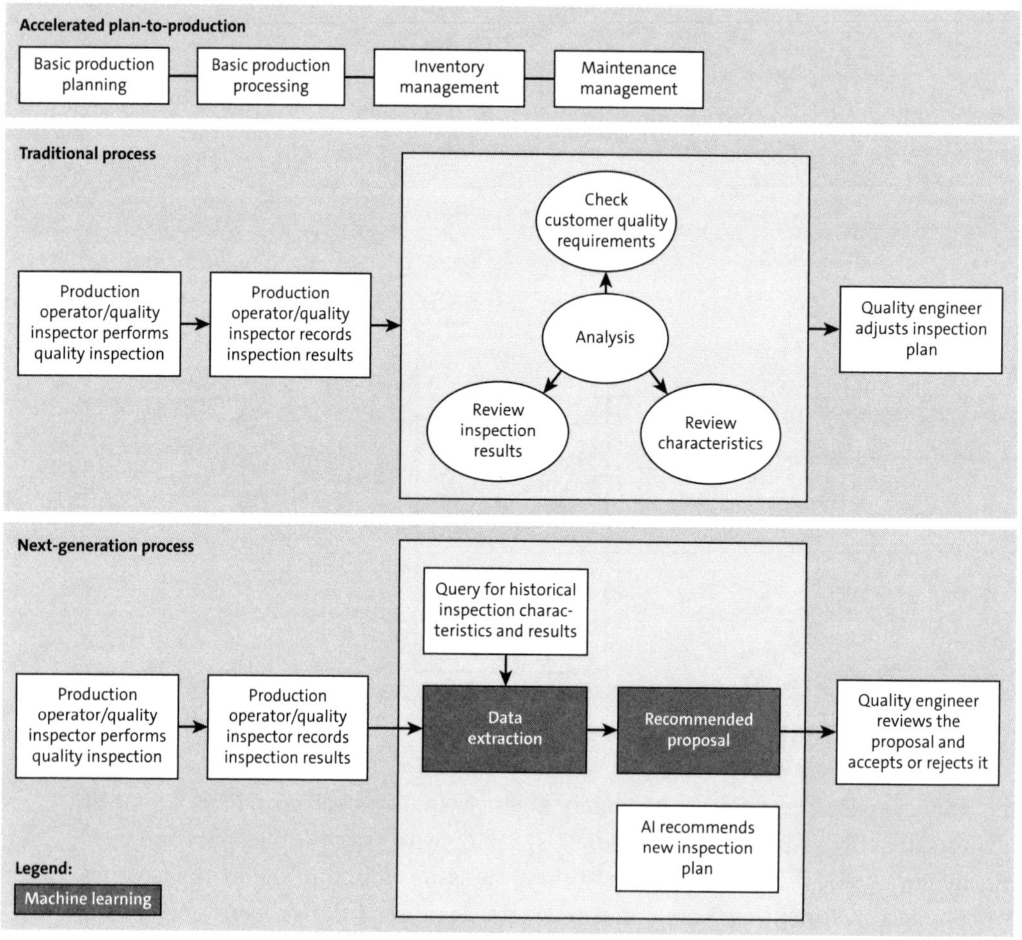

Figure 6.56 Optimize Inspection plans

In a traditional process, the production operators and quality inspectors need to review the quality characteristics and the inspection results. Later, they check the various customer quality requirements, and eventually the quality engineer adjusts the detailed inspection plans. The proposed solution leverages machine learning algorithms and the historical characteristics of the inspections plans and the results, then proposes how the inspection plans can be adjusted. This can greatly reduce the manual effort needed from quality engineers for reviewing all the inspection plans. Because this is in the evaluation phase, the architecture decisions are not yet finalized for which approach would be considered: embedded, side-by-side, hybrid, or explorative analytics.

Defect Recording

We discussed defect processing and defect recording at length in Section 6.5.3 in the context of the inventory and supply chain lines of business. Some of this functionality is applicable to the component manufacturing industry as well. Figure 6.57 shows the scenario that is built for this context that we'll discuss in this section.

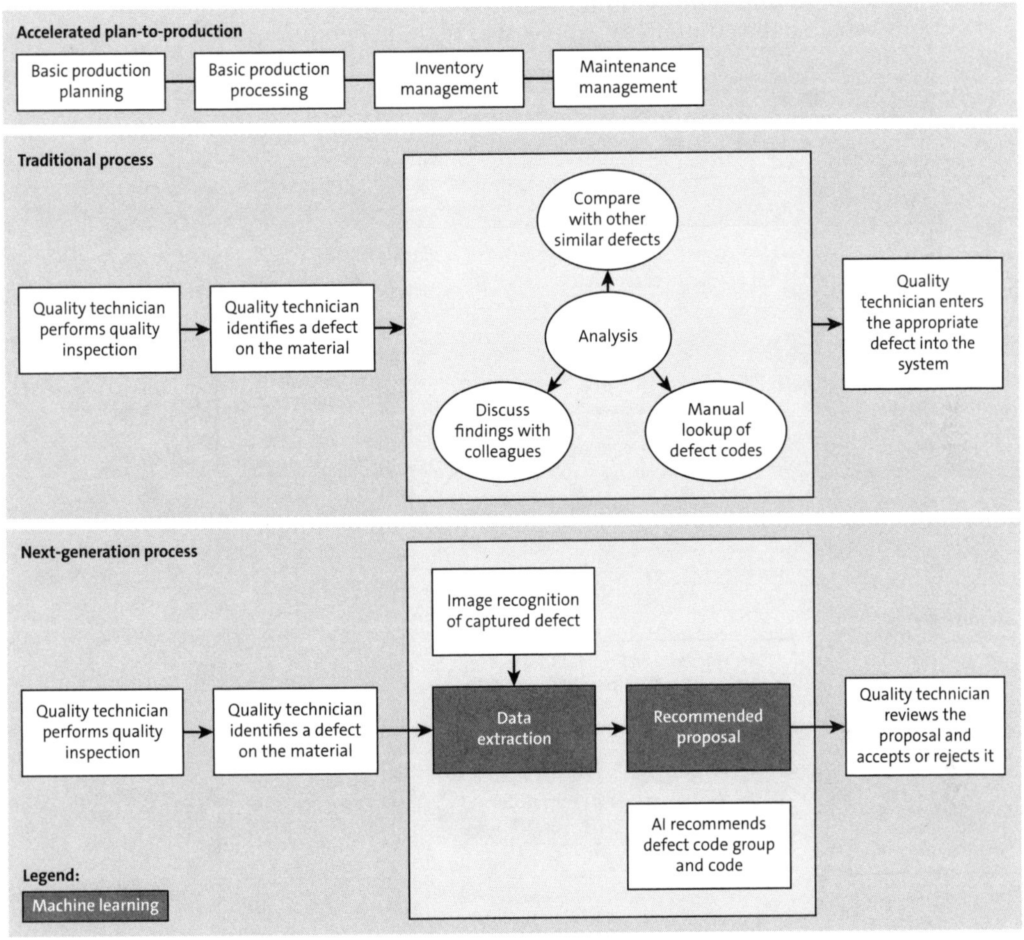

Figure 6.57 Defect Recording

The machine learning service is embedded into the SAP S/4HANA business process with streamlined defect code proposals based on the historic trends, so quality technicians can spend more quality time finding many other defects. In Section 6.5.3, we discussed a similar use case in the inventory and supply chain lines of business, but focused on text recognition while doing defect code proposals. In the component manufacturing industry, there will be some overlaps, but here the plan is to focus on the image recognition part and the machine learning algorithms to propose defect code and defect code groups based on other similar defects. This is again in the pre-evaluation phase and subject to change with more updates.

Analysis of Backflush Issues

In this section, we'll focus on the requirements for analyzing the generated list of errors in material movement transactions and helping the inventory analyst to address them quickly and easily. Figure 6.58 shows how machine learning can provide recommendations based on the different categorizations of the generated errors.

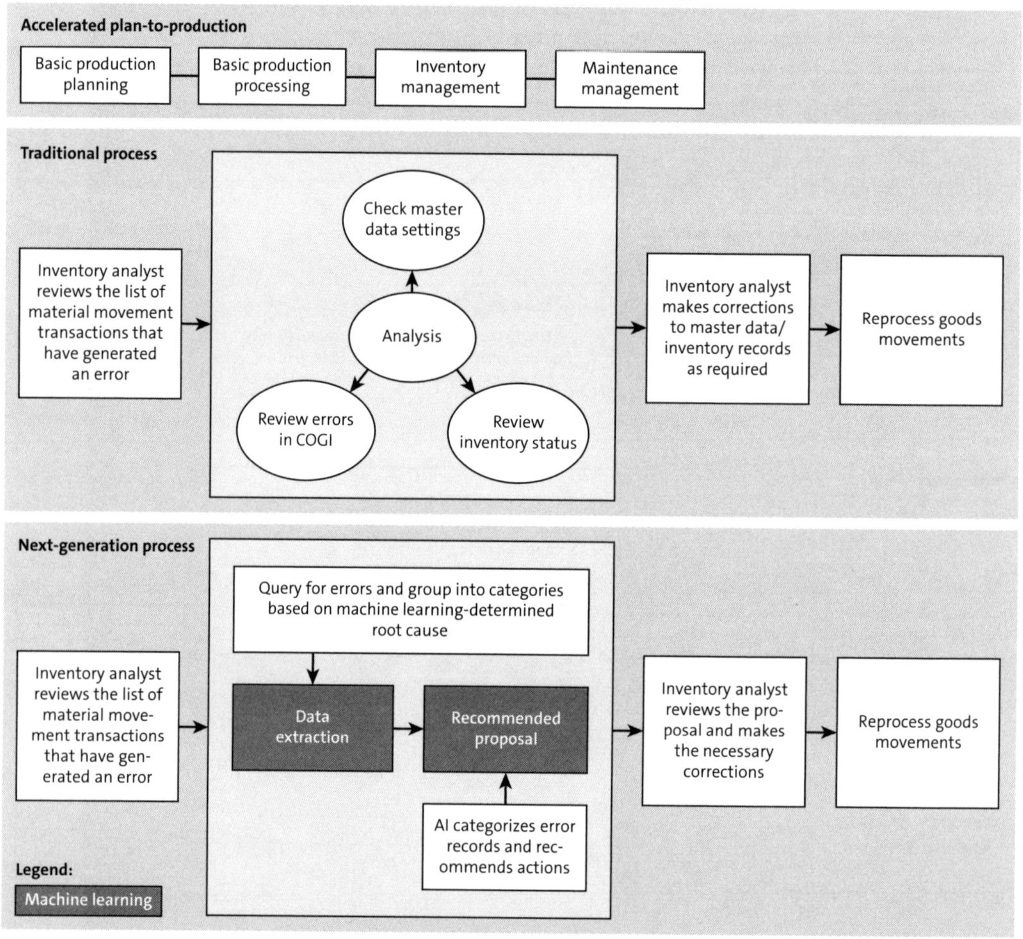

Figure 6.58 Analysis of Backflush Issues

In a traditional process, the inventory analyst reviews the list of all the material movement transactions manually and identifies the error-prone material movements. This involves checking the master data settings, reviewing the inventory status, and checking errors in controlling of goods issued (COGI). When there are issues in the automatic goods movement, the error handling provides the details of the errors encountered. This involves a lengthy process of corrections to the master data and the inventory records as required.

The proposed solution leverages a machine learning service that does a data extraction of the material movements and groups them into different categories based on the historic analysis and past goods movement. The machine learning algorithm after categorizing these generated errors provides recommendations that can be utilized by the inventory analyst and makes corrections to the master data and updates the inventory as needed. A lot of time can be saved by the inventory analyst in this process because the machine learning algorithm provides some suggestions that could be helpful without reviewing all the errors manually. This can save a lot of costs over time.

Predict Output Batch

Now, we'll focus on the requirements for predicting the output batch requirements for the production planner to issue production orders.

Figure 6.59 shows how machine learning can improve the production planning process, which we'll unpack in this section.

In a traditional process, the production planner reviews the output batch characteristics to schedule upcoming production orders by checking the documentation for these batch combinations, reviewing various reports, and talking to the warehouse managers. Eventually the production planner prepares and issues the production orders. This takes up a lot of time and a lot of costs are involved.

The proposed solution leverages a machine learning service that can do a lot of data extraction for these output batches and historical analysis of the batch characteristics from the various locations. A lot of the documentation for these batches and the historical reports can be reviewed, and the machine learning algorithm categorizes these under different batch combinations and proposes suggestions. These suggestions can be picked up by the production planner, who can make appropriate changes and issue production orders. This can reduce the time lag in creation of the production orders based on the various output batches and the combinations that arrive from various warehouses and plant locations.

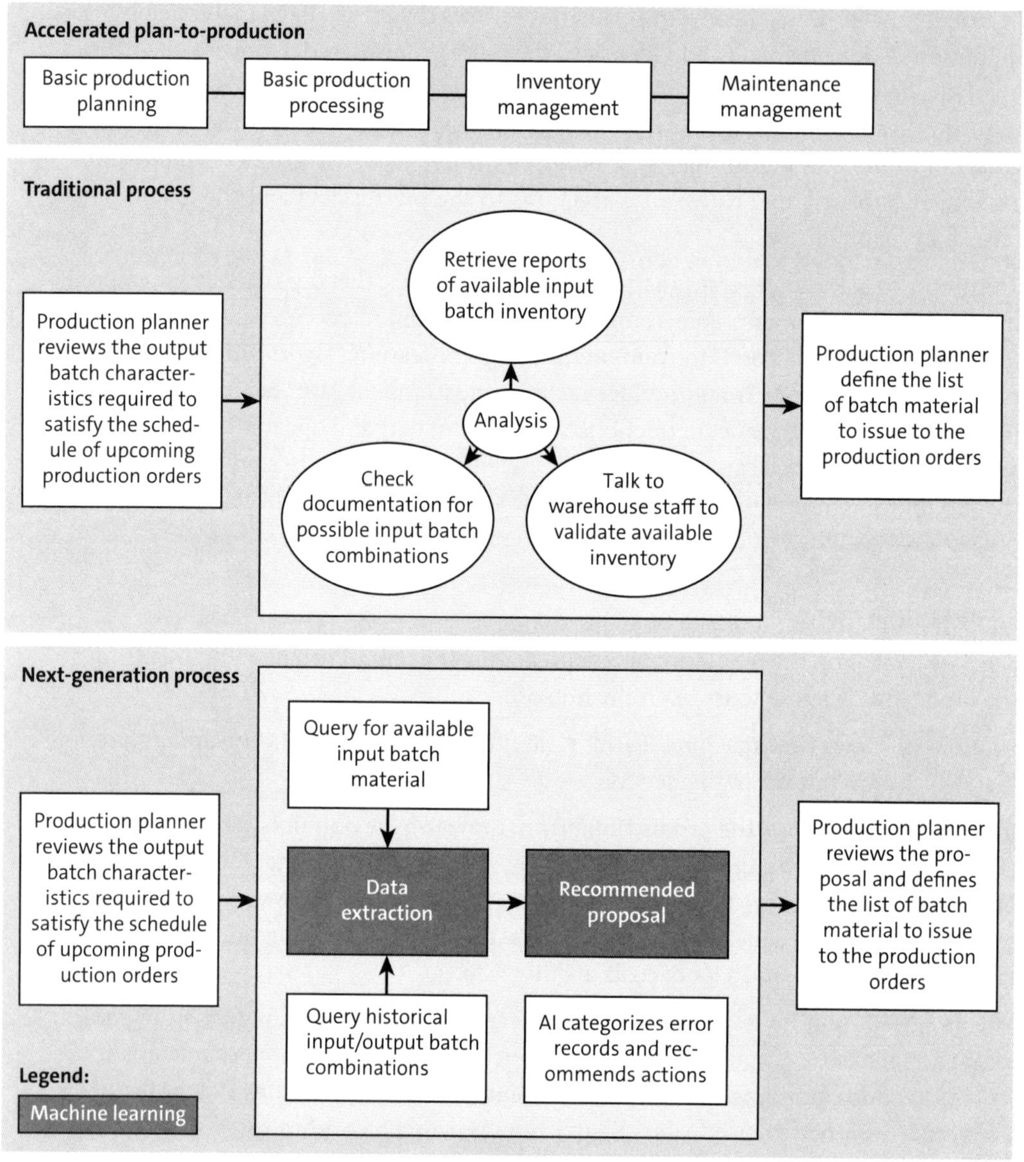

Figure 6.59 Predict Output Batch

6.8.3 Retail

Let's now discuss the machine learning functionality that is built for the retail industry. We'll look into some of the use cases centered on demand forecasting, product promotion, affinity analysis, and so on, plus the customer activity repository, retail sales insights, and more. The retail managers and sales representatives will benefit immensely from machine learning and predictive modeling functionality in retail management solutions. Identifying the demand for fast-moving items and promoting the slow-moving items is very important while working in the retail space.

As part of the machine learning solutions in the retail industry, we'll focus on a few apps that are available as part of the SAP Customer Activity Repository bundle:

- SAP Customer Activity Repository is a platform to centralize customer and transactional data.
- SAP Merchandise Planning consumes data for SAP Customer Activity Repository and performs end-to-end planning for merchandise sales, margin, and inventory.
- SAP Assortment Planning maximizes the regular sale price and optimizes based on the location clusters.
- SAP Promotion Management is an application for creating retail offers and advertising them to customers.
- SAP Allocation Management identifies preseason allocations of merchandise and allocates it to stores.

The SAP Customer Activity Repository bundle and its applications are actively configured to work with SAP S/4HANA Retail.

Let's now look at some of the options that have already been delivered.

Unified Demand Forecast: SAP Customer Activity Repository

In this section, we'll focus on the machine learning service built around predicting the unified demand forecast (UDF) embedded in SAP Customer Activity Repository. SAP Customer Activity Repository is an SAP application that helps to extract, collect, cleanse, and organize customer and transactional information for easy consumption. SAP S/4HANA Retail for merchandise management is available as the backend for SAP Customer Activity Repository, so here we're discussing the machine learning services built into the UDF feature of SAP Customer Activity Repository.

The UDF learns the impact of various demand-influencing factors on historical sales. Demand-influencing factors are conditions that have an impact on the consumer demand, such as promotions, like buy one, get one free. Seasonality, trend, weighting of promotion effects, day effects, and price elasticity are additional demand-influencing factors. The UDF machine learning service creates a predictive model by balancing a priori information with historical data, identifying which are preprocessed with outlier and out-of-stock corrections. In the case of a lack of historic information for the actual product location, a priori information can reference products, hierarchical priors (from product and location hierarchy), and global (static) priors.

When aggregate data is used to understand historical demand-influencing factors, you get a model that's able to capture some aspects of customer demand that aren't visible at the product location level. With this feature, you can take advantage of these aggregated insights and apply them to improve modeling and forecasting on the product location level.

The business value and benefits include the following:

- Robust, precise, and automatic forecasting for reducing out-of-stock and overstock situations.
- UDF gives detailed results for full transparency, the baseline forecast, and the impact of each demand-influencing factor and the forecast confidence index.
- UDF serves planning applications with statistical forecasts for assortment planning, promotion management, allocation management, and SAP Forecasting and Replenishment on supply chain management.

Configuration Guide

For configuration steps, refer to the SAP Help Portal at *https://help.sap.com* and search for "Unified Demand Forecast".

Store Clustering: SAP Customer Activity Repository

In this section, we'll focus on the machine learning service built around store clustering embedded in SAP Customer Activity Repository.

Creating location cluster sets that group locations into clusters based on selected clustering criteria such as KPIs helps retail and store managers identify which stores are doing better business and which are not performing well. Analyzing the existing location cluster set data in a list or graphical format can help improve the performance in a location cluster. The classification analysis machine learning algorithm helps propose an optimal number of clusters within a given range, and then additional analysis can be done on these clusters.

Store clusters are important in business for easy planning, configuration, and automatic data analysis and visualization.

The business value is in doing the following (which helps retail business managers):

- Size curve analysis on the location cluster level
- Assortment planning on the location cluster level
- Allocation management with parameter maintenance and KPI calculation via store clusters

Configuration Guide

For configuration steps, refer to the SAP Help Portal at *https://help.sap.com* and search for "Store Clustering".

Assortment Lists with Planning Rules and Request Proposals

In this section, we'll focus on the machine learning service built around assortment lists with request proposals for retail businesses. This is embedded with the SAP Assortment Planning application available as part of the SAP Customer Activity Repository bundle.

Planning rules are used as guidance when manually determining product statuses and module assignments. These rules are considered by the My Assortment Lists application, and the machine learning algorithm proposes the products to keep in the assortment list and which modules to assign the products to. If there is an option plan, rules for options are automatically created based on the option plan (or they can be recreated from option plans). The feature uses linear programming in the SAP HANA Optimization Function Library. This library provides complex optimization functions that leverage the core SAP HANA features.

The business value and the benefits include the following:

- Decision support is provided for the assortment planner.
- The resulting assortment list is optimized in the sense that it maximizes sales revenue or gross margin and simultaneously aims to fulfill all rules, whether manually created from the option plan or budgeted.

> **Configuration Guide**
>
> For configuration steps, refer to the SAP Help Portal at *https://help.sap.com* and search for "Assortment List".

Assortment Planning: Prepack Optimization

In this section, we'll focus on the machine learning service built around assortment planning with prepack optimization for retail businesses.

Typically, prepacks are customized bundles made available so that customers can save time and money by choosing the best possible combinations of products. These prepacks are designed based on the best match for product demand across different suites of products and locations. It's a very time-consuming process for merchandise and assortment planners to make prepack decisions; the optimizer machine learning algorithm should be able to help in creating the right prepacks based on consumer demand. The optimizer machine learning algorithm is a technique that iteratively compares the different combinations until a satisfactory combination is identified.

The business value and benefits include the following:

- Enhanced inventories shipped to stores
- Time and payroll saved versus doing this manually

- Put away units calculated automatically while determining prepacks on initial allocation
- Money saved by selecting the smallest number of prepack boxes while still fulfilling demand

Configuration Guide

For configuration steps, refer to the SAP Help Portal at *https://help.sap.com* and search for "Assortment Planning".

Promotion Management Affinity Analysis

In this section, we'll focus on the machine learning service built around promotion management and affinity analysis for retail businesses. The promotion management and affinity analysis focus on identifying product relationships, determining how products sell together, and deciding how promotions can be offered and managed effectively.

When considering how to manage products that interact with each other, it's important to understand how an action such as an offer on one product affects the sales of another. Understanding these interactions can help you make smarter and more profitable decisions with the machine learning service. The affinity analytics machine learning service in Demand Data Foundation (DDF) allows you to identify such relationships between your products based on transaction log data available in SAP Customer Activity Repository. DDF is a module in SAP Customer Activity Repository that focuses on analysis, modeling, and forecasting services based on the available customer and transaction data.

The business value and benefits include the following:

- Easily determine affinities by running a dedicated report.
- Explore the affinity relationships for all products within an offer.
- Explore these relationships for all products even outside of the offer.

Configuration Guide

For configuration steps, refer to the SAP Help Portal at *https://help.sap.com* and search for "Promotion Management".

SAP Sales Insights for Retail

In this section, we'll focus on the machine learning service built around the sales analysis and insights into the retail businesses.

Millions of transaction records are generated every day by many retailers, and almost every business decision is open to statistical evaluation. The SAP Sales Insights for Retail machine learning service can help retailers to become more analytical enterprises by providing a business-driven analytics methodology with the necessary tools. SAP Sales Insights for Retail typically helps identify and analyze your point-of-sale data at a market-basket level. SAP S/4HANA for retail is available as the backend for the SAP Sales Insights for Retail application. The machine learning service is embedded into the SAP Sales Insights for Retail application. Most retailers would agree that that there is potential to improve business decisions by learning more systematically from data analysis. But typical roadblocks are as follows:

- Lack of vision to look for the right information in the available data
- Lack of tools and technology for complex analysis on mass data
- Lack of time from business users

The business value and benefits include the following:

- Promotions that help better understand the retail sales' true impact and optimize the promotion strategy
- Assortment of retail products to align better with the customer behavior
- Performance of stores and analysis by doing root cause analysis ahead of time and reacting more effectively
- New product introductions to reduce the time needed for in-store tests
- Cost and effort reduction for data analysis

Configuration Guide

For configuration steps, refer to the SAP Help Portal at *https://help.sap.com* and search for "Sales Insights for Retail".

Product Similarity Scoring

In this section, we'll focus on the machine learning service built around comparative scoring for similar products in retail businesses for better positioning and promotion management.

Figure 6.60 gives an overview of product similarity scoring. The product similarity scoring feature is again a module in the SAP Customer Activity Repository, which helps when analyzing and identifying similar products from a batch of products and thus helps with assortment planning.

Many planning processes, such as assortment planning and replenishment, need appropriate reference information from the past for planning new products. Maintaining reference products manually is tedious and error-prone. The machine learning service for product similarity scoring finds the most similar products to a given one to solve this problem. The machine learning service is embedded into the product similarity scoring functionality, which helps analyze a batch of products to choose a similar product for assortment purposes.

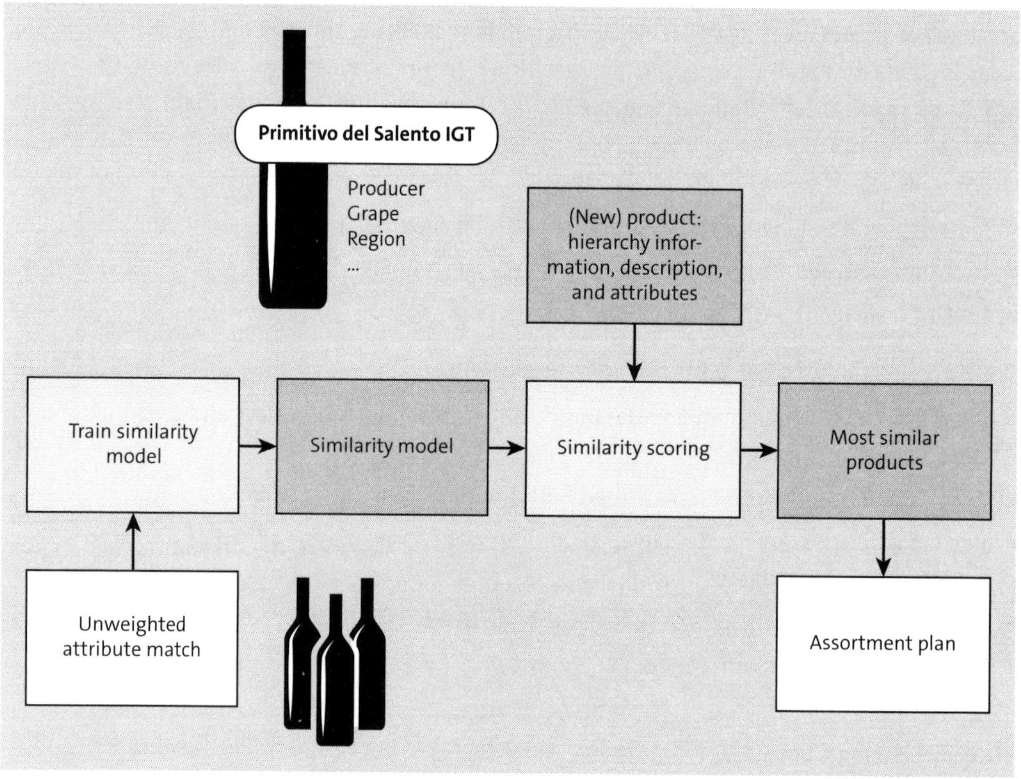

Figure 6.60 Product Similarity Scoring

It makes use of structured product attributes like brand, price category, color, fabric, and/or other textual information provided as sort of a description.

The business value and benefits include the following:

- Supportive service for assortment planning
- Finding suitable products in your initial assortment with suitable other products
- Finding good candidates for reference products when creating placeholder products

Configuration Guide

For configuration steps, refer to the SAP Help Portal at *https://help.sap.com* and search for "Product Similarity Scoring".

Cross Product Allocation

In this section, we'll focus on the machine learning service built around the cross product allocation with SAP Allocation Management. Cross product allocation helps in identifying the store demand with an optimum mix of the cross products and allocating accordingly to the stores.

Figure 6.61 gives an overview of cross product allocation. Notice how the limited availability of a product is handled by optimizing the product mix and allocating accordingly.

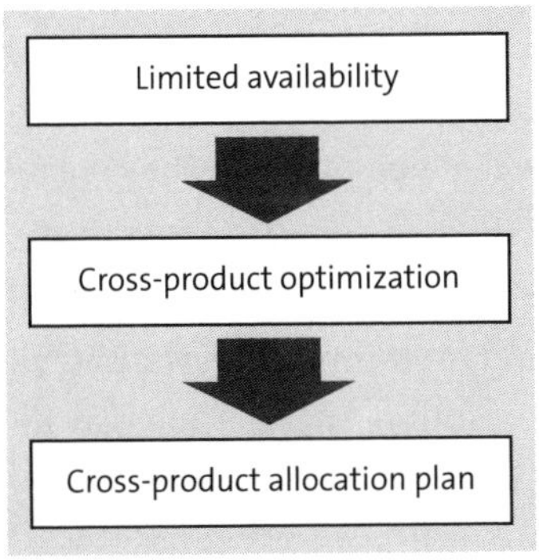

Figure 6.61 Cross Production Allocation

In the case of very limited product availability in a distribution center, store allocation is a very cumbersome task if lacking automation and intelligence. This situation is common for outlet businesses, which are based on but not limited to returns from full-price stores. Often, there are similar products or the same product in another distribution center, all of which together could fulfill the total store demand.

Typically, when there is a similar demand from the stores, the product mix is provided to the stores with a minimized mixture of the products. This helps all stores maintain the different product mix for positioning to customers and then order accordingly. This helps all stores to maintain the different product mix for positioning to customers. The stores can then order more of these product combinations based on demand. This helps keep allocation management across the stores for similar products seamless. The machine learning service is embedded into the cross product allocation function, where it identifies these similar demands across a store and assigns the stores with a minimal mix of the products so that each store is fully satisfied. The cross product allocation machine learning service uses the SAP HANA Optimization Function Library.

The business value and benefits include the following:

- Higher margins through enhanced service levels and product presentation in stores
- Higher efficiency through ease of use and time savings for business users, coupled with high transparency into the optimization results

Configuration Guide

For configuration steps, refer to the SAP Help Portal at *https://help.sap.com* and search for "Cross Product Allocation".

Promotion Analyzer

In this section, we'll focus on the machine learning service built around product promotions and deal analysis in the retail businesses.

Promotions are an extremely important aspect of most retailers' business. For many retailers, a large portion of their sales come from promoted items. Despite the importance of promotions, it's very hard to quantify their true business impact. They create a number of very different effects, both positive and negative.

The proposed solution leverages a machine learning service to analyze the total business case of past promotions in order to better understand their impact on the overall business and to optimize the promotion strategy. The Promotion Analysis app is available with SAP Integrated Business Planning for Supply Chain (SAP IBP) to help businesses do promotions analysis. SAP S/4HANA is integrated with the SAP IBP to get all the master data and transactional data. The machine learning service is embedded into the Promotion Analysis app. Based on point-of-sale data, the solution computes various effects for each historic promotion.

These effects, which add up to the total business impact of the promotions, include the following:

- Hoarding
- Cannibalization
- Increased consumption of the promoted product
- Net new customers for the store or the promoted product
- Drag-along effects
- Cost of promotion

Configuration Guide

For configuration steps, refer to the SAP Help Portal at *https://help.sap.com* and search for "Promotion Analyzer".

Upcoming Ideas from Retail Industry

The retail industry, whether in-store or online, is very vibrant, and there are a lot of customer requirements that help consumers better their daily routines. The competition in the retail industry space is also very tough, but leveraging machine learning services helps you better understand customer sentiments and behaviors to keep you ahead of your competition. Some of the new ideas currently in the evaluation and planning phases are as follows:

- **Moving products with radio frequency identification (RFID)**
  The machine learning service would help move products from the backroom to the sales floor in an automated and easy way. The business value is that the service would allow you to distinguish the stock in the back room and the stock on the sales floor, needing only minimal manual effort to post the movement and achieve high product availability on the sales floor.
- **Capacity allocation in promotion-driven allocation**
  The machine learning service helps create promotional products that are available for short periods only to attract customers, especially for discounter and supermarket chains. Allocation can optimize special store areas by predicting store inventory at the start of an offer, determining free capacity, and balancing product pressure and range of coverage across stores with data intelligence. The business value is that the service would provide better-looking promotional store areas, better balancing of free capacity across stores, higher sales, and more satisfied customers.
- **Cost-optimal ordering in store replenishment**
  The machine learning service can help find cost-optimal order quantities for automated replenishment of retail stores considering different cost aspects like costs for write-offs, costs for lost revenue, handling costs, and storage costs. The business value is that intelligent store replenishment increases stores' margins by ordering products in a cost-optimal manner, reducing required manpower, and increasing process quality.
- **Smart pack sizes in allocation**
  The machine learning service can help retailers use logistical units for buying at vendors and distributing to stores, such as cartons, layers, or pallets. To serve store demands in the best possible way, smart pack size optimization helps order and distribute alternative logistic units in the right quantity for the right store. The business value is better demand fulfillment in stores through more adapted pack sizes.
- **Intelligent store management with retail in-store product location microservice**
  The machine learning service can support customers in finding products in the store or processing individual shopping lists. Associates are supported in processing click and collect orders and their routes through the store are optimized when ordering, counting or labeling products. The business value is in the convenience for the shoppers, increased efficiency of store services, and more.
- **Predicted store inventory**
  The machine learning service helps in automatically detecting products in the retail store with the wrong inventory info and eventually predicting the deviation from the actual physical inventory. The business value is reduced write-offs or overstocks caused by wrong orders from the wrong inventory numbers, improved data quality for the retailer, and so on.

- **Store video monitoring for operations department**
  The machine learning service helps store and analyze video content from the store video cameras by adding value for district managers, marketing departments, loss prevention departments, and so on. The business value is found in additional tools for operations departments that can help optimize their work and increase return on investment (ROI).
- **Improved consumer demand forecast with localized real-world data**
  The machine learning service can help improve retail demand not only via promotions but also via external factors like weather market trends, traffic, and more. The business value is improved demand forecasting.
- **Close the loop on in-store RFID processes**
  The machine learning service can help with centralized merchandizing, improved footprint in the store, and so on. The business value is in store process automation, fraud protection, and single item tracking.

6.8.4 Utilities

In this section, we'll discuss the machine learning functionality built around the utilities industry. We'll look into some of the use cases around smart metering, energy consumption, and predictive maintenance that help the utility companies, quality professionals, and sustainability officers to a great extent. The utility companies will benefit immensely from machine learning and predictive modeling functionality built into these everyday solutions for consumers. In the utility industry, it's always very important to analyze and predict how energy is being consumed and how you can help the planet become more greener or sustainable.

Let's now look into some of the ideas that have already been delivered.

Smart Meter Data Analytics

In this section, we'll focus on the data analytics built around smart metering for utilities companies.

You can analyze transactional smart meter data in combination with master data from various sources, including SAP S/4HANA systems, and predict interesting insights for utilities companies. The prediction model embedded into the SAP Cloud for Energy solution provides interesting insights for SAP S/4HANA Utilities customers.

The business value and benefits include the following:

- Converting large volumes of data into meaningful business intelligence
- Understanding how your customers use energy by any dimension and any level of granularity instantly
- Turning deep predictive insights into actions

Configuration Guide

For configuration steps, refer to the SAP Help Portal at *https://help.sap.com* and search for "Cloud for Energy".

Implausible Smart Meter Readings

In this section, we'll focus on the machine learning service built around the implausible meter readings for utilities companies.

Figure 6.62 shows the machine learning service built for this process, which we'll unpack further in this section.

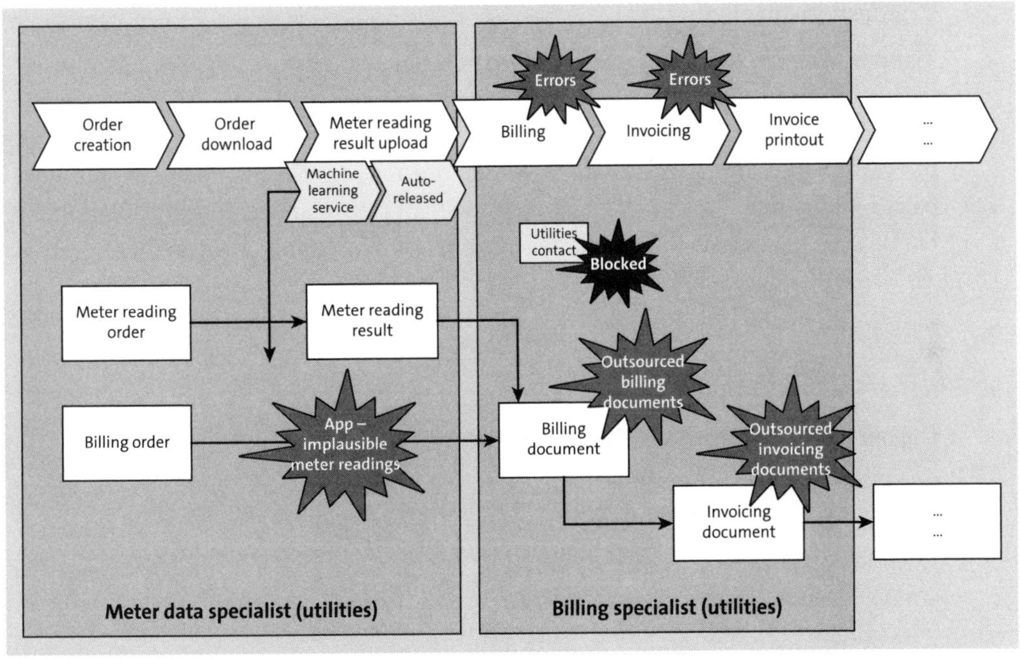

Figure 6.62 Implausible Meter Readings

This machine learning service helps to resolve implausible meter readings and removes the need for agents to solve the exceptions manually. The regression algorithm understands the agents' past behavior and automatically suggests a release of all noncritical exceptions for an efficient reduction in the amount of manual work. The algorithms from PAL/APL are leveraged to build the machine learning service in a side-by-side model on SAP Cloud Platform for consumption by the SAP S/4HANA business processes.

The business value and benefits include the following:

- Enhanced speed in resolving exceptions for an improved cash flow
- Optimizing current manual processes
- Cost savings

Configuration Guide

For configuration steps, refer to the SAP Help Portal at *https://help.sap.com* and search for "Implausible Meter Readings".

Predictive Maintenance for Energy Providers

In this section, we'll focus on the machine learning service built around predictive maintenance for energy providers.

Manufacturers of today's high-tech equipment for sustainable energy production, like wind turbine manufacturers, require modern systems that combine intelligent machine health prediction models and predictive maintenance strategies with a seamless integration into robust maintenance scheduling and logistics planning systems for repair optimization, spare parts planning, and repair crew management. This is an example of how integrated intelligent machine and maintenance management systems work together by leveraging the SAP S/4HANA business processes with SAP Data Intelligence. The algorithms from PAL/APL with R integration are leveraged to build the machine learning service in a side-by-side model on SAP Cloud Platform for consumption by the SAP S/4HANA business processes.

The business value and benefits include the following:

- 2-D and 3-D line charting to help technicians and operations with a visual way to check their sensor data and history
- Machine learning algorithms written in R to provide prediction models for assets' health status
- Maintenance order integration with SAP ERP
- Easy navigation from one central dashboard to manage many different types of assets

Configuration Guide

For configuration steps, refer to the SAP Help Portal at *https://help.sap.com* and search for "Predictive Maintenance".

Predictive Maintenance for Efficient Supply of Electricity

In this section, we'll focus on the machine learning service built as a predictive maintenance for efficient supply of electricity add-on for utilities companies such as energy providers.

By leveraging this machine learning service, you can provide insights into the health of critical utilities assets and can support better decision-making for maintenance and investment decisions for energy providers. Oil quality analysis for power transformers uses industry best practices approaches and international standards. The Utilities Linear Asset Analytics tool in the SAP Predictive Maintenance and Service solution provides analysis for underground residential distribution assets, overhead lines, and poles. The algorithms from PAL/APL with R integration are leveraged to build the add-on machine learning service as a side-by-side model on SAP Cloud Platform for consumption by the SAP S/4HANA business processes.

The business value and benefits include the following:

- Better asset availability, reduced maintenance costs, and improved customer experience
- A single platform for all types of assets from all vendors, avoiding siloed applications for different asset classes
- Evaluation of IoT data collected from power transformers, applied to machine learning techniques to predict the future health and resulting maintenance needs

Configuration Guide

For configuration steps, refer to the SAP Help Portal at *https://help.sap.com* and search for "PDMS add-on for utilities".

Smart Master Data Quality Improvements

In this section, we'll focus on the machine learning service built around the master data improvements.

Improving the quality of master data by completing missing technical data such as the material or construction year for all assets with high accuracy is crucial for any company. This machine learning approach replaces the traditional purely rule-based solutions for managing the quality of asset data solutions, which are time-consuming, are unable to detect unknown patterns, and require tremendous expert involvement. The solution combines internal business knowledge with unparalleled data mining expertise to rapidly improve data quality, leveraging data mining and statistical programming techniques to identify patterns, extract regularities from the data itself, and complete missing asset data automatically. The algorithms from PAL/APL are leveraged to build the machine learning service using the side-by-side model on SAP Cloud Platform for consumption by the SAP S/4HANA business processes.

The business value and benefits include the following:

- Quickly improve data quality at a lower cost
- Reduce asset maintenance costs by 1-2%

- Meet regulatory and governance requirements
- Better and more efficient planning for asset investments
- Streamlined data usage across the business for more informed decision-making

Configuration Guide

For configuration steps, refer to the SAP Help Portal at *https://help.sap.com* and search for "Smart Master Data Quality".

Outsorted Billing Documents

In this section, we'll focus on the machine learning service built around the outsorted billing documents for utilities companies.

The regression machine learning algorithms using the side-by-side model are applied to the process and resolve out-sorted billing documents for utilities companies. There is a growing need for agents to solve these exceptions manually. Application of machine learning to understand agents' past behavior and to either automatically release or suggest a release of all noncritical exceptions can lead to an efficient reduction in the amount of manual work.

The business values and benefits are as follows:

- Enhanced speed in resolving exceptions for an improved cash flow
- Optimization of current manual processes
- Cost savings

Configuration Guide

For configuration steps, refer to the SAP Help Portal at *https://help.sap.com* and search for "Outsorted Billing Documents".

Customer Profitability Analysis

In this section, we'll focus on the machine learning service built around the profitability analysis for utilities subscribers.

The machine learning service is built to understand the utilities subscribers and their behaviors by performing various kinds of data science techniques to identify the outlier detection, supervised and unsupervised clustering, and delta analysis. This gives an understanding of the margin drivers with the subscribers and helps clarify patterns. The algorithms from PAL and APL are leveraged here to build the machine learning service using the side-by-side model that is consumed from SAP Cloud Platform.

The business value and benefits include the following:

- Helping utilities to transition from revenue assurance to margin assistance
- Detecting changes in key figures like profit growth over time
- Detecting possible reasons for changes or other changes that correlate with the changes in key figures over time

Configuration Guide

For configuration steps, refer to the SAP Help Portal at *https://help.sap.com* and search for "Customer Profitability Analysis in Utilities".

6.8.5 Consumer Products

In this section, we'll focus on the machine learning functionality that is built around the consumer products industry. We'll look into some of the use cases for demand forecasting, products promotion management, affinity analysis, trade management, and demand signal management. The sales teams will benefit hugely from machine learning and predictive modeling functionality in the consumer packaged goods (CPG) industry.

Let's now look at some of the ideas that have already been delivered.

SAP Trade Management Forecasting

First, we'll focus on the machine learning service built around SAP Trade Management, leveraging the time series algorithms. SAP Trade Management is built for managing trade promotions, customer information, and other trade management analytics. The machine learning service extracts additional data from SAP S/4HANA while building the predictive models using time series data.

Figure 6.63 shows the machine learning service built for this process, which we'll explain in this section.

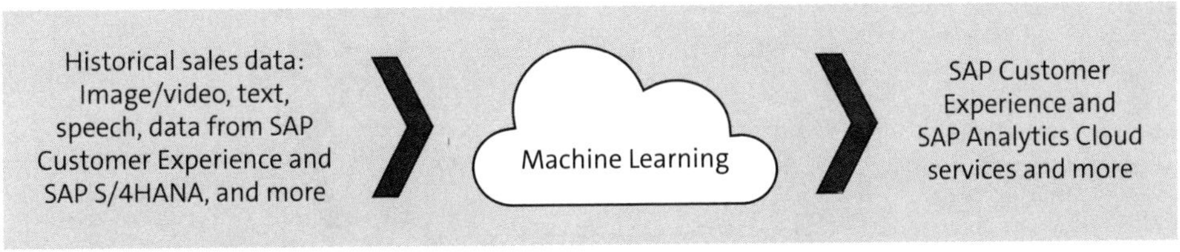

Figure 6.63 SAP Trade Management Forecasting

SAP Trade Management forecasting automates the understanding and optimization of baseline, price, lift, tactic lift, seasonality, holiday, and cannibalization volumes. Sales teams are empowered through machine learning to determine the right promotion attributes, from price to merchandizing tactics and associated investments. The algorithms from PAL/APL are leveraged to build the machine learning service, using the side-by-side model on SAP Cloud Platform for consumption by the SAP S/4HANA business processes.

The business value and benefits include the following:

- Better and faster sales planning
- Accurate total, baseline, and uplift volumes
- Steering investments to the right activities
- Improved plan accuracy and cycle times
- Integrated with SAP Trade Management for reduced TCO and time to value
- Allows focusing on business rather than debates about the baseline
- Empowers sales teams to focus on strategic tasks and selling

Configuration Guide

For configuration steps, refer to the SAP Help Portal at *https://help.sap.com* and search for "SAP Trade Management"

SAP Demand Signal Management Data Quality

Next, we'll focus on the machine learning service built around SAP Demand Signal Management, leveraging the classification algorithms. SAP Demand Signal Management holds large volumes of internal master data and transaction data along with external point-of-sale data, including market research data, to provide detailed reports on and analysis of products and their demands. It leverages SAP S/4HANA for extracting the internal master data.

Figure 6.64 shows the machine learning service built for this process, which we'll unpack in this section.

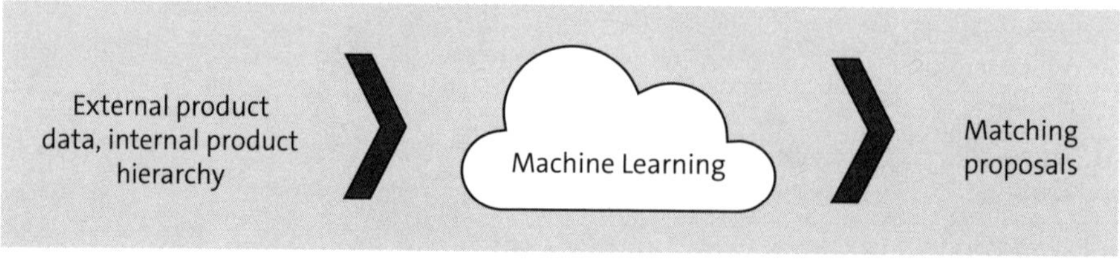

Figure 6.64 Demand Signal Management: Data Quality

The machine learning service automatically classifies internal and external products in the product hierarchy. If products are mapped to SAP S/4HANA products, they can inherit classifications like category, subcategory, and segment from the SAP S/4HANA record, but the customer also receives competitor products, which need to be classified and mapped into product hierarchies. The algorithms from PAL/APL are leveraged to build the machine learning service, using the side-by-side model on SAP Cloud Platform for consumption by the SAP S/4HANA business processes.

The business values and benefits include the following:

- Automatically learn from master data analysts' actions and apply them to future classifications
- Enable substantial reduction of labor cost per data source and faster resolution
- Help automation shift effort toward value-adding processes
- Improve product classification time and quality
- Integrate with SAP Trade Management for reduced TCO and time to value
- Allow shared services to scale as the business grows
- Empower analysts to focus on strategic tasks and service quality

Configuration Guide

For configuration steps, refer to the SAP Help Portal at *https://help.sap.com* and search for "SAP Demand Signal Management".

Claims Management in Trade Promotion Management

Now, we'll focus on the machine learning service built around trade claims, remittances, invoices, and trade promotions. This is the claims management functionality in the Trade Promotions Management SAP Fiori app, which leverages SAP S/4HANA as a backend for customer, transactional, and master data.

Figure 6.65 shows the machine learning service built for this process, which we'll unpack in this section.

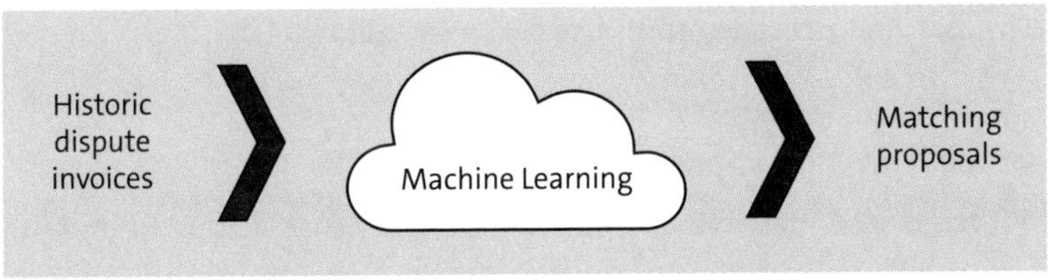

Figure 6.65 Trade Claims

The machine learning service automatically matches incoming invoices and deductions with open items and trade promotions or credit memos. It boosts the efficiency of the traditionally very labor-intensive validation process, being executed with nearly no user interaction. The application takes advantage of machine learning to enable an effortless setup and to increase automatic matching rates. The algorithms from the TensorFlow library are leveraged to build the machine learning service using the side-by-side model on SAP Cloud Platform for consumption by the SAP S/4HANA business processes.

The business values and benefits include the following:

- Automatically learn from claims analysts' actions and apply those to future claims
- Enable substantial reduction of labor cost per claim and faster resolution
- Help automation shift efforts towards better service quality
- Improve claim resolution time and quality
- Empower claims analysts to focus on strategic tasks and service quality
- Integrate with SAP Trade Management for reduced TCO and time to value
- Allow shared services to scale as the business grows

Configuration Guide

For configuration steps, refer to the SAP Help Portal at *https://help.sap.com* and search for "SAP Trade Claims".

Upcoming Ideas from the Consumer Products Industry

One key idea in the evaluation and planning phases is related to omnichannel promotion and pricing: intelligent planning and pricing of the consumer products with omnichannel promotions.

Manufacturers struggle with consistent pricing for customers and shoppers via the different sales channels that may or may not be promotion-related. The machine learning service could be designed in such a way that all pricing elements agreed on with the customer need to be reflected in the pricing during the order-to-cash process regardless of where the order has been entered. The business value would be proper pricing regardless of which sales channels customers use for order taking.

6.8.6 Insurance

In this section, we focus on the machine learning functionality that is built around the insurance industry. We'll look into some of the use cases around fraud management, claims management, online service request automation, automated underwriting of claims, and so on. Insurance companies will benefit immensely for machine learning and predictive modeling functionality for faster management of insurance cases.

Let's now look into some of the ideas that have already been delivered.

Predict Claims Severity

In this section, we'll focus on the machine learning service built around insurance for claims management.

Figure 6.66 shows the machine learning service built for this process. As you can see, the captured claim facts and vehicle damage are applied, taking into consideration the additional historic claims and payout data, which helps predict the total payout in a particular claim scenario.

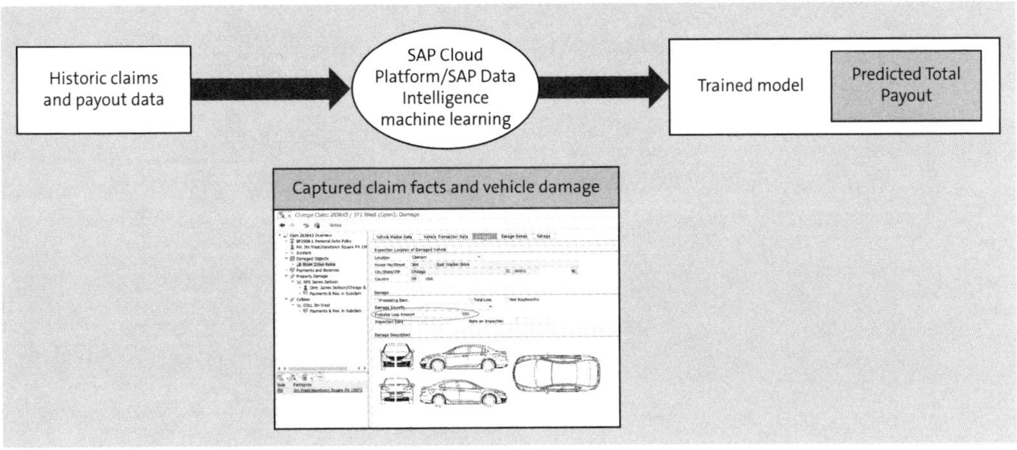

Figure 6.66 Predict Claims Severity

The machine learning service helps to accurately predict the severity of a claim and enables insurers to deploy appropriate strategies to optimize the claim handling process and enhance the customer experience. The predicted severity of a claim could directly influence crucial aspects of the claims process, including assignment of the right claim handler, setting accurate reserves, and identifying claims for straight-through processing. This machine learning service is built into the SAP S/4HANA Insurance for claims management functionality and leverages a side-by-side model.

The business values and benefits include the following:

- Expand claim handling expertise on the right cases
- Improved financials by setting more accurate reserves
- Identify high-exposure cases early in the claims process
- Identify claims for straight-through processing

> **Configuration Guide**
>
> For configuration steps, refer to the SAP Help Portal at *https://help.sap.com* and search for "SAP S/4HANA Insurance for Claims Management".

Email to Record

In this section, we'll focus on the machine learning service built around service request automation.

Figure 6.67 shows the machine learning service built for this process.

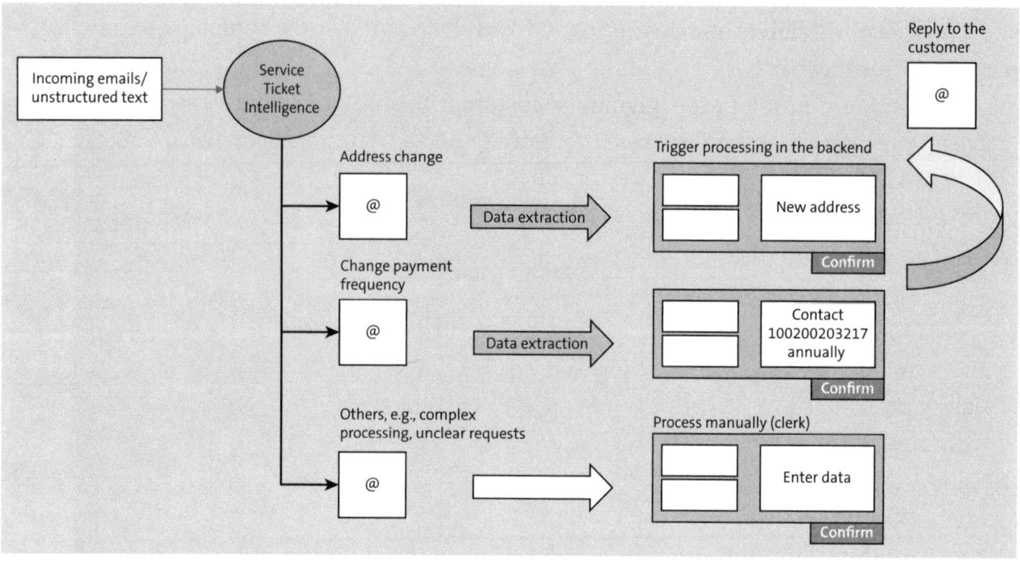

Figure 6.67 Service Request Automation

Service Ticket Intelligence is a generic machine learning service that provides a solution for processing emails by automating their categorization, extraction, and labeling of relevant information and triggering suitable follow-up processes, like changing payment details or executing business transactions on enforced policies with data derived from the email. This generic machine learning service that leverages the side-by-side model triggers the processing of the backend functions and extracts the information needed to update the insurance customer records.

The business value and benefits include the following:

- Automation of email processing
- Reduces response time for consumers
- Frees clerks to focus on complex tasks
- Reduces resources needed

Configuration Guide

For configuration steps, refer to the SAP Help Portal at *https://help.sap.com* and search for "SAP Service Ticket Intelligence".

Upcoming Ideas in the Insurance Industry

The insurance industry relies heavily on customer data, historic information about incidents, and payment data to calculate risks. Machine learning helps to increase operational efficiency, improve customer service, and detect fraud faster. Some of the new ideas in the evaluation and planning phases are as follows:

- **Customer profile from insurance**
  This machine learning idea could support agents to derive a customer profile based on a customer's life situation, giving the customer the option to adjust their coverage according to their preferences. The profile is used to find the best personalized insurance product recommendations. The machine learning algorithm leverages the collected customer data and creates a customer profile. The business value is more competitive offerings and better personalized recommendations based on sales experience for similar customers.
- **From paper policy to record (as a service)**
  The machine learning idea is to provide a solution for direct insurers (and potentially portals like Check24) to convert image-based policy documents into a structured and potentially standardized data format. The solution could be exposed as a ready-to-use API trained and hosted by SAP and charged based on the usage (software-as-a-service [SaaS] offering). The business value is less effort involved (just taking photos), better data quality and no implementation costs for insurers, and industry standards easily being defined in this domain with this machine learning service.
- **Robot underwriting supported by machine leaning**
  The machine learning idea is to provide a solution to support system underwriting decisions based on data instead of rules. After being trained, the machine learning algorithm evaluates data related to the applicant (e.g., existing policies, claims) and related to the application (e.g., coverages involved) in order to make or propose an underwriting decision. The business value is reduced effort for underwriting rule maintenance, underwriters being able to focus on the most relevant decisions, and fewer underwriting resources needed.
- **Robot pricing**
  An expected claim value would be supported by machine leaning. This is a machine learning idea to calculate premiums based on historical data instead of rules. After being trained, the algorithm evaluates data related to the applicant (e.g., age or job), data related to the application (e.g., coverages, insured sum), and the rates of potential competitors in order to determine a price. The business benefits are reduced effort to adjust pricing conditions (training instead of programming), more flexibility, and adjustments to prices to optimize earnings.
- **Invoice inspector**
  Analyze incoming invoices for savings potential. This is a machine learning idea to support employees like claim adjusters when inspecting incoming invoices by predicting

their cost-reduction potential. The business benefits include increased effectivity of invoice verification and securing the highest cost-reduction potential.

- **Predict next actions**
  This is a machine learning idea to support a clerk and propose one or a sequence of next actions in the system learned from historical insurance claims processing data. The business value is to improve the performance of processing a claim, improve usability, and increase automation.
- **Predict optional processing of a claim**
  This is an idea to support the clerk and predict optional claim processing steps recommended based on historic claims experiences. The business value is support for unexperienced claim handlers and reduced time efforts.
- **Detect anomaly**
  Support clerks when capturing data in the system. Learn from historical data, letting the system detect suspicious entries and raise a warning. The business value is error prevention.

6.8.7 Telecommunications

In this section, we'll focus on the machine learning functionality that is built around the telecommunications industry. We'll look into some of the use cases around fraud prevention, customer profitability analysis, and more. Telecom companies benefit in a huge way from machine learning and predictive modeling functionality for easier handling of their subscribers.

Let's now look at some of the use cases that have already been delivered.

Customer Profitability Analysis

In this section, we'll focus on the machine learning service built around customer profitability analysis with outlier detection and clustering for the telecom business.

Profitability analysis of telecom subscribers includes margin drivers that help increase the margin revenue. The machine learning service uses a side-by-side model, leveraging algorithms such as outlier detection, supervised and unsupervised clustering, and delta analysis. It runs on SAP Cloud Platform. SAP S/4HANA is the backend data source that connects the billing information and sales order data to perform profitability analysis.

By leveraging the machine learning service, there are quite a few benefits to glean:

- Helping telecom transition from revenue assurance to margin assurance
- Detect margin leakages of subscriptions based on various machine learning algorithms
- Segmentation of customers based on margins using supervised and unsupervised clustering techniques

- Detecting patterns of subscriptions based on the usage of services
- Detecting possible reasons for changes, or other changes that correlate with the changes in key figures over time

Configuration Guide

For configuration steps, refer to the SAP Help Portal at *https://help.sap.com* and search for "Telco Customer Profitability Analytics".

Payment Fraud Prevention

In this section, we'll focus on the machine learning service built around payment fraud prevention for the telecom business.

By leveraging the machine learning service with blockchain technology, the suppliers put bank account information into the blockchain, which helps identify and handle payment fraud easily. Of course, it's the supplier's responsibility to have the correct data in the blockchain. While the blockchain technology uses the secure account information, the regression machine learning algorithms help analyze the various factors for identifying and preventing potential fraud. The side-by-side machine learning models are used for creating the machine learning service that runs on SAP Cloud Platform. With easy-to-access information (restricted read access) for customers and authentic bank account verification, this helps prevent any payment fraud.

The machine learning service with blockchain technology provides many benefits, including the following:

- Long-term vision of leveraging master data as a service
- Increased process efficiency via timely updates and separated responsibilities
- No central authority but rather joint usage of general supplier data possible
- Multiparty collaboration on a single source of truth, enabling efficiency gains and transparency for all involved parties

Global International Mobile Equipment Identity Storage and Services

Next, we'll focus on the machine learning service built around the theft of mobile devices.

Figure 6.68 shows the blockchain service with machine learning built for this process, which we'll unpack in this section.

The blockchain service with machine learning capabilities helps to deny stolen phones access to the mobile network. While the blockchain service keeps stolen phones out of the network and provides no access, the machine learning service helps to augment this by analyzing the access activity on the stolen phones with the help of International Mobile Equipment Identities (IMEIs). The multiparty capabilities of blockchain enable mobile operators, smartphone OEMs, public authorities, insurance companies, and others to share data in a secure and transparent way to dry out the black market and consequently secure consumers and businesses against the theft of mobile phones and other wireless devices.

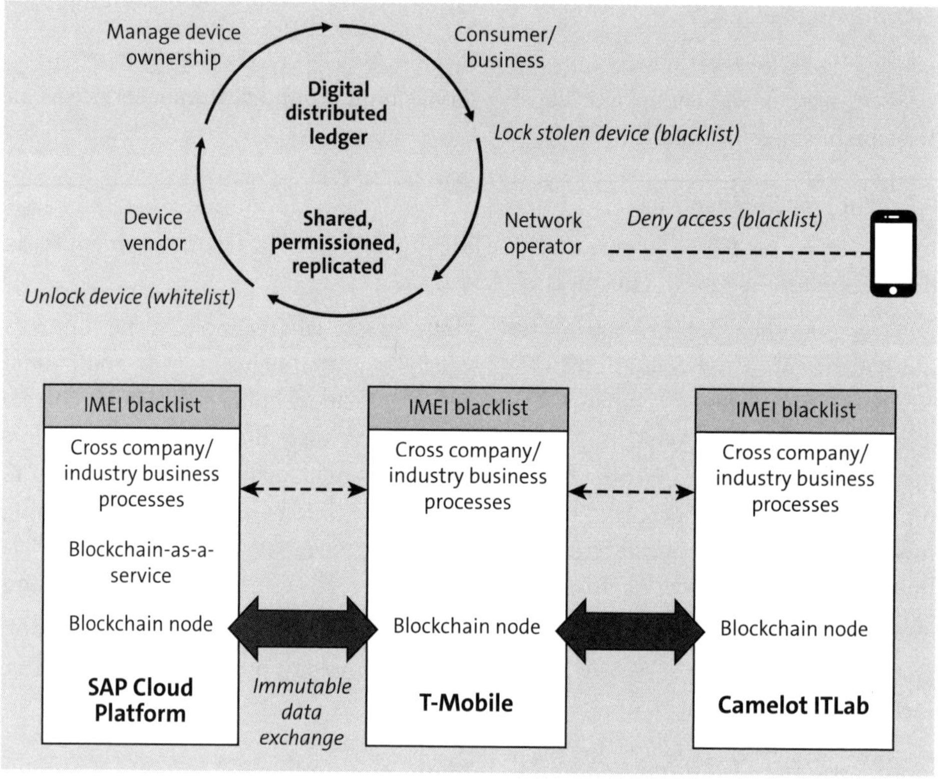

Figure 6.68 Global International Mobile Equipment Identity Storage and Services

Mobile devices and the personal information stored on them are consistently among the highest priority targets for thieves. Studies estimate that by now, one in ten smartphone owners in the United States have been the victims of phone theft. This is a major challenge faced by mobile operators.

Blockchain-based lists of IMIEs are accessible by network operators, hardware vendors, and so on. (The IMIE is a code that uniquely identifies a mobile device.) A connection can be established between the blockchain and mobile network to deny stolen phones access to the mobile network, which can be a typical administrative functionality for all network members.

The business benefits include the following:

- Global sharing of blacklisting information across participants brings transparency and mutually assured trust
- Secures consumers and businesses against the theft of mobile phones

- Mobile operators sharing blacklist information with multiple participants, like insurance companies and large customers
- Efficiency gains through removal of data silos
- Potential for the development of new products and services

Customer Service Chatbot

Now, we'll focus on the machine learning service built around customer service in the telecom business network.

Figure 6.69 shows the machine learning service built for this process, which we'll unpack in this section.

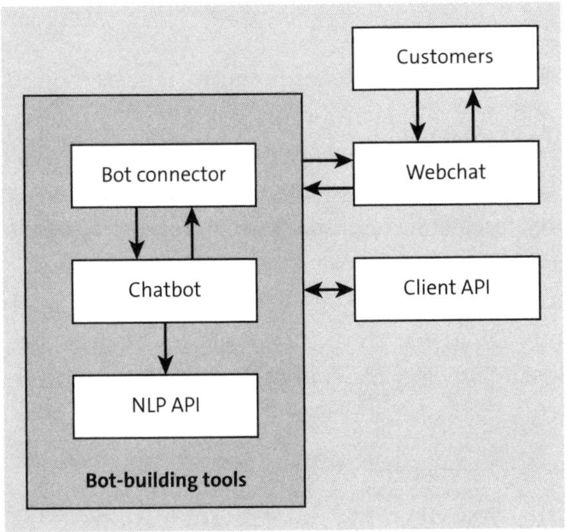

Figure 6.69 Customer Service Chatbot

The customer service chatbot is an entry point for customer requests. It first understands customer requests and either directs them to the correct department (presales or customer service) or manages their inquiry independently. It can be broken down into three parts: customer dispatch, automated conversation, and standard answers. The customer service chatbot built here has a mix of natural language processing (NLP) and machine learning capabilities. While the NLP functionality reads the webchat and interprets it, the machine learning service helps to divert the customer requests accordingly.

The typical business benefits include the following:

- Reduces the length of support conversations
- Ensures customers are talking to the correct agent on the first call
- Frees up a lot of time for agents so that they can focus on added-value tasks such as presales or personalized client support

6.8.8 Banking

Let's now discuss the machine learning functionality that is built around the banking industry. We'll look into some of the use cases around loan management, customer analytics, customer behavior, and financial services. Bank managers and sales representatives will benefit hugely from machine learning and predictive modeling functionality in the banking solutions. They will be able to identify the customer behavior and promote the right solutions accordingly.

Let's now look at some of the ideas that have already been delivered.

Efficient Processing of Misdirected Payments

In this section, we'll focus on the machine learning service built around processing misdirected payments in the banking industry.

The Intelligent Account Finder for Payments app proposes potential target accounts for incoming payments without valid account identification. Intelligent Account Finder for Payments is an app in SAP Transactional Banking for SAP S/4HANA, which leverages machine learning to identify misdirected payments using the k-nearest neighbors algorithm in a side-by-side model. Such misdirected payments typically occur during processing of international, non-single unit payments area (SEPA) payments due to missing standards. This machine learning service compares specific attributes of the incoming payment with productive customer accounts and their related data. It calculates a similarity score, which indicates the confidence in the proposal.

The business benefits include the following:

- Reduced cost and time for manual processing of misdirected payments
- Increased straight-through processing of payments by faster determination of a target account for misdirected payments
- Consumable by different payment exception handling applications (e.g., within a payment transaction system or account management system) with SAP Transactional Banking for SAP S/4HANA

Configuration Guide

For configuration steps, refer to the SAP Help Portal at *https://help.sap.com* and search for "Intelligent Account Finder".

Banking Customer Analytics

In this section, we'll focus on the machine learning service built around customer analytics in the banking domain.

Personal contacts with customers are decreasing as customers more often prefer to interact with a bank via online channels, particularly for the simple things that represent the majority of customer-bank interactions. Retail banks face the problem that they lose information about customer needs and valuable insights to design and position competitive offerings for the right customers. The machine learning service provides detailed analysis of the customer behavior based on a wide variety of customer interaction data. The regression machine learning algorithm uses a side-by-side model by leveraging customer profiles, customer online behavior and other banking transactional data, and so on.

The business benefits include using behavioral scores, which increases the success rates of campaigns and sales activities significantly. At the same time, with machine learning, the automated analytics approach accelerates the analytical process and allows more tailor-made and agile campaigns.

Configuration Guide

For configuration steps, refer to the SAP Help Portal at *https://help.sap.com* and search for "Banking Customer Analytics".

Upcoming Ideas from the Banking Industry

Retail banking is changing a lot nowadays, with varying competitions between banks to provide online services: mortgage banking, credit banking, loan banking, and so on. There is always a hunger for banks to attract more customers, and at the same time customers are becoming more tech-savvy in choosing better and smarter services. Some of the new ideas that are in the evaluation and planning phases are as follows:

- **Tailor-made offers for loan rollover contracts**
  Intelligent rollover supports the bank in providing tailor-made offers to borrowers at the end of a fixed interest period of loan contracts. Up to the time of writing (summer 2020), rollover offers are created via a mass run (i.e., by means of tables with variants and modifications) or via a single individual rollover. Machine learning algorithms could be included in the mass run in order to derive new conditions for the rollover contract: by clustering the borrower, analyzing his payment behavior in the past (number of dunning cases, etc.), and so on. The business value would be improved quality of mass run results, potentially lower number of individual offers, higher customer satisfaction through individual rollover offers, reduced risks by individual customer analysis and conditioning, and reduced time effort by considering the individual customer situation in rollover mass runs.
- **Bonded loans management**
  Bonded loans combine the economic advantages of capital markets instruments with the favorable legal requirements of loans. Distributed ledger technology (DLT) provides the possibility to significantly simplify and speed up the processes of bonded loan issuing paperless service while allowing clients to constantly track

issued loans. The business value would be a reduced processing duration of placement and servicing transactions for bonded loan contract execution, increased attractiveness for investors by enhancing insight and transparency, and fungibility in bonded loans operations and constellations via a trust-providing platform.

- **Intelligent decisions based on predicted account behavior (intelligent overdraft decisions)**
  In commercial banking, regular peaks of critical but high-volume payments like salary for all employees at month end or payments to suppliers may lead to overdrafts on the ordering party's side. Banks need to decide whether to allow the overdraft—make an intelligent overdraft decision based on predicted account data—or support the bank clerk in his overdraft decisions by prediction of credit payments and or predicted account balance in the near future. Machine learning algorithms allow you to use large volumes of historical data (balances, payments) to predict future account behavior. The business value would be more information available for personal customer interaction and well-founded decisions with predicted balances, and turnovers give additional insights for intelligent overdraft decisions, in addition to reducing the effort of manual research into customer account behavior.
- **Banking financial services data platform**
  Introduce a data management strategy that allows you to easily access new information needs and opportunities, act at the speed of business, and engage in data-driven innovation, which allows you to leverage existing data management siloes. Additionally, acknowledge the strengths of different data management tools and technologies, including open-source ones, and introduce a roadmap for simplification and reduction of data redundancies. The business value includes the following:
  - Enable access to a consolidated dataset across the enterprise through a simplified data platform architecture supporting extraction from and federated data access to diverse data sources.
  - Provide a standardized and business-driven financial services data model with full integration into SAP S/4HANA, ensuring extensibility and upgradeability.
  - Directly run high-performance analytical scenarios (e.g., finance, risk, regulatory, customer) on a finely granular level of data in the platform without replication and reconciliation efforts to enable new intelligent business processes.

6.8.9 High-Tech

In this section, we focus on the machine learning functionality built around the high-tech industry. We'll look into some of the use cases around high-tech equipment manufacturing, mass production, and assembly yield lines. The production planners, quality inspectors, asset managers, and others in the high-tech line of manufacturing will greatly benefit from the machine learning and predictive modeling functionality.

Note

The use cases discussed here for the high-tech industry are in the planning and evaluation phases and subject to change.

Here are some of the ideas from the high-tech industry that are interesting to review:

- **Factory data brought to light**
  Sometimes a customer remaining in the mass production market for a longer time (during rough economic times) might experience a decline in revenue and fear disappearing from the market. Changing from mass production to lot sizes means a tremendous change in the customer's way of working, hence initiating the digital backbone program to enter into this new market as soon as possible. The digital backbone should be the single source of truth for customers' factories located all around the world, increasing the visibility from the shop to the top floor. The business value is in global track and trace, an asset information network, and IoT discovery.
- **Quality management analytics**
  A quality risk predictive analysis solution improves the yield of assembly lines, reduces the risk of material batches, and also brings data analysis capabilities to a new level. The business benefits include the following:
  - Improved accuracy and efficiency of quality management by disentangling the influence of the quality or raw materials, the assembly process, and the test procedure itself.
  - Enabling production line managers to monitor quality in real time and intervene early.
  - Treating uncertainty and error sources within a self-containing statistical framework to get reliable estimates for informed business decisions.
- **Quality first**
  High-tech equipment manufacturers produce large numbers of consumer products by combining and assembling many components and subparts into the finished product. Any quality deficiencies in this multilevel supply chain or anomalies in test equipment leading to malfunctioning products can pose huge network and recall risks, including significant brand damage lawsuits. The quality first predictive analytics solution could help identify these quality deficiencies ahead of time so that appropriate measures can be taken to counter them. The business benefits include the following:
  - Identification of potential material batches with quality risks as early as possible.
  - Assisting the test engineers in making decisions to replace materials or to discontinue production.
  - Predicting and identifying the test equipment that has failed or potential problems and anomalies.

- **SAP Supply Base Optimization**
  This machine learning solution is targeted at discrete manufacturing enterprises dealing with lots of purchased parts, cross products, cross bill of materials (BOMs), price fluctuations, outsourced and/or in-house manufacturing, etc. This is an intelligent decision support application that uses market benchmark data and predictive analytics to determine negotiation price targets, and it provides a rule-based optimization and simulation framework to make business award decisions against these targets. The business benefits include the following:
  - Streamline the data collection, negotiation, and decision support process to enable commodity managers to spend more time on high-impact negotiations.
  - Advanced tools and insights identify high-value opportunities and negotiate better business award decisions.
  - Overall reduction in direct material spend increases bottom-line earnings.
- **Material tracing: component validation for quality inspector**
  At an original equipment manufacturer (OEM), a quality inspector needs to verify the originality of the high-value components (e.g., processors). Birth and lifecycle certificates for individual components are stored in a public blockchain, which is fed by suppliers. The quality inspector of the OEM rechecks the bill of material component against the blockchain certificates and can verify the originality. After verifying the originality, the status of the certificates can be changed. The business value is prevention of counterfeiting high-value components.

6.8.10 Sports and Entertainment

In this section, we focus on the machine learning functionality built around the sports and entertainment industry. We'll look at some of the use cases around sports teams coaching management with data analytics. Sports coaches and team management will benefit immensely from machine learning and predictive modeling functionality built into sports apps.

Here are some of the use cases from the sports and entertainment industry that are interesting to review:

- **NHL in-arena coaching**
  The In-Arena Coaching app is a next-generation mobile application. This application deployed on iPad will help support National Hockey League (NHL) team coaches by providing them the information they need on the bench. The NHL In-Arena Coaching tool can be used on-bench by the 31 NHL coaching staff during live NHL games. This tool will be showcased by NHL and SAP as a means to introduce technology/data analytics via the Apple iPad during a live NHL game in an area in which that technology has not been previously used. This application can be used for an in-game summary, face-off display, shoot-out planner, time on ice, shot chart, and chance definition.

- **SAP Digital Boardroom integrating a big data 360 view on business and player data**
  The SAP SportsOne application blends external sports data sources such as Opta and Prozone and delivers a boardroom experience by using SAP Analytics Cloud as the frontend. A holistic view of players, fans, and businesses is provided in real time, managing 500,000 visitors per year. It also provides awareness levels of capacity and sales during match days and beyond. The business benefits include the following:
  - Gain real-time insights into performance KPIs
  - Total transparency
  - Instant data-driven insights
  - Simplified boardroom processes
- **Biometric bridge**
  Sailing team AkzoNobel, together with Foundry Performance, coinnovated with SAP to build an IoT platform to gather biometric data suited for offshore sailing and to provide insights based on the data to the team using machine learning. Insights into the fatigue level are balanced across the team, including how many calories someone burns in a day or the preferred sleep conditions that might lead to better strategical decisions and improve overall performance. Differences between teams are very small, and a 1% performance improvement may very well win the race. The business benefits include the following:
  - Provide insights based on data science
  - Better strategical decisions and improvements in overall performance
  - Avoid deep or accumulated fatigue for faster recovery, keeping crew optionally fit until the end of the race
  - Balance out fatigue across the team
  - Review to create analytical reports and have an overview of the last leg

6.8.11 Public Services

In this section, we'll focus on the machine learning ideas and functionalities that can help the public sector domain. We'll look into some of the use cases around public sector services and agencies, along with the citizens involved in the process.

Handling interagency transactions and citizen services, public sector administrative officials can greatly benefit from machine learning and predictive modeling functionality to structure and better serve citizens. Here are some of the ideas from the public sector domain that are interesting to review:

- **Flood prediction and emergency management**
  Imagine a tool that will predict when a waterbody such as river or a bayou will overflow in advance. Think of the leverage it provides, with the ability to make evacuation decisions sooner, taking advanced precautions and warning people ahead of time, or

initiating pooling of the necessary resources well before a natural calamity or an adverse weather event actually hits. The business benefits include the following:

- Examine historical data combined with information from the ground to predict and act fast. This feeds early-warning systems for better response times and event optimization to avoid disaster or to support disaster recovery.
- Predict impending overflows.
- Analyze weather events and thus find patterns for a statistical understanding of the frequency and damage magnitude of potential events.

- **Behavioral insights: tax services with machine learning**
  Allow proactive intervention strategies such as personalized payment plans customized to an individual's circumstances and needs to prevent them from becoming debtors based on historical data and machine learning models. The business benefits include the following:
  - Get empowered and more engaged employees by using the machine learning algorithms
  - Potential to bring revenue sooner
  - Potential to reduce the amount of money owed to a tax department
  - Easier, simpler, and better experience for land-tax payers

6.9 Summary

In this chapter, we focused on the business implementation of the predictive analytics and machine learning scenarios leveraging different approaches. We have sorted them based on different lines of business and industries rather than by implementation approach.

We started by providing some examples of the end-to-end processes and then briefly discussed how to access configuration instructions for your use case. Later we continued with the different machine learning and predictive analytics use case functionalities implemented in the lines of business, from finance to R&D/engineering, explaining the focus of these use cases and how they have been implemented.

Then we focused on explaining the use cases in different industries, discussing the use cases that have already been delivered and a few new ideas that are in the planning and evaluation phases. We didn't expand on some of these use cases because they're still in the evaluation phase as of the time of writing (summer 2020—and amid the impact of COVID-19, as well!).

In the next chapter, we'll give a brief overview of the services on the SAP Cloud Platform.

Chapter 7
Services on SAP Cloud Platform

In this chapter, we'll widen our lens to view the full scope of SAP Cloud Platform services related to machine learning that are available to help your business realize the vision of the intelligent enterprise.

We've discussed how machine learning and predictive analytics services are implemented technically and in various business scenarios. Now it's time to consider other SAP Cloud Platform services that are interwoven with machine learning, from SAP Data Intelligence to Internet of Things (IoT) and blockchain. Let's start with a rundown of the trends to consider in the context of SAP Cloud Platform.

7.1 Key Trends and Capabilities

Let's look into some key trends, customer needs, and value propositions in the context of an intelligent enterprise and see how SAP Cloud Platform can help address them.

Figure 7.1 gives a brief overview of SAP Cloud Platform and how it's leveraged in the context of other SAP applications.

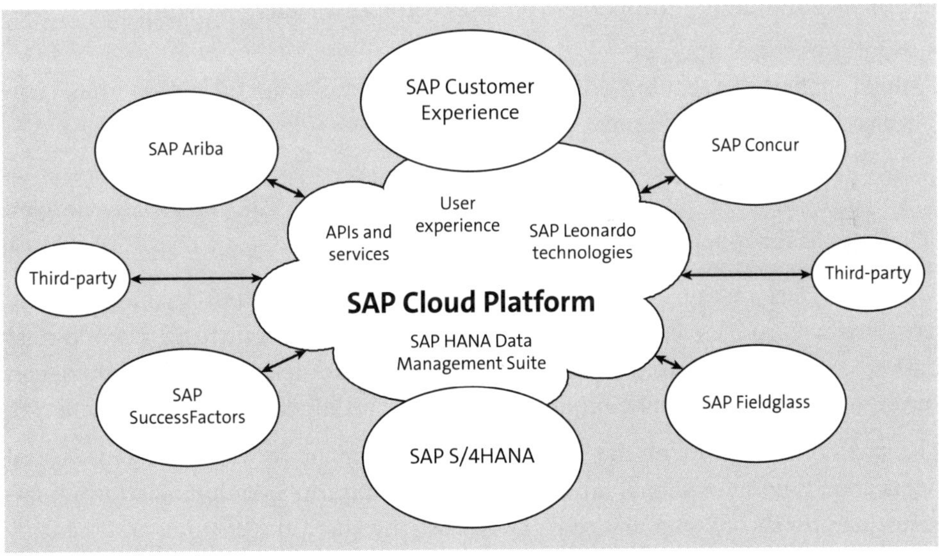

Figure 7.1 SAP Cloud Platform

Disruption is the buzzword of note, with changes happening in all spheres of life. With the current global pandemic, disruption is also quickening and shaping the next economy. Making business processes more intelligent with artificial intelligence and machine learning, adding IoT services, and leveraging blockchain and real-time analytics are becoming some of the essential customer needs in the present day to redefine and reinvent businesses. Interacting with business systems using conversational UIs and consumer-grade, user-centric experiences while deriving actionable insights from big data in real-time appeals to enterprise customers in a big way. Enterprises always prefer solutions that can excel in hybrid IT landscapes by integrating and combining on-premise business systems and cloud solutions with differentiating technologies. In general, businesses expect a faster time to market and personalized, collaborative, mobile-enabled cloud applications to meet their priorities, taking advantage of innovations such as IoT, big data, machine learning, blockchain, and real-time analytics. SAP Cloud Platform can address these collective needs.

Using open standards and capitalizing on existing investments, some of the key capabilities of SAP Cloud Platform are as follows:

- **Cloud scalability**
  Deliver scalable applications with a free choice of underlying infrastructure provider.
- **Business services**
  Leverage a fast-growing set of smart business services that deliver prepackaged content and domain-specific functionality.
- **Integration and security**
  Safely integrate on-premise and cloud applications, IoT, social networks, collaboration tools, business networks, and diverse customer interaction channels.
- **Runtimes and containers**
  Build applications that run on any browser and device, using the programming language/models of your choice.
- **Tools and SDKs**
  Empower developers with high-productivity tools and SDKs to rapidly develop enterprise-grade apps with a delightful user experience.

SAP Cloud Platform and its underlying infrastructure hold various certifications and attestations. It runs in a secured and certified environment consisting of world-class data centers, advanced network security, reliable data backup, and built-in compliance, including integrity and confidentiality, with 99% availability of the infrastructure.

The additional support of the Kubernetes environment provides container-based applications and services that are built and deployed on an SAP Cloud Platform infrastructure provider of your choice without worrying about the creation and management of the underlying container clusters. Added flexibility is available to run any workload (stateful, stateless) in a secure, enterprise-grade cluster environment.

SAP Cloud Platform supports the ABAP environment, and the development and runtime environment for cloud applications is written in ABAP. The environment is based on a whitelisted set of APIs and optimized for SAP HANA. This helps to easily develop SAP Fiori apps that are optimized for SAP HANA and ready to deploy enterprise-ready business applications to enable the ABAP community to transition to the cloud. In addition, you can utilize existing ABAP know-how in the cloud, extend ABAP-based products such as SAP S/4HANA Cloud, transform ABAP-based custom code for the cloud, and use the service ecosystem of SAP Cloud Platform.

7

Figure 7.2 gives an overview of the different services available from SAP Cloud Platform.

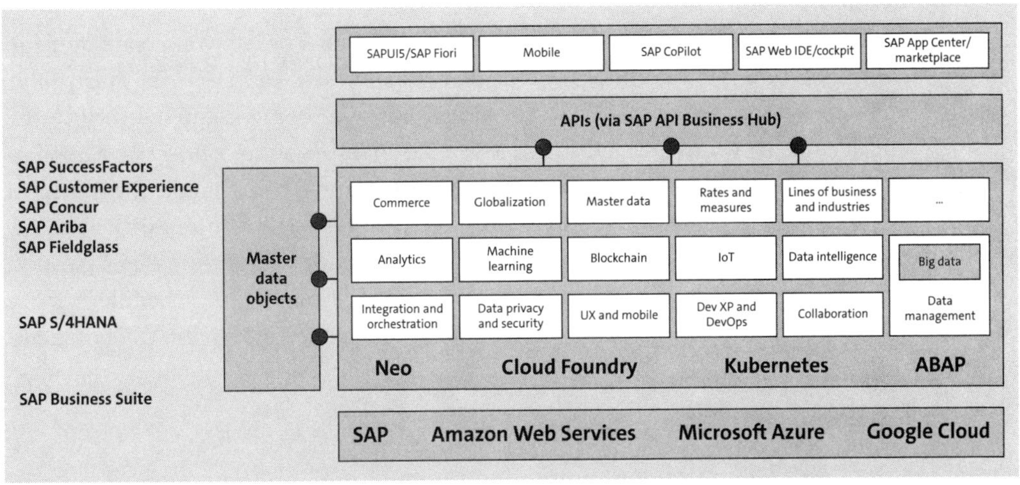

Figure 7.2 SAP Cloud Platform Services

Let's now look at the SAP Cloud Platform services in little more detail:

- **Analytics**
  Analytics plays a key role, providing insights based on data with interactive results and helping with reporting for the different functions. Tracking organizational KPIs in real time with up-to-the-minute updates using interactive dashboards is the key. Analytics functionality is typically leveraged with SAP S/4HANA's embedded SAP Fiori apps, the SAP Cloud Platform SAP Fiori apps, and SAP Analytics Cloud. This is discussed at length in Chapter 1, Section 1.4.
- **Data management**
  ObjectStore is a service that enables Cloud Foundry applications to obtain storage space in the underlying infrastructure-as-a-service (IaaS), with access to that storage space. It provides access to a distributed and consistently available object storage space that Cloud Foundry applications can use to store their objects. The storage space can be an S3 bucket on AWS, a container on Azure storage, or a bucket on Google Cloud Storage. This also supports a cross-consumption model that includes

making the service compliant with the Open Service Broker API (OSBAPI) by adapting it with a service manager. The Open Service Broker API enables software vendors, software-as-a-service (SaaS) providers, and developers to provide backing services for workloads based on cloud-native platforms such as Cloud Foundry and Kubernetes.

PostgreSQL on SAP Cloud Platform offers an object-relational database management system with many advanced features, such as user-defined types, table inheritance, foreign key referential integrity, and multiversion concurrency control.

Redis-as-a-service is an open-source, advanced, in-memory key-value store with on-disk persistence, which can be used as a cache, database, and message broker.

- **DevOps**
  SAP Cloud Platform Continuous Integration and Delivery lets you create and run continuous integration/continuous delivery pipelines that automatically test, build, and deploy your code changes to speed up your development and delivery cycles. The agent activation for the Dynatrace service enables you to connect Java applications to your Dynatrace SaaS monitoring environment. It helps handle the transport of application development content and application-specific content (in a SaaS scenario) between SAP Cloud Platform subaccounts with the cloud-based transport management service of SAP Cloud Platform. In addition, the job scheduler service allows applications to define and manage jobs that run on either one-time or recurring schedules.
- **Data privacy and security**
  Authorization management support for SAP Cloud Platform and business users within different environments, including configurable authorization assignments to users during runtime and the like, are handled with data privacy services. This also extends support for groups and role collections. In-depth authorization management for custom roles with advanced UI enhancements is also available as a service. Additional platform APIs for authorization, trust, and audit log OAuth client management are supported.

  Some of the key aspects of the security services in SAP Cloud Platform include single sign-on to protect customer data, securing mobile and IoT scenarios, secure integration with customers' corporate user directories, propagation of logged users, secure storage of confidential data, configuration of own identity providers for business and platform users, and usage of fine-grained authorizations through custom platform roles.
- **Commercialization**
  The commercialization aspects of SAP Cloud Platform include decoupling procurement from use, usage transparency, flexible adoption, scale-up cycles, and self-service access, available with a minimum price guarantee.
- **Globalization**
  The SAP Localization Hub tax service automates indirect local tax determination

and calculation in a single source to determine and compute tax. The SAP Translation Hub is an additional SAP Cloud Platform service that provides access to translation capabilities across multiple products and languages.

- **Collaboration**
  Collaboration is one of the key aspects of SAP Cloud Platform and is discussed in the context of SAP Jam. Delivering functionality that further improves the customer experience and ease of use is a key priority for SAP Jam collaboration. Some innovations include the following:
  - Enhancing group messages to enable features such as *share to* for feed events and document attachment sharing
  - Speeding adoption with in-application user guides and expansion of quantitative and qualitative feature data usage
  - Delivering support for running SAP Jam collaboration on Amazon Web Services
  - Easing mobile adoption with easier activation and an enhanced user interface (UI)
- **Blockchain**
  The blockchain service helps developers to build a holistic solution on blockchain integrated with the SAP landscape. The entire lifecycle is supported from creating service instances, building secure networks, configuring channels to deployment, and exploration of chaincode. We'll discuss the blockchain services on SAP Cloud Platform in more detail in Section 7.5.
- **Document management**
  With this service, you can build API- and UI-based document management capabilities for your business applications to organize your enterprise content. A repository for storing and managing your business documents and attachments is provided. However, your own content management system can also be integrated with and embedded into business applications.
- **Developer experience**
  Let's briefly look at the developer tools in SAP Cloud Platform. Some of the application programming model capabilities of the SAP Cloud platform include the following:
  - Using CDS views for data and service modeling, consuming and extending SAP S/4HANA services and cloud solutions, leveraging a preferred programming language with databases and tools, and support for SAP Web IDE and local development tools.
  - Support of a Java VDM to access APIs in SAP S/4HANA and abstractions for simple integration of SAP Cloud Platform services.
  - Easy access to advanced OData functionality and SOAP services and the SAP S/4HANA Cloud SDK for JavaScript.
  - Zero installation or upgrades, rapid app development, integration with various data sources, integrated Git source control, and extensible and modular architecture.

- Full-stack application development, extension and deployment, UI, business logic, and database modeling.
- Support for SAP technologies (SAP Fiori, SAP Internet of Things, SAP HANA business data platform, etc.) and open-source technologies (JavaScript, OData, Java, node.js).

- **Integration**

  Some of the key integration capabilities include open API support and a tight connection to SAP API Business Hub. The framework interface handles support between SAP Process Orchestration and SAP Application Interface Framework. Additional intelligent content management and governance capabilities with community collaboration and role-based model sharing enhance the intelligent interface extension approach.

  SAP API Business Hub accelerates development of apps, extensions, and integrations with high-quality APIs. APIs are basically digital building blocks for modern apps and open integrations. API management typically includes full lifecycle management of your APIs with support for SAP Data Intelligence flows and CDS views and monetization of partner content via integration with SAP App Center. SAP API Management will offer highly scalable and resilient architecture to manage your APIs and microservices. SAP Cloud Platform enterprise messaging connects applications, services, and systems so that they can interact with each other through messages and enterprise events. It enables you to decouple application logic, develop microservices, and support event-driven architectures.

  SAP Cloud Platform Integration seamlessly connects cloud applications with other cloud and on-premise apps, both from SAP and from third-party providers. Integration with and access to open-source message broker RabbitMQ is also available for asynchronous communication between applications.

- **Machine learning**

  With SAP Data Intelligence, a machine learning platform is provided that enables scalable training and inference serving. Because SAP HANA is also included in SAP Cloud Platform, the corresponding machine learning libraries (APL, PAL) can be used. These capabilities were discussed throughout the book. In Section 7.3, we'll discuss reusable machine learning services, which can be incorporated into business applications.

- **Internet of Things**

  This service represents a network of physical objects known as *things* that are used to collect and exchange data. IoT enables setting up a thing model and provides a collection of REST-based as well as OData-based services to store and retrieve data efficiently for a thing, a company, and service personnel. We'll discuss the IoT services available on SAP Cloud Platform in Section 7.4.

- **Output management**

  With this service, you manage business documents such as printouts, emails, and alerts, and configure print forms. It includes features like email templates, native integration into business applications, and support for print forms using SAP Cloud Platform Forms by Adobe. In particular, you can send documents to the print service and then print out documents using a physical printer while you monitor the printing status.

- **Master data**
  Master data services include delivering functionality that adds value for existing customers through improved data and simplified API interactions. There are additional verification endpoints available while refining the licensing model for customers that are using multiple SAP Cloud Platform solutions or are interested in larger transaction volumes. Some of the key capabilities included address cleansing: parsing, correction, validation, global enhancement, geocoding, reverse geocoding, and so on.
- **Mobile**
  SAP Cloud Platform mobile services provide cloud services, SDKs, and tools that accelerate the development of mobile apps with powerful enterprise capabilities. This includes accelerating development of native apps (coded and metadata-driven) with new tools and SDKs for iOS and Android. Enterprise-grade security, offline data synchronization, and flexible cross-platform push notifications with connectivity for SAP and third-party backend systems are also supported. The benefits include delivering a wide variety of mobile apps quickly with powerful SDKs and incorporating tools to offer enterprise mobile features such as security and supportability, which are available offline for all the apps.
- **User experience and user management**
  Portals empower users to become more productive by providing easy and smart access to the right content and helping to enhance processes anywhere, anytime. Some of the key capabilities include building branded websites and SAP Fiori 2.0 launchpad sites by leveraging packaged content for SAP Business Suite and SAP S/4HANA, integrating SAP Fiori, SAPUI5, and on-premise SAP applications. Some of the benefits include driving user experience (UX) innovations and gaining agility via cloud services.

  SAP Fiori provides SAP S/4HANA and SAP Business Suite customers with a simple approach to adopt the SAP Fiori UX by leveraging SAP Fiori prepackaged content and services on SAP Cloud Platform while keeping business data on-premise.

  SAPUI5 is an HTML5 framework designed to build cross-platform, responsive, enterprise-grade applications. It's the leading UI technology of SAP's UX strategy, with major parts of the framework being open-sourced under the OpenUI5 product.

  The UI theme designer is a browser-based tool that allows you to create your own themes to adapt the visual appearance of applications. It's a single tool for theming and branding SAP's key UIs.

SAP Cloud Platform's Identity Authentication service provides authentication, web-based single sign-on, and user management as a SaaS. It's based on open industry-standard protocols such as SAML, OAuth, and SCIML. It also offers a comprehensive, low-cost approach to identity lifecycle management in the cloud.

- **Orchestration**
  SAP Cloud Platform business rules is a platform service that enables business key users and application developers to author and change business logic without rewriting an application. One of its key capabilities is enhanced decision table authoring. The benefits include building rule-based applications on SAP Cloud Platform that can easily extend standard cloud application workflows. SAP Cloud Platform workflow service can be used to orchestrate a sequence of tasks across different people and organizations.
- **Runtimes and containers**
  The SAP Cloud Platform application runtime provides a reserved runtime capacity for applications to run in the Cloud Foundry environment, allowing you to choose from a variety of programming languages and runtimes.
- **Rates and measures**
  This feature allows you to handle currencies, exchange rates, and currency conversions. For business applications supporting globalization and localization for different regions and countries, this service is required to manage the various currencies based on configuration.

7.2 SAP Data Intelligence

In this section, we focus briefly on SAP Data Intelligence and its services in the context of SAP Cloud Platform.

SAP Data Intelligence is a data warehouse-as-a-service (DWaaS) designed for business and IT users as an integrated and fully managed service. SAP Data Warehouse Cloud will extend its data modeling, wrangling, and connectivity capabilities to give the business analyst instant access to data. It's suitable for SAP and non-SAP customers alike.

SAP Data Intelligence provides one platform that combines individual SAP cloud services to deliver a scalable, integrated solution embedded in SAP Cloud Platform to serve various use cases, such as SAP Business Warehouse (SAP BW), SQL data warehouses, and agile analytical data marts. It provides one tool set with collaboration options for all tasks related to building data lakes, data marts, and data warehouses and analytical solutions for different personas (e.g., the business user, modeler, operator, etc.). It provides a central metadata catalog enabling data lineage, data quality, and profiling of datasets. The central connection management is leveraged to connect various data sources, both on-premise and in the cloud. The central repository enables versioning and lifecycle management processes.

The flexibility aspect of SAP Data Intelligence helps to deliver out-of-the-box integration with SAP and non-SAP systems, as well as to support structured and unstructured datatypes. It also offers data preparation capabilities for different user types, with options to switch from easy to advanced mode and to empower businesses to easily extend global data warehouse models.

SAP Data Intelligence is business-ready due to delivering a content library with packages of end-to-end business scenarios for specific industries and lines of business, supporting the intelligent enterprise. It is fully managed by SAP, so customers can innovate faster and leverage the latest and greatest capabilities.

The three key competitive advantages of SAP Data Intelligence are as follows:

1. Seamlessly connect all data sources with lineage and clear governance.
2. Provide powerful pipelines and notebooks across SAP, hyperscalers, and other open-source tools.
3. Availability of pre-engineered features and pipelines for SAP solutions on an open marketplace.

7.3 Machine Learning

This book is predominantly focused on machine learning services with SAP S/4HANA—but now, let's briefly touch on the machine learning services available as part of SAP Cloud Platform. As discussed in earlier chapters, you can enhance the business processes with intelligent algorithms and enterprise-grade easy-to-use machine learning services by creating customized machine learning services for images, videos, and text based on your own data without requiring machine learning expertise. Bringing your own model and deploying and serving custom machine learning models is the norm. The image-processing services that are available can help detect objects in pictures or classify and match images based on similarity by recognizing humans and faces while extracting text from document images. Natural language processing (NLP) services help to match text documents based on similarity or topic and keyword extraction. Easy-to-consume APIs are targeted at nonexperts in machine learning and offer simple integration in the cloud with on-premise, mobile apps and workflows via REST APIs.

New capabilities in the areas of data science and machine learning model lifecycle management, with new machine learning services in the area of text analysis, speech recognition, and master data matching, are also available. Data scientist support is provided with Jupyter Notebook and reproducible experimentation, text service analysis, speech recognition, master data matching, and so on.

Overall, these offerings help reduce time for development and maintenance of machine learning-powered applications to unlock knowledge from unstructured data (text and speech) or complex data schemas. Democratization of machine learning is attained via simple-to-use APIs.

In addition, SAP Cloud Platform provides reusable machine learning facilities with SAP AI Business Services. SAP AI Business Services offers strategic machine learning capabilities that automate and optimize processes and enrich the customer experience across the intelligent suite. Compared to full-fledged business applications, SAP AI Business Services can be consumed easily across the entire business and from a more generic viewpoint. They solve a problem that can be found in multiple business processes and are not specific to only one application.

As illustrated in Figure 7.3, SAP AI Business Services provide business applications on top of and integrated with other intelligent SAP offerings, such as SAP Conversational AI, SAP Intelligent Robotic Process Automation (SAP Intelligent RPA), and SAP Data Intelligence, in the form of open and flexible building blocks.

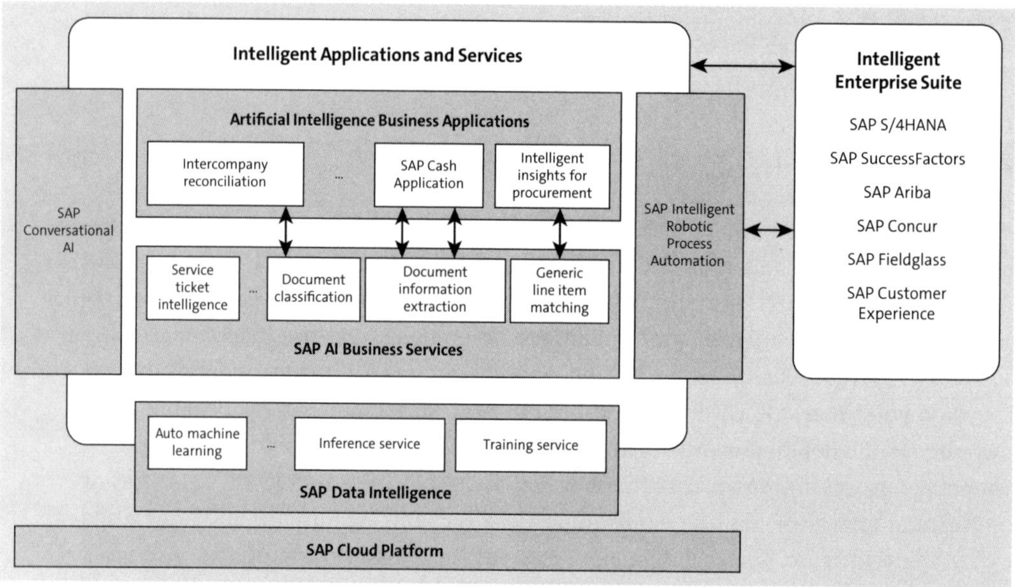

Figure 7.3 SAP AI Business Services

The portfolio of SAP AI Business Services currently covers the following ready-to-use machine learning services, which are exposed via REST-based APIs (we discussed these services at length in Chapter 6):

- **Document Classification**
  Many companies today still struggle with low process efficiency across their organization due to extensive paper- and document-based communication. This service classifies business documents into user-defined categories using machine learning. It analyzes PDF documents and proposes classifications based on previous customized classification models. To do so, the service includes training and inference capabilities

to fit a model using a custom dataset. With customized classification models, the service can be used in a wide range of business scenarios. Organizations achieve higher quality and become more efficient when handling documents in critical business processes like, for example, enterprise mail inbox processing, contract management, and invoice processing.

- **Document Information Extraction**
  This service uses machine learning to automatically extract information for documents. It helps to process various documents that have content in headers and tables. Organizations can use the extracted information, for example, to automatically process payables, invoices, or payment notes while making sure that invoices and payables match. The service requires a document file as input and returns the information from header fields and tables in JSON format. To extract information, the document file is submitted to optical character recognition, and a JSON file with the extracted information is provided, which the user can query. With this service, businesses can finally process more documents efficiently with fewer errors and reduced costs.
- **Data Attribute Recommendation**
  This service targets organizations struggling with the consistency and accuracy of their master data. This service applies machine learning to classify data records. It helps to classify entities such as products, stores, and users into multiple classes, using free text, numbers, and categories as input. Data Attribute Recommendation can be used, for example, to enrich missing attributes of a data record and classify incoming product information. Data records can be easily matched and classified, making more efficient and higher-quality master data management a reality while speeding up the data creation or cleaning process. The service consists of a data manager for training data, a model manager for lifecycle management, and inference for classifying the records.
- **Service Ticket Intelligence**
  This service helps organizations handle large volumes of service interactions. With Service Ticket Intelligence, incoming customers' service tickets are automatically classified into categories and routed to the right agent. The agent is then provided with recommended solutions to improve operational efficiency. This service leverages deep learning neural networks trained on large amounts of historical data, such as that existing in customer relationship management applications. The model recognizes the semantics of unstructured ticket messages, classifies the tickets into their most likely categories, and suggests solutions or knowledge base articles from similar previously answered tickets for the agent. The model accuracy increases over time as more and more service tickets are processed and user feedback is incorporated. With the Service Ticket Intelligence service, response times are accelerated, digital interactions are scaled, the volume of messages from omnichannel customer service is tackled, customer satisfaction is increased, and total cost of ownership (TCO) is reduced.

- **Invoice Object Recommendation**
  This service supports the accounts payable department in its decision-making through the prediction of correct account assignments. For that, the service includes training and inference capabilities to fit a model using a custom dataset. Historic invoice data recommends general ledger accounts and cost centers for incoming invoices without a purchase order reference by leveraging this data for model training. The inference results actually enhance financial reporting by minimizing barriers due to discrepancies in accounts and budgets, and they reduce repetitive and redundant tasks such as the need to search for general ledger accounts and correcting wrongly assigned accounts. The service enables efficient and fast processing of invoices, increases quality and compliance of general ledger accounts and cost centers, and reduces manual and repetitive work, allowing the team members to focus on more relevant tasks.
- **Business Entity Recognition**
  This is a machine learning business service that supports detecting and highlighting any given type of named entity in unstructured text in predefined categories. Business Entity Recognition can be used, for example, to automatically extract the context from incoming emails with invoice inquiries, automating the recurring tasks of answering queries about the status and payment of invoices. Thus, the service helps to increase overall document processing productivity, increase efficiency, reduce errors through the ability to detect named entities in text entities, and increase the quality and compliance of text document processes.

SAP AI Business Services have already proven successful with customers across various industries and LoBs. Customers and partners were able to improve efficiency and automation, empower employees, save value-adding time, and improve data quality.

7.4 Internet of Things

Let's briefly discuss IoT services and functionality as part of SAP Cloud Platform with SAP Internet of Things. This was very briefly discussed in Chapter 3 in the context of the intelligent enterprise and how machine learning services are enhanced.

You can connect, monitor, and manage devices to bring IoT data into SAP Cloud Platform applications and business processes, unleashing an ecosystem of innovation. SAP Internet of Things connects devices to SAP Cloud Platform to further manage and configure the data transmission from and to these devices as requested by IoT applications. Support for OData access is available to the IoT core API, yielding improved standard access into business applications and web-based development tools such as SAP Cloud Platform's SAP Web IDE. The IoT cockpit based on the SAP Fiori UX is the UI of the IoT service.

Some of the key capabilities include lifecycle management at scale for device sensors and gateways, from onboarding to decommissioning. SAP Internet of Things also helps to collect and process sensor data at scale and store it on SAP Cloud Platform, feeding asset-related information to business processes. The benefits include enabling integration from other business applications with the IoT core components via standardized OData access. Real-time data consumption is possible using a WebSocket and standard integration with SAP Data Intelligence. Device location visualization is also available via an integrated map view. The processing services configure data forwarding to databases in SAP Cloud Platform, wherein messages might be persisted in a database or forwarded to a message broker, an event stream processor, a document repository, or any other system.

7.5 Blockchain

SAP strongly believes in the open-source ecosystem and community efforts to advance blockchain technology and the ecosystem. SAP is a premium member of Hyperledger, which it sees as the most promising open-source consortium at the moment. SAP offers technical and business expertise with its ecosystem advantage.

The standard mantra says that innovation does not happen in isolation. There is always an existing customer situation, established business process, and classic system infrastructure that must be considered. SAP Cloud Platform is the innovation and extension platform that is also now offering blockchain capabilities. On the enabling layer, there is the technical integration that is necessary to have connectivity between different participants. This is where blockchain will play a major role in collaborative, peer-to-peer scenarios. The key element is to understand that with blockchain, we cannot expect a homogenous infrastructure and technology stack. The whole concept needs to be open to support blockchain technology running on different platforms and nevertheless be able to interact with each other. Then there is the business process layer, allowing a semantical integration of cross-company activities and process flows. This is the level on which SAP wants to bring together classic/existing SAP solutions with the new capabilities of the emerging blockchain ecosystem.

SAP follows an open and protocol-agnostic approach, integrated into the blockchain ecosystem. Its ambition is to run multinodes on multiclouds and multiplatforms. Therefore, SAP is supporting different blockchain technologies. It's aware that this space is in constant evolution and change is very fast paced. SAP therefore doesn't enforce a technology or platform decision but offers different options to its customers. Whenever new technologies emerge and gain relevant market traction, SAP offers further solutions, as provided in the past for the following:

- **MultiChain**
  Proven production readiness with enterprise-grade features. A lightweight opportunity to get started with scenarios that don't require application logic on blockchain.
- **Hyperledger Fabric**
  A blockchain framework implementation allowing components such as consensus and membership services to be plug-and-play.
- **Quorum**
  An enterprise variant of the Ethereum blockchain technology.

Typically, decentralized ledgers don't stop at the perimeter of individual data centers. Therefore, SAP is offering options to accommodate different infrastructure requirements that its customers are faced with, such as the following:

- **External node**
  Single blockchain nodes that are integrated into a bigger network but remain in other locations (for example, integrating an on-premise node to an SAP-provisioned blockchain network).
- **Multicloud**
  A blockchain node that is provisioned by SAP and run within the SAP Cloud Platform environment that connects with or integrates into a multipartner blockchain network across different operators. Components are hosted and operated by different cloud vendors.
- **Connect your own network**
  Connect the SAP stack (business processes, applications, blockchain services) to a multipartner blockchain network across different operators.

7.6 Summary

In this chapter, we discussed a high-level overview of SAP Cloud Platform and the services available as part of the SAP Cloud Platform, with a focus on machine learning-relevant trends and capabilities.

In the next (and final) chapter, we'll focus on the roadmap, upcoming features, and some suggested readings.

Chapter 8
The Road Ahead and Further Learning

In this chapter, we'll discuss the road ahead for machine learning in SAP S/4HANA, as well as resources for further information to continue your journey with machine learning and SAP.

We've reached the end of our exploration of machine learning implementation with SAP S/4HANA. However, the path continues with upcoming features and additional resources to discover. We'll take a look at the machine learning and predictive analytics roadmap in this chapter and point you in the direction of relevant blogs and resources to further support your learning.

8.1 Upcoming Features and Functionality

There are quite a few new and upcoming feature functionalities that are going to improve and advance the way predictive analytics and machine learning are built with SAP S/4HANA processes. This includes the new framework feature enhancements for the intelligent scenario lifecycle management (ISLM) framework. A backlog of requirements, including UX improvements, identity access management with cloud enablement, and more, are also in the pipeline. Additional setup requirements, including ISLM communication and advanced extensibility, better metrics support, and batch interference, are also in the evaluation phase.

> **Disclaimer**
>
> This information is all subject to change and is not a commitment or legal obligation to deliver any material, code, or functionality.

We'll walk through the major roadmap topics in this section.

8.1.1 Embedded Predictive Models

Throughout this book, we have worked with the next version of the Predictive Analytics integrator (PAi), now called the ISLM framework, and we discussed some of its architecture features and functionality in Chapter 4 and Chapter 5. This functionality can help customers build intelligent scenarios leveraging SAP HANA Predictive Analysis Library (PAL)/Automated Predictive Library (APL) algorithms or open-source R, TensorFlow,

Scikit-Learn, or Python libraries and manage the predictive models embedded into the SAP S/4HANA processes in a much more intuitive way. This functionality can be extended as needed.

Support for the creation of new intelligent scenarios using the ISLM technology for PAL/APL and SAP Data Intelligence will be possible in the future, as well as the migration of the current embedded PAi intelligent scenarios into the new ISLM framework. New SAP S/4HANA intelligent scenarios that leverage both PAL and APL with the ISLM framework are also planned. The support for all of these features is planned for both on-premise and cloud versions of SAP S/4HANA.

Feature upgrades with online inference, activation, and deactivation and triggered training with extension parameters including technical configurations and the like are in the pipeline for future releases. Additional features providing better metrics support with explainability of the predictive models embedded into the business processes are in the development phase.

Embedding of PAL and APL algorithms into the SAP S/4HANA business processes is available as of the time of writing (summer 2020). These embedded algorithms can be extended by building predictive models using the SAP Analytics Cloud smart predict functionality. You can also publish these predictive models built by smart predict back into the SAP S/4HANA business processes leveraging the ISLM framework. Future functionality is planned to embed the SAP Analytics Cloud predictive storyboards that are currently available as independent SAP Analytics Cloud stories. This would help customers and partners by offering an additional way of leveraging their predictive models that are available as SAP Analytics Cloud stories. Currently, individual business intelligence (BI) dashboards and stories are available as embedded SAP Analytics Cloud stories in SAP S/4HANA as tiles on the SAP Fiori launchpad. We should soon see predictive stories from SAP Analytics Cloud embedded into SAP S/4HANA.

8.1.2 Machine Learning Models on SAP Cloud Platform

Migration of all the released SAP Leonardo Machine Learning Foundation use cases to the new SAP Data Intelligence platform is planned. Currently the use cases built with the old SAP Leonardo Machine Learning Foundation framework are hosted as machine learning services on SAP Cloud Platform and configured to be consumed by SAP S/4HANA business processes. These machine learning services will be streamlined to be run directly on SAP Data Intelligence (which runs on SAP Cloud Platform). Upgrades to the current machine learning services are also planned that involve using the new SAP Data Intelligence tools and services in conjunction with connectivity to leverage the Jupyter Notebook framework environments.

New use cases from the different lines of businesses are in the planning phase and will soon be developed on SAP Data Intelligence. This will also enable leveraging the new Jupyter Notebook environments to build machine learning services that can be easily consumed not only by SAP S/4HANA business processes but also by external and third-party apps. This will help create new SAP Data Intelligence scenarios to be leveraged by SAP S/4HANA.

Support for multimodel intelligent scenarios in SAP Data Intelligence is planned, which will help to create more hybrid scenarios for use by the SAP S/4HANA business processes.

Features including prerequisite checks and backend implementation, as well as historical checks, are planned. Additional features including dark inferencing, model degradation, model comparison, and A/B testing are in the evaluation phase. This will help create many more qualitative and performance improvements to the machine learning models.

Many more proofs of concept (PoCs) and use cases are in the planning mode, which include cluster creation in the industry cloud and will lead to ideal follow-up action planning. As discussed in Chapter 6, there are many more use case scenarios planned in industries such as retail, insurance and banking, and so on.

8.1.3 SAP Analytics Cloud

SAP Analytics Cloud is going through a major facelift and improvement on many levels, from performance improvements to live connection access to many SAP S/4HANA processes. This should help open up the SAP S/4HANA business processes to quickly adapt and build more standard extensions to the current embedded predictive models and new predictive models. Let's walk through some of the planned features for SAP Analytics Cloud:

- **Augmented analytics**
  The feature allows users of the SAP Analytics Cloud solution to generate time series forecasting with day-based time dimensions.
- **Collaborative enterprise planning**
  This functionality makes sure that any changes made in the script editor for advanced formulas will translate to and be available within the visual editor.
- **Data actions monitoring**
  This capability enables users to see and track all triggered data actions, such as date and time, execution status, runtime, or initiator.
- **Enterprise reporting**
  This is the ability to create and share highly formatted and paginated reports manually or dynamically over the scheduler—for example, to expand the size of tables dynamically or extend support for paginated output exports.
- **Data locking**
  This feature allows users to allocate remaining amounts over unlocked cells. It simplifies several types of forecasting in which the original, targeted full-year amount needs to be spread over the open periods of the forecast.

8.1.4 Extensions of Existing Approaches

As of the time of writing (summer 2020), extensions to the released embedded predictive models and the machine learning models are available for the on-premise versions but not completely for the cloud versions. Additional extensions for the embedded predictive models and machine learning models will be possible on SAP S/4HANA Cloud for all approaches (embedded, side-by-side machine learning with SAP Data Intelligence, and side-by-side predictive analytics with SAP Analytics Cloud).

8.2 Blogs for Continuous Information

Blogging has long been one of the easier ways to share and communicate continuous information from experts and knowledge gurus. There are a few blogs that might interest you and that you might want to follow for additional information and updates. Alternatively, the regular SAP Community updates with SAPPHIRE and TechEd update resources will be of great use. You can find a great number of helpful blogs at *https://blogs.sap.com*.

There are some blogs specific to machine learning and predictive analytics that we would like to recommend:

- *http://s-prs.co/v515804*
  An interesting blog series about leveraging predictive intelligence in the context of SAP S/4HANA.
- *http://s-prs.co/v515805*
  A blog series on PAi and SAP S/4HANA explaining some of the nuances of predictive modeling in the context of SAP S/4HANA and SAP Analytics Cloud.
- *http://s-prs.co/v515806*
  An additional blog about predictive automation in the context of SAP Analytics Cloud.
- *http://s-prs.co/v515807*
  Another machine learning blog series that is focused for managers and consultants.
- *http://s-prs.co/v515808*
  This blog gives a quick overview for getting started with SAP S/4HANA embedded analytics.
- *http://s-prs.co/v515809*
  This blog explains how to implement machine learning scenarios based on SAP Data Intelligence. For this, a linear regression model is created with Python and trained and consumed with REST APIs so that the reader becomes familiar with the core components of SAP Data Intelligence.
- *http://s-prs.co/v515810*
  This blog describes how to extract application data from SAP S/4HANA based on CDS

extraction views. The concept and the necessary annotations and tools are explained so that the reader is enabled to reuse the existing CDS extraction views to defining their own extractors.

8.3 Summary

In this chapter, we highlighted upcoming features and functionality for SAP S/4HANA, SAP Data Intelligence (on SAP Cloud Platform), and SAP Analytics Cloud in the context of predictive analytics and machine learning.

In our journey forward with machine learning and future innovation, it's possible to create a world where AI benefits humanity, in which machine learning and automation frees people from dangerous, repetitive tasks and affords us the freedom to focus on activities that only humans can perform with empathy, moral judgement, and passion. The business technology platform is the foundation that powers and enables our vision. It is the technology that enables us to create value for our customers.

The utilization of machine learning is to improve and automate businesses led by the power of our imaginations. This is ultimately what intelligent enterprises are all about. Beyond the actual technology is the human mind; the goal of all these tools is to help you create value for your business.

We wish you happy reading, developing, deploying, and predicting the future!

The Authors

Dr. Siar Sarferaz is a chief software architect at SAP. In this role, he drives digital transformation by focusing on artificial intelligence and predictive analytics. He began his career as a method researcher at Siemens AG, before moving to SAP, where he has now worked for more than 20 years, holding various positions. He is the lead architect for machine learning implementation in SAP S/4HANA and is in charge of all concepts for infusing intelligence into business processes. He studied computer science and philosophy and holds a Ph.D. in computer science.

Raghu Banda is a senior director of AI product strategy at SAP Labs, where he is responsible for infusing AI technologies in SAP S/4HANA. He began his career as a software developer and architect in India before moving to the US in 1997. He joined with SAP in 2001 and worked in various roles such as engineering development, customer support and implementations, product marketing, and product management. He has worked with predictive analytics and machine learning since SAP entered this arena in 2012. He holds a Bachelor of Science in computer science and engineering and will soon graduate from the prestigious INSEAD business school. He is the lead product manager for leveraging machine learning into SAP S/4HANA.

Index

B

C

E

F

G

Q

R

S

T

U

V

W